Electromagnetic Field Theory

H. Parthasarathy

TECHSAR™
NEW DELHI

Published and Printed by
Techsar Pvt. Ltd.
4435–36/7, Ansari Road, Daryaganj
New Delhi-110002 (India)
E-mail: tech_info@techsarworld.com
Website: www.techsarworld.com

ISBN 978-81-88237-09-8

Reprint 2025

This book is dedicated to
My Parents

PREFACE

This is an introductory book to the subject of classical electromagnetic field theory but simultaneously carries the reader to a level from where he will be able to read more advanced textbooks on the subject like *Jackson's Electrodynamics*. The material for this book arose out of a one semester course on electromagnetic theory which the author delivered to fourth semester undergraduate students of electronics and communication engineering at the Netaji Subhas Institute of Technology. The author's chief research interests have been on problems related to the formulation of boundary value problems in electromagnetics in terms of partial differential equations and their solution by numerical methods. Such problems include computing the capacitance between two conducting surfaces of arbitrary shape, propagation modes for waves inside a waveguide of arbitrary cross section etc. Although the book does not put much emphasis on numerical methods, it does tell the reader how to formulate such problems starting from the basic Maxwell equations. Several projects have been guided by the author on such numerical problems in electromagnetic theory and the theoretical methods used in these projects have been discussed extensively in this book. The book differs from most of the engineering textbooks on the subject in the style in which it has been written laying more emphasis on physical principles rather than on engineering applications. For example the topic of relativistic transformation of the fields is not to be found in most of the engineering textbooks. A number and variety of solved problems have also been incorporated in each chapter to illustrate the concepts discussed. For example, in the chapter on Coulomb's law and electrostatics, several problems related to the solution of the Laplace equation under different kinds of boundary conditions have been included. The exercises in each chapter are also supported with hints so that the student attempting to solve these does not get discouraged.

The first chapter deals with vector analysis. Different kinds of co-ordinate systems commonly used in electromagnetic theory are discussed, i.e. the Cartesian, the cylindrical and the spherical polar systems. Methods of transforming scalar and vector fields from one system to another are discussed. The chapter discusses the notion of line, surface and volume integrals connecting these to the flux of a vector field through a closed surface and the line integral of a vector field around a closed loop, concepts that play an important role in the formulation of the basic equations of electrodynamics. The chapter discusses the important theorems of vector calculus namely Gauss' and Stokes' theorems which play an important role in transforming the Maxwell equations from integral form to differential form. Expressions for the gradient, curl and divergence in the various co-ordinate systems are also obtained.

The second chapter deals with Coulomb's law and electrostatics. The basic experimental findings of Coulomb are discussed using which and the principle of superposition the expression for the electric field due to a distribution of charges is obtained. In the course, the notion of electric field as an entity permeating space and causing a force to be exerted on a charge is introduced. The electric field is shown to have the irrotational property using which the concept of electrostatic potential functions is derived, which not only helps in computing the electric field in various electrostatic problems but also has a useful physical interpretation. The energy of a system of charges is derived from the concept of the potential. Computation of the potential for various kinds of electrostatic problems using line integrals is performed. The multipole expansion for the potential function is derived which is a useful approximation especially for determining fields at a great distance from the charge distribution. We then derive Gauss' law of electrostatics starting from Coulomb's law. This is an integral law that relates the flux of the electric field out of a closed surface to the charge enclosed. The complete equations of electrostatics are then formulated as a boundary value problem for the potential function using the Poisson equation. Green's function for the solution of the potential problems under various kinds of boundaries is obtained in the solved examples. The solution of Laplace's equation for electrostatic problems where no external charge is present is also illustrated through examples. Finally, the idea of the polarization vector to describe the electric field inside a dielectric medium is discussed. How the equations of electrostatics can be formulated in terms of the polarization vector and electric displacement vector for a dielectric is given a detailed discussion. Boundary conditions for the tangential and normal components of the electric field at the interface of two media are also treated. This chapter forms a self contained treatment of all important problems of electrostatics and some engineering applications like computation of the capacitance between two conducting surfaces are also discussed. After reading this chapter, the student should be in a position to write his own software for solving electrostatic field problems, plotting the electric lines of force due to any distribution of charges and computing the capacitance using numerical methods between any pair of conducting surfaces.

The third chapter introduces the basic equations of magnetostatics, namely Ampere's law in conjunction with the no magnetic monopole condition. Ampere's formulation of his law based on experimental findings is stated and its formulation in differential form using Stokes' theorem is deduced. Illustrations of Ampere's law using elementary current configurations is performed. The no magnetic monopole condition is then stated in the form of an integral theorem, the physical interpretation of which is that the magnetic field lines are all closed. The differential form of this law is derived in the form of the vanishing divergence of the magnetic field. The no monopole condition leads to the definition of the magnetic vector potential, for which, we derive the Poisson equation after imposing the Coulomb gauge. The solution to this equation in terms of the usual Green's function used in electrostatics is accomplished and from this, we derive the Biot-Savart law for the magnetic field produced by a static current distribution. A variety of problems are solved using the Biot Savart law. The problems include magnetic field by a circular current loop, by a straight wire of infinite length, by a small loop of arbitrary shape, by a current distribution at distances great from the distribution. The idea of a magnetized material and the magnetization vector is also introduced. Finally, we talk about the force exerted by a magnetic field on a current carrying conductor. This force equation is derived from the Lorentz force equation of a magnetic field upon a charge in motion. This force equation is used for example in the computation of the torque exerted by the magnetic field on the coil of a motor.

The fourth chapter introduces Faraday's law of induction which is also called sometimes as Lenz's law of induction. This relates a changing magnetic flux through a loop to the induced electromotive force along the loop. Many problems related to the motion of wire loops through a magnetic field are discussed. The differential version of Faraday's law is also discussed.

The fifth chapter completes the story of electromagnetism in the form of the four Maxwell equations. It is a summary of Gauss' law, Ampere's law with the displacement current correction term, the Faraday law and the no magnetic monopole condition. The correction to Ampere's law stems from the fact that it violates the principles of charge conservation for non-steady currents. The Maxwell equations in free space are then discussed and from them, we derive the wave equations for the electric and magnetic field. These show that electromagnetic waves propagate at the speed of light in vacuum. The experimental confirmation of this fact was obtained by Heinrich Hertz one hundred years after they were theoretically predicted by Maxwell. The complete solution to the Maxwell equations in terms of retarded potentials is also discussed here. To obtain these solutions, we need to introduce the electrostatic scalar and magnetic vector potential. A variety of applications of the Maxwell equations are also described in this chapter. These applications include proving Snell's laws of reflection and refraction for waves at a boundary separating two dielectrics, obtaining formulae for the reflection and transmission coefficients when an electromagnetic wave is incident upon a boundary, describing the propagation of electromagnetic waves inside rectangular and circular waveguides and, exploring voltage and current waves in a transmission line. A problem on the radiation due to an infinitesimal current element is also described. Two nonlinear problems in electromagnetic theory are mentioned at the end of the chapter. These problems arise in situations when the properties of the medium are field dependent. They've been mentioned because of their importance in ionospheric communication as well as in nolinear optics. The chapter also contains a separate section on relativistic transformation of the fields. These formulas involve describing the relationship between the fields in two reference frames moving with uniform velocity relative to each other.

The sixth chapter is an introduction to quantum mechanics. This chapter has been included because of the important theoretical and practical connections between electromagnetic field theory and quantum theory. The connection arises in problems dealing with the motion of an electron in a magnetic field, and quantum electrodynamics. The subject of quantum electrodynamics has not been touched upon here because of the technical complexity, but the reader of this chapter will be in a position to start understanding this subject involving quantization of the radiation field wherein the field is regarded as an ensemble of harmonic oscillator and also treats some tunnelling problems which are of importance in quantum electronics. The Schrodinger and Heisenberg pictures of quantum theory are discussed here and a section on the simulation of a quantum system is also included. This chapter has however omitted the discussion of the hydrogen spectrum which is available in all the standard references on quantum mechanics. The author felt that for understanding the basic principles of quantum theory, it is enough to understand the harmonic oscillator. Finally, a problem on the splitting of the energy levels of an atom placed in a uniform magnetic field is considered. This is known in the literature as the Zeeman splitting.

H. Parthasarathy

CONTENTS

1

CHAPTER

BASICS OF VECTOR ANALYSIS

Contents

1. Definition of three dimensional vectors.
2. The unit vectors of a cartesian system $\hat{x}$, $\hat{y}$, $\hat{z}$.
3. The scalar product of vectors and computation of the distance between two points specified by vectors.
4. The cylindrical co-ordinate system. Expression for the unit vectors in terms of cartesian co-ordinates, vector fields.
5. The spherical polar co-ordinate system.
6. Area and volume measures in the different systems.
7. Line integrals and flux of a vector field.
8. Gradient, divergence and curl operations and their physical meanings.
9. Gauss' and Stokes' theorems and their corollaries.
10. Computation of gradient, divergence and curl in the different co-ordinate systems.
11. Green's theorems.

INTRODUCTION

Vector analysis plays a fundamental role in describing several physical phenomena using quantitative, i.e. mathematical analysis. Its applications include mechanics of systems of particles and electromagnetism. A vector is a physical quantity that has both a magnitude and a direction in space. Of course, we can from a purely mathematical viewpoint, consider also vectors in n dimensional Euclidean and curvilinear space, but for our purposes, three dimensional vectors will suffice. Either we specify a vector in terms of its magnitude and direction or in terms of its three components with respect to a reference frame. The kind of co-ordinate system used to specify a vector is also important. Thus the same vector can be described in a variety of ways by introducing different kinds of co-ordinate systems. For our purposes, we shall use three kinds of systems, namely the Cartesian, the cylindrical and the spherical polar co-ordinate system. In mechanics, the kinds of vectors that we come across are displacement, velocity, acceleration, momentum and force. Angular momentum and torque are pseudo vectors in the strict sense of the word. A vector undergoes a definite linear transformation law under rotations and under reflections. Under a reflection, the sign of the vector under-goes a change. This holds for position and velocity and acceleration and force but not for angular momentum or torque. Hence the latter are pseudo vectors. Further under rotations, angular momentum and torque do not undergo the same transformation as do position and velocity, the reason being that angular momentum and torque are obtained as the cross product of two vectors. We shall assume herein that the reader is familiar with the familiar operations of scalar or dot product, the vector or cross product and the scalar and vector triple products built out of the former two. The equations of motion in mechanics are vector equations, i.e. we express the velocity vector as the derivative of the position vector, the acceleration vector as the derivative of the velocity vector, the acceleration as being proportional to the force with mass as the proportionality constant, or equivalently, rate of change of momentum vector to be the force vector and rate of change of angular momentum vector to be the torque vector. Each such vector equation is equivalent to three scalar equations for the three components of the vector. The specific choice of the reference frame and co-ordinate system employed is problem dependent and should be such that the symmetries of the problem are utilized to the maximum enabling an easy solution. In electromagnetic theory, we come across the notion of a vector field, i.e. a vector at each point in space and perhaps each vector at each point in space may vary with time. Such fields are the electric and magnetic fields. Strictly speaking, from the standpoint of relativity, the electric and magnetic fields are components of an antisymmetric tensor of rank four. The electromagnetic potentials, namely, the scalar electric potential and the magnetic vector potential, are components of a four vector, i.e. a quantity that transforms like time and space under the Lorentz transformation. However, we shall for our purposes regard the electric field as a vector and magnetic field also as a vector as each of these is specified by three components. At each point in space and time, we specify the electric and magnetic fields and hence these form vector fields. Scalar fields are scalar quantitites at each point in space and time like the electric potential. Strictly speaking, the electric potential is not a scalar field because it is not invariant under the Lorentz transformation. In this chapter, we shall introduce the different kinds of co-ordinate systems, explain how to calculate the length of a curve in each of these systems, how to calculate the integral of a vector field along a curve, how to calculate the flux of a vector field out of a surface and then introduce the important operations and theorems of vector calculus, namely, the calculus that deals with the partial derivatives of these vector fields, how using the vector operations the circulation of a vector field around a closed curve can be related to the flux of a derived vector field through the area enclosed by the curve and how the flux of a vector field through a closed surface bounding a volume can be related to the volume integral of a derived vector field over the volume bounded by the surface. These are the classic Gauss' and Stokes' theorems. We derive using the Gauss' theorems and the Green's theorems. The operations of vector calculus namely gradient, divergence and curl are very important in the study of electromagnetism because they enable us to express the empirical laws of Gauss, Faraday and Ampere in the language of partial differential equations. These equations are the Maxwell equations and their solution can be derived using the standard mathematical tool of the Green's function. This chapter will also introduce you to how the operations of vector calculus can be expressed in different co-ordinate systems. This is very important because each problem in electromagnetic theory will usually have a certain kind of symmetry like rectangular, cylindrical, spherical, azimuthal, etc. and by appropriate choice of the co-ordinate system, we can obtain the solution to the problem.

1. Definition of three-dimensional vectors

A vector in three-dimensional Euclidean space is specified by a magnitude and a direction. Depending upon the type of co-ordinate system chosen, we can specify a vector in different ways. Examples of vectors include the displacement of an object specified by stating the distance of the object from a given reference point and the direction in which the overall displacement has taken place relative to a set of axes erected at the reference point, the velocity of an object in space at a given time, specified by the magnitude of the velocity, that is the speed of the object and the direction in which the object is moving at that time instant, the electric field at a given point in space at a given time specified by specifying the strength of the field and its direction of polarization. A physical quantity that is specified by only a magnitude but no direction is called a scalar. Examples include pressure at a given point in the atmosphere and at a given time, temperature at a given point in space at a given time, intensity of sound coming from the vocal chords of a person, brightness at a given point in a room, weight of a person, height of a building, etc. In general, to specify a vector in three-dimensional Euclidean space one number is needed for specifying its magnitude and two numbers for specifying its direction, making an overall of three numbers required for completely specifying the vector.

2. The unit vectors of a Cartesian system

The Cartesian co-ordinate system is specified by choosing a point in space and calling it the origin of our reference frame. From this point, three mutually perpendicular lines are drawn and label-ed counter-clockwise as the x, y and z axes respectively. A unit distance along the x axis is called the vector $\hat{x}$, i.e. the unit vector in the x direction. A unit distance along the y and z axes are likewise label-ed as $\hat{y}$ and $\hat{z}$. Given any point P in space, we calculate its projections along the x, y, z axes and denote these projections by (x, y, z) respectively. Then the point P is expressed as $x\hat{x} + y\hat{y} + z\hat{z}$. This is also denoted by the vector OP, where O is the orgin. We say that the vector OP is obtained by moving a distance x along the $\hat{x}$ direction, a distance y along the $\hat{y}$ direction and a distance z along the $\hat{z}$ direction. When it is understood that the co-ordinate system being employed is Cartesian, we may say that $P = OP = (x, y, z)$. x, y, z are called the Cartesian components of the vector OP.

3. The scalar product of vectors and computation of the distance between two points specified by vectors

Given vectors $P = (x, y, z)$ and $Q = (x', y', z')$ relative to the same co-ordinate system, the distance between the two points in space corresponding to these two vectors is by Pythagoras' theorem

$$|P-Q| = ((x-x')^2 + (y-y')^2 + (z-z')^2)^{1/2}$$

In particular, the distance of the point P from the origin is

$$|OP| = |P| = |P-O| = (x^2 + y^2 + z^2)^{1/2}$$

Note that the coordinates of O are (0, 0, 0). Given two vectors $OP = (x, y, z)$ and $OQ = (x', y', z')$, we want to calculate the angle between them. Consider the triangle OPQ. From elementary triangle geometry, if O denotes the angle between OP and OQ, then

$$|OP|^2 + |OQ|^2 - 2|OP||OQ|\cdot\cos\theta = |PQ|^2 = |P-Q|^2$$

or

$$x^2 + y^2 + z^2 + x'^2 + y'^2 + z'^2 - 2\sqrt{x^2 + y^2 + z^2}\sqrt{x'^2 + y'^2 + z'^2} \cdot \cos\theta$$
$$= (x - x')^2 + (y - y')^2 + (z - z')^2$$

Simplification of this identity gives

$$\cos\theta = \frac{xx' + yy' + zz'}{\sqrt{x^2 + y^2 + z^2}\sqrt{x'^2 + y'^2 + z'^2}}$$

We denote the quantity $xx' + yy' + zz'$ by (P,Q) or $P.Q$ and call it the scalar or dot product between the vectors P and Q. This quantity is invariant under rotations of the co-ordinate system. We may thus write

$$P.Q = |P||Q|.\cos\theta$$

From this equation, we immediately arrive at the Cauchy-Schwarz inequality: $|P.Q| \le |P||Q|$. Suppose $P_1, \dots, P_n$ are vectors specified by the Cartesian components (x_i, y_i, z_i), $i = 1, 2, \dots, n$ respectively. Then, the Cartesian components of $\sum_{i=1}^{n} P_i are (\sum_{i=1}^{n} x_i, \sum_{i=1}^{n} y_i, \sum_{i=1}^{n} z_i)$. Equivalently,

$$P_i = x_i\hat{x} + y_i\hat{y} + z_i\hat{z}, i = 1, 2, \dots, n$$

so that,

$$\sum_{i=1}^{n} P_i = \left(\sum x_i\right)\hat{x} + \left(\sum y_i\right)\hat{y} + \left(\sum z_i\right)\hat{z}$$

We have in addition,

$$|\sum P_i|^2 = \sum |P_i|^2 + 2 \sum_{1 \le i < j \le n} P_i.P_j$$

Hence, if θ_{ij} is the angle between P_i and P_j, then

$$|\sum P_i|^2 = \sum |P_i|^2 + 2 \sum_{1 \le i < j \le n} |P_i||P_j|\cos\theta_{ij}$$

The formula for the sum of vectors follows the parallelogram law of addition. Equivalently, vectors may be added according to the diagram shown in Fig. 1.1.

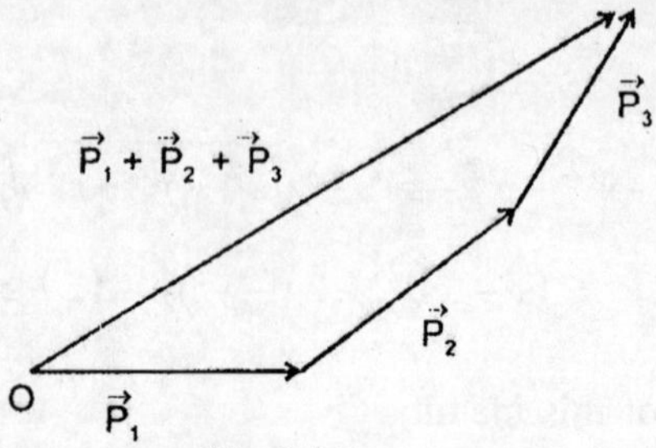

Fig. 1.1

4. The cylindrical co-ordinate system

Cylindrical co-ordinates (ρ, ϕ, z) of a point P in space defined by its Cartesian components (x, y, z) are given by

$$\rho = \sqrt{x^2 + y^2}, \phi = \tan^{-1}(y/x), z = z$$

or equivalently,

$$x = \rho.\cos\phi, \ y = \rho.\sin\phi, \ z = z$$

ρ is the distance of the point P from the $\hat{z}$ axis and ϕ is the angle between the OXZ plane and the plane formed by the vectors $\hat{z}$ and P, i.e. the OZP plane. This is shown in Fig. 1.2. The point P can be expressed as

$$P = \mathbf{r} = \rho.\cos\phi\,\hat{x} + \rho.\sin\phi\,\hat{y} + z\hat{z}$$

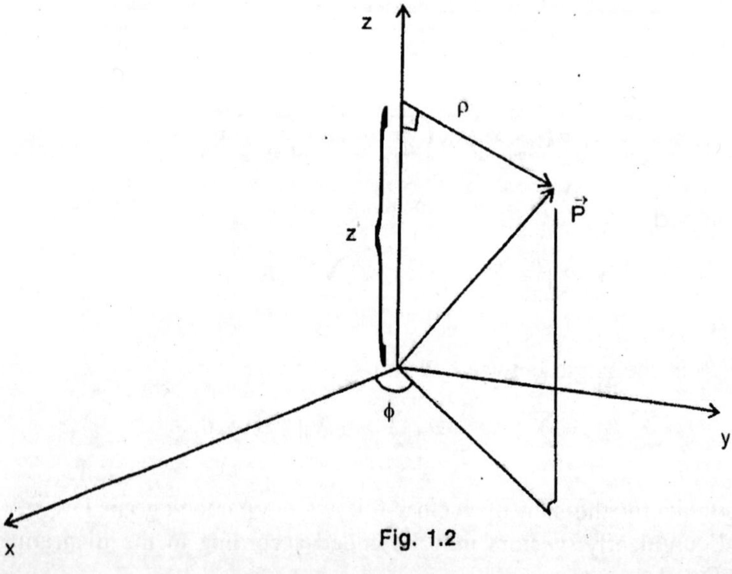

Fig. 1.2

The unit vectors along the ρ, ϕ, z directions are calculated as

$$\hat{\rho} = \frac{\dfrac{\partial \mathbf{r}}{\partial \rho}}{\left|\dfrac{\partial \mathbf{r}}{\partial \rho}\right|} = \hat{x}.\cos\phi + \hat{y}.\sin\phi$$

$$\hat{\phi} = \frac{\dfrac{\partial \mathbf{r}}{\partial \phi}}{\left|\dfrac{\partial \mathbf{r}}{\partial \phi}\right|} = -\hat{x}.\sin\phi + \hat{y}.\cos\phi$$

Note that while differentiating **r**, we take $\hat{x}, \hat{y}, \hat{z}$ as constant. This means that for any vector

$$\psi(\alpha,\beta,\gamma) = \psi_1(\alpha,\beta,\gamma)\hat{x} + \psi_2(\alpha,\beta,\gamma)\hat{y} + \psi_3(\alpha,\beta,\gamma)\hat{z}$$

where α, β, γ are arbitrary parameters, we have

$$\frac{\partial \psi}{\partial \alpha} = \hat{x}.\frac{\partial \psi_1}{\partial \alpha} + \hat{y}\frac{\partial \psi_2}{\partial \alpha} + \hat{z}\frac{\partial \psi_3}{\partial \alpha}$$

with analogous expressions holding for the partial derivatives with respect to β and γ. The notion of a constant co-ordinate surface plays a fundamental role *i* applications. For the Cartesian system, $x = x_0$ is the surface plane parallel to the *OYZ* plane and at a distance x_0 from it. $y = y_0$ and $z = z_0$ are likewise planes parallel to the *OXZ* and *OXY* planes at distances y_0 and z_0 respectively from these planes. The surface $\rho = \rho_0$ is a cylindrical surface with axis as the z axiz and of radius ρ_0. $\phi = \phi_0$ is a plane passing through the z axis and making an angle ϕ_0 with the *OXZ* plane. $z = z_0$ has been discussed above. The constant ρ and constant ϕ surfaces have been shown in Fig. 1.3 and Fig. 1.4 respectively. We can express $\hat{x}$ and $\hat{y}$ also in terms of $\hat{\rho}$ and $\hat{\phi}$ by inverting the expressions that give $\hat{\rho}$ and $\hat{\phi}$ in terms of $\hat{x}$ and $\hat{y}$:

$$\begin{pmatrix} \hat{\rho} \\ \hat{\phi} \end{pmatrix} = \begin{pmatrix} \cos\phi & \sin\phi \\ -\sin\phi & \cos\phi \end{pmatrix} \begin{pmatrix} \hat{x} \\ \hat{y} \end{pmatrix}$$

which on inversion give

$$\begin{pmatrix} \hat{x} \\ \hat{y} \end{pmatrix} = \begin{pmatrix} \cos\phi & -\sin\phi \\ \sin\phi & \cos\phi \end{pmatrix} \begin{pmatrix} \hat{\rho} \\ \hat{\phi} \end{pmatrix}$$

or

$$\hat{x} = \hat{\rho}.\cos\phi - \hat{\phi}.\sin\phi$$

$$\hat{y} = \hat{\rho}.\sin\phi + \hat{\phi}.\cos\phi$$

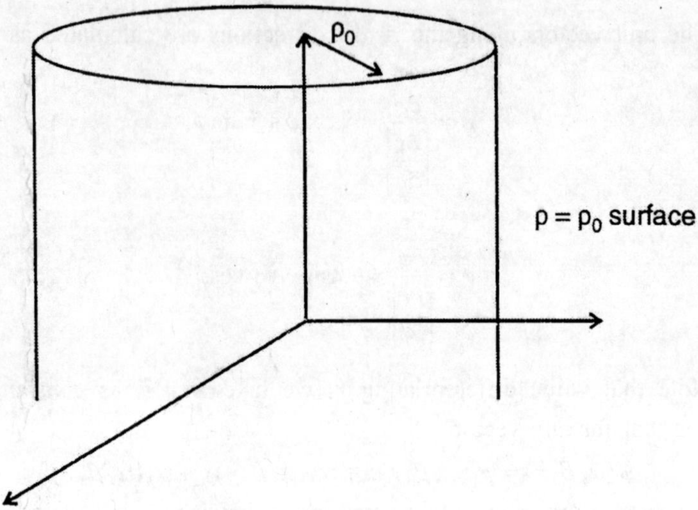

Fig. 1.3

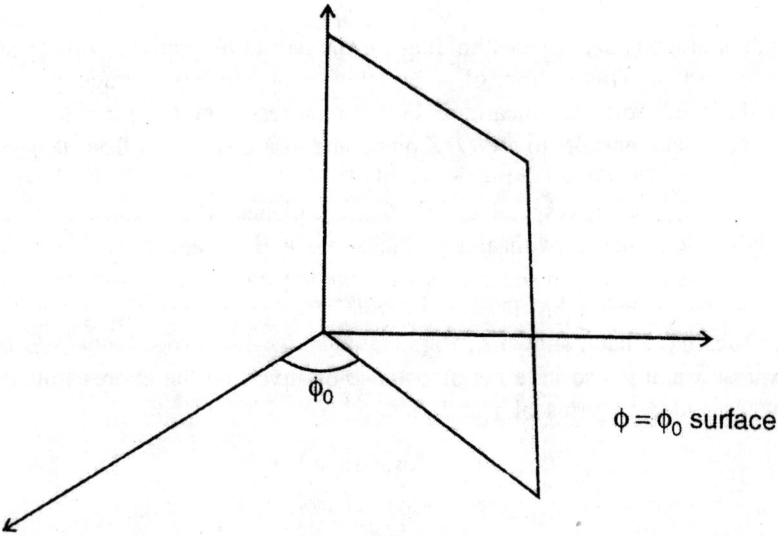

Fig. 1.4

A vector field is specified by defining a vector at each point in space. If **F** (x, y, z) is such a field, then in terms of Cartesian co-ordinates,

$$\mathbf{F}(x,y,z)=F_x(x,y,z)\hat{x}+F_y(x,y,z)\hat{y}+F_z(x,y,z)\hat{z}$$

Here the vector field is obtained by specifying the three Cartesian components at each point (x, y, z) in space. The vector field may also be specified using the cylindrical co-ordinate system

$$\mathbf{G}(\rho,\phi,z) = G_\rho(\rho,\phi,z)\hat{\rho} + G_\phi(\rho,\phi,z)\hat{\phi} + G_z(\rho,\phi,z)\hat{z}$$

The vector field **F** expressed above in terms of Cartesian co-ordinates can be expressed in cylindrical co-ordinates by expressing x, y, z in terms of (ρ, ϕ, z) and $\hat{x}, \hat{y}, \hat{z}$ in terms of $\rho, \phi, \hat{z}$. Thus,

$$\mathbf{F} = F_x(\rho.\cos\phi, \rho.\sin\phi, z)(\hat{\rho}.\cos\phi - \hat{\phi}.\sin\phi) + F_y(\rho.\cos\phi, \rho.\sin\phi, z)$$

$$(\hat{\rho}.\sin\phi + \hat{\phi}.\cos\phi)F_z(\rho.\cos\phi, \rho.\sin\phi, z)\hat{z}$$

Comparing coefficients of $\hat{\rho}$, $\hat{\phi}$ and $\hat{z}$, we get

$$G_\rho(\rho,\phi,z) = F_x(\rho.\cos\phi, \rho.\sin\phi, z).\cos\phi + F_y(\rho.\cos\phi, \rho.\sin\phi, z).\sin\phi$$

$$G_\phi(\rho,\phi,z) = F_y(\rho.\cos\phi, \rho.\sin\phi, z).\cos\phi - F_x(\rho.\cos\phi, \rho.\sin\phi, z).\sin\phi$$

$$G_z(\rho,\phi,z) = F_z(\rho.\cos\phi, \rho.\sin\phi, z)$$

Likewise, given G in terms of cylindrical co-ordinates, we can express it in terms of the Cartesian co-ordinate system:

$$G = G_\rho(\sqrt{x^2 + y^2}, \tan^{-1}y/x, z)(\hat{x}\cos\phi + \hat{y}\sin\phi)$$

$$+ G_\phi(\sqrt{x^2 + y^2}, \tan^{-1}y/x, z)(-\hat{x}.\sin\phi + \hat{y}.\cos\phi)$$

$$+ G_z(\sqrt{x^2 + y^2}, \tan^{-1}y/x, z)\hat{z}$$

$$= \mathbf{F}(x, y, z)$$

where

$$F_x(x,y,z) = \cos\phi.G_\rho - \sin\phi.G_\varphi$$

$$= \frac{x}{\sqrt{x^2 + y^2}} G_\rho(\sqrt{x^2 + y^2}, \tan^{-1}y/x, z) - \frac{y}{\sqrt{x^2 + y^2}} G_\phi(\sqrt{x^2 + y^2}, \tan^{-1}y/x, z)$$

$$F_y(x,y,z) = \sin\phi.G_\rho + \cos\phi.G_\phi$$

$$= \frac{y}{\sqrt{x^2+y^2}} G_\rho(\sqrt{x^2+y^2}, \tan^{-1} y/x, z) + \frac{x}{\sqrt{x^2+y^2}} G_\phi(\sqrt{x^2+y^2}, \tan^{-1} y/x, z)$$

$$F_z(x,y,z) = G_z(\sqrt{x^2+y^2}, \tan^{-1} y/x, z)$$

5. The spherical polar co-ordinate system

Here, we parametrize a point P in space by three co-ordinates (r, θ, ϕ), where r is the radial distance of P from the origin, θ the angle made by OP and the z axis. Let Q be the point of intersection of the perpendicular dropped from P onto the XY plane. Then ϕ is the angle between the planes OXZ and OPQ. The angle θ is called the elevation angle and the angle ϕ is the azimuth. For a fixed r, different values of θ correspond to different latitudes on the surface of a sphere of radius r, while different values of ϕ correspond to different longitudes. Figure 1.5 shows the spherical polar co-ordinates of the point in space. The length of the projection OQ of OP onto the XY plane is $OQ = OP.\sin\theta = r.\sin\theta$. Thus, the X and Y co-ordinates of P are $x = OQ.\cos\phi$, $y = OQ.\sin\phi$. The length of the perpendicular PQ from P onto the XY plane determines the z co-ordinate of P. It is given by $PQ = OP.\cos\theta = r.\cos\theta$. We thus obtain the following expressions for the (x, y, z) co-ordinates of P:

$$x = r.\sin\theta.\cos\phi, \quad y = r.\sin\theta.\sin\phi, \quad z = r.\cos\theta$$

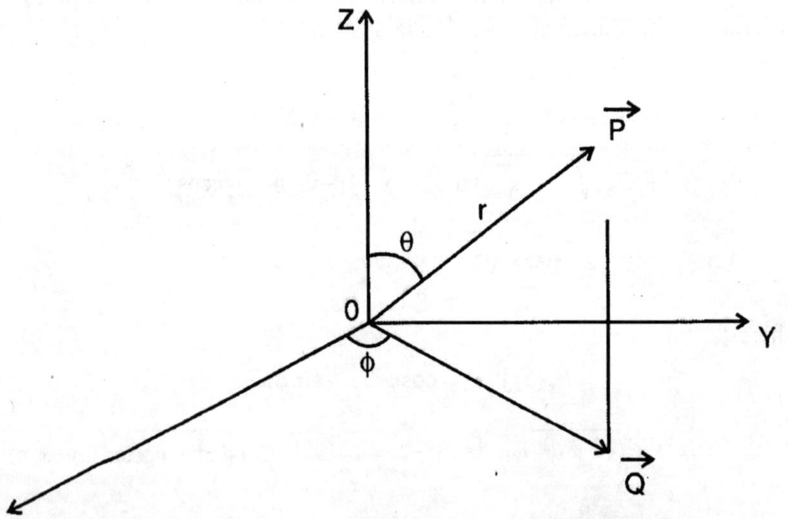

Fig. 1.5

The inverse relations determine (r, θ, ϕ), the polar co-ordinates of the point in terms of the Cartesian co-ordinates:

$$r = \sqrt{x^2 + y^2 + z^2},\ \theta = \cos^{-1} z/r, \phi = \tan^{-1} y/x$$

The point P can be expressed using vectorial notation as

$$\mathbf{r} = OP = P = r.\cos\phi.\sin\theta\,\hat{x} + r.\sin\phi.\sin\theta.\hat{y} + r.\cos\theta\,\hat{z}$$

Moving P in the direction r keeping θ and ϕ fixed gives the unit vector $\hat{r}$. Likewise moving P in the direction θ keeping r and ϕ fixed gives $\hat{\theta}$ and finally, moving P in the direction ϕ keeping r, θ fixed gives $\hat{\phi}$. Equivalently since the spherical polar system is orthogonal, $\hat{r}$, $\hat{\theta}$, $\hat{\phi}$ are respectively perpendicular to the surfaces r = const, θ = const, and ϕ = const that pass through the given point $\mathbf{r}_0$. The co-ordinate surface $r = r_0$ is a sphere of radius r_0 with centre at the origin. The co-ordinate surface $\theta = \theta_0$ is a cone with apex at the origin, $\hat{z}$ axis as the axis of the cone and with the cone angle being $2\theta_0$. Finally, the co-ordinate surface $\phi = \phi_0$ is a plane passing through the $\hat{z}$ axis and making an angle ϕ_0 with the xz plane. These surfaces are depicted in Fig. 1.6a-c. We compute the unit vectors $\hat{r}$, $\hat{\theta}$, $\hat{\phi}$:

$$\hat{r} = \frac{\dfrac{\partial \mathbf{r}}{\partial r}}{\left|\dfrac{\partial \mathbf{r}}{\partial r}\right|} = \hat{x}.\cos\phi.\sin\theta + \hat{y}.\sin\phi.\sin\theta + \hat{z}.\cos\theta$$

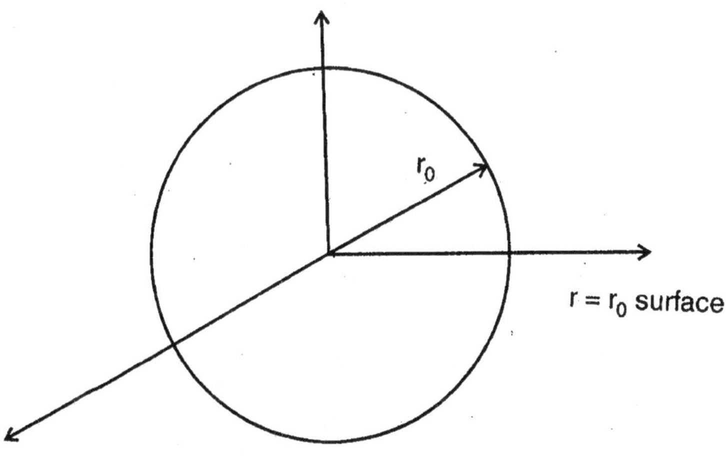

Fig. 1.6a

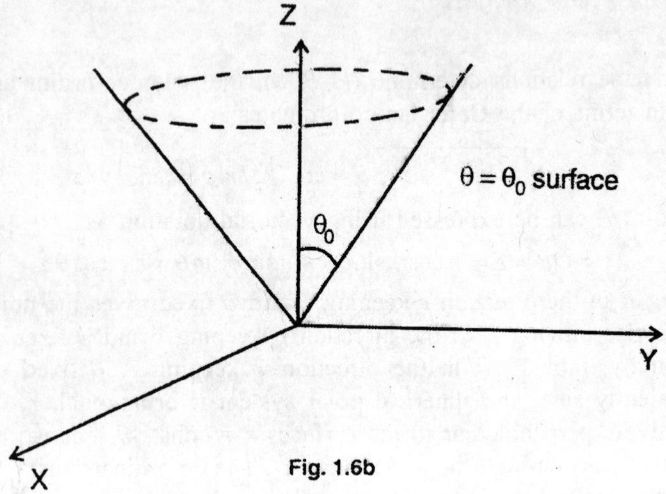

Fig. 1.6b

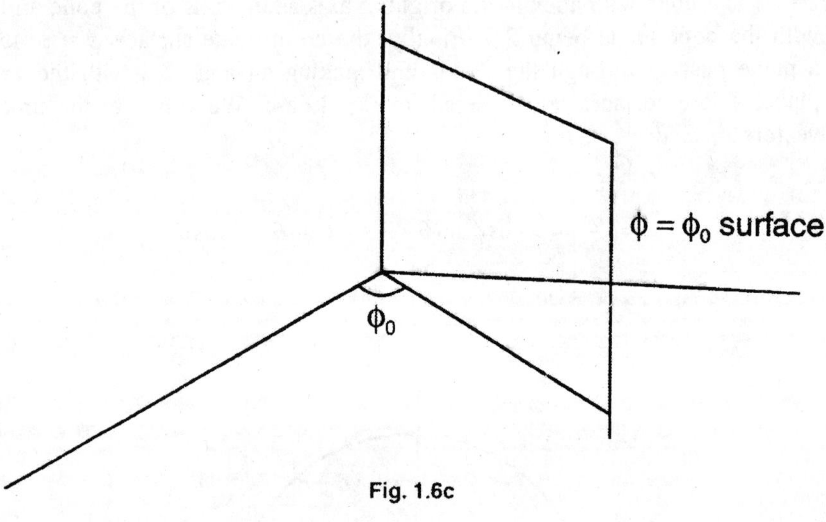

Fig. 1.6c

$$\hat{\theta} = \frac{\dfrac{\partial \mathbf{r}}{\partial \theta}}{\left|\dfrac{\partial \mathbf{r}}{\partial \theta}\right|} = \hat{x}.\cos\phi.\cos\theta + \hat{y}.\sin\phi.\cos\theta - \hat{z}.\sin\theta$$

$$\hat{\phi} = \frac{\dfrac{\partial \mathbf{r}}{\partial \phi}}{\left|\dfrac{\partial \mathbf{r}}{\partial \phi}\right|} = -\hat{x}.\sin\phi + \hat{y}.\cos\phi$$

We note that at each point (r, θ, ϕ), the unit vectors $(\hat{r}, \hat{\theta}, \hat{\phi})$ define a right handed orthogonal system of axes. This is also true for $(\hat{\rho}, \hat{\phi}, \hat{z})$ in the cylindrical system. The vectors $(\hat{r}, \hat{\theta}, \hat{\phi})$ and $(\hat{\rho}, \hat{\phi}, \hat{z})$ however, vary from point to point unlike the Cartesian vectors $(\hat{x}, \hat{y}, \hat{z})$. This follows from their expressions as coefficients of $\hat{x}, \hat{y}, \hat{z}$ depending on (θ, ϕ) and ϕ respectively. We can also express $(\hat{x}, \hat{y}, \hat{z})$ in terms of $(\hat{r}, \hat{\theta}, \hat{\phi})$:

$$\hat{x} = (\hat{x}.\hat{r})\hat{r} + (\hat{x}.\hat{\theta})\hat{\theta} + (\hat{x}.\hat{\phi})\hat{\phi}$$
$$\hat{y} = (\hat{y}.\hat{r})\hat{r} + (\hat{y}.\hat{\theta})\hat{\theta} + (\hat{y}.\hat{\phi})\hat{\phi}$$
$$\hat{z} = (\hat{z}.\hat{r})\hat{r} + (\hat{z}.\hat{\theta})\hat{\theta} + (\hat{z}.\hat{\phi})\hat{\phi}$$

Equivalently, by inverting the following equations noting that the coefficient matrix is orthogonal:

$$\begin{pmatrix} \hat{r} \\ \hat{\theta} \\ \hat{\phi} \end{pmatrix} = \begin{pmatrix} \cos\phi.\sin\theta & \sin\phi.\sin\theta & \cos\theta \\ \cos\phi.\cos\theta & \sin\phi.\cos\theta & -\sin\theta \\ -\sin\phi & \cos\phi & 0 \end{pmatrix} \begin{pmatrix} \hat{x} \\ \hat{y} \\ \hat{z} \end{pmatrix}$$

so that

$$\begin{pmatrix} \hat{x} \\ \hat{y} \\ \hat{z} \end{pmatrix} = \begin{pmatrix} \cos\phi.\sin\theta & \cos\phi.\cos\theta & -\sin\phi \\ \sin\phi.\sin\theta & \sin\phi.\cos\theta & \cos\phi \\ \cos\theta & -\sin\theta & 0 \end{pmatrix} \begin{pmatrix} \hat{r} \\ \hat{\theta} \\ \hat{\phi} \end{pmatrix}$$

The latter coefficient matrix is obtained by simply transposing the former. Given a vector field **F** (x, y, z) in terms of the Cartesian system, it is expressed in terms of the polar system by replacing $x, y, z, \hat{x}, \hat{y}, \hat{z}$ by their corresponding expressions in terms of $r, \theta, \phi, \hat{r}, \hat{\theta}, \hat{\phi}$:

$$\mathbf{F} = F_x(r\cos\phi\sin\theta, r.\sin\phi.\sin\theta, r.\cos\theta)(\hat{r}.\cos\phi.\sin\theta + \hat{\theta}.\cos\phi.\cos\theta - \hat{\phi}.\sin\phi)$$

$$+ F_y(r.\cos\phi\sin\theta, r\sin\phi\sin\theta, r.\cos\theta).(\hat{r}.\sin\phi.\sin\theta + \hat{\theta}.\sin\phi.\cos\theta + \hat{\phi}.\cos\phi$$

$$+ F_z(r.\cos\phi.\sin\theta, r.\sin\phi.\sin\theta, r.\cos\theta).(\hat{r}.\cos\theta - \hat{\theta}.\sin\theta$$

We leave the problem of expressing **F** as

$$G_r(r,\theta,\phi)\hat{r} + G_\theta(r,\theta,\phi)\hat{\theta} + G_\phi(r,\theta,\phi)\hat{\phi}$$

to the student.

6. Area and volume measures in the different systems

The area measure on the XY, YZ and ZX planes are respectively given by $dxdy$, $dydz$ and $dzdx$. Thus, for example, if we want to calculate the integral of a function $\phi(x, y)$ over the triangular region $0 \leq x \leq 1$, $0 \leq y \leq 1 - x$ of the XY plane shown in Fig. 1.7, we would compute it as

$$\int_0^1 dx \int_0^{1-x} \phi(x,y)dy$$

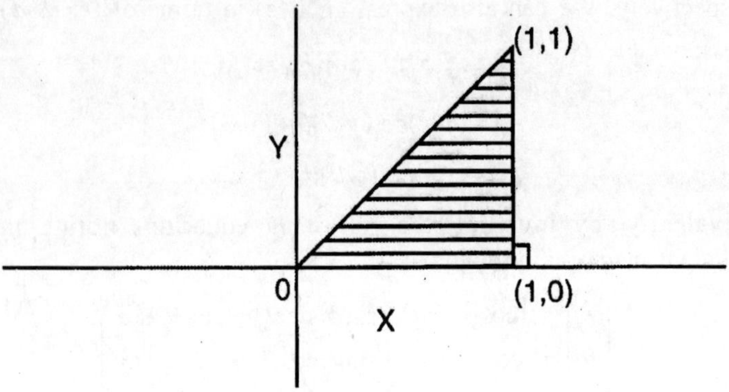

Fig. 1.7

Suppose we are given a surface in Euclidean space specified by two parameters u,v, i.e. a point on the surface is

$$\mathbf{r}(u,v) = x(u,v)\hat{x} + y(u,v)\hat{y} + z(u,v)\hat{z}$$

then the area measure on this surface is

$$dS(u,v) = \left|\frac{\partial \mathbf{r}}{\partial u} \times \frac{\partial \mathbf{r}}{\partial v}\right| du dv \ .$$

This is seen via the diagram of Fig. 1.8. Let $\mathbf{r}(u, v)$ be a point on the surface. Changing u to $u + du$ keeping v fixed takes us to the point

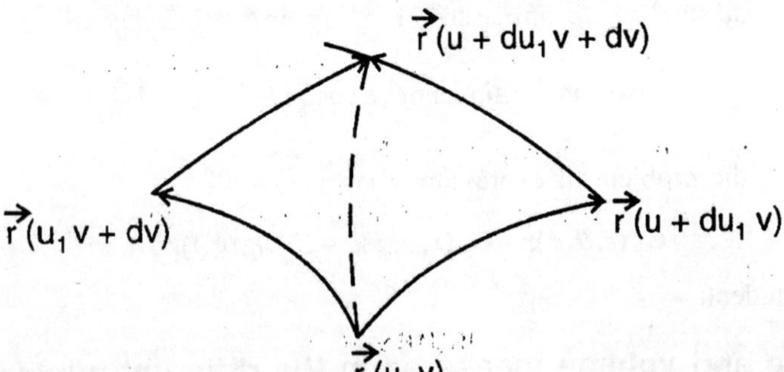

Fig. 1.8

$\mathbf{r}\,(u + du, v)$. On the other hand, changing v to $v + dv$ keeping u fixed takes us to the point $\mathbf{r}\,(u, v + dv)$. The parallelogram subtended by these two displacement vectors has the area $dS\,(u, v)$ given by the above formula. Given a scalar function $\phi\,(\mathbf{r})$, we would thus compute the surface integral $\int \phi dS$ of ϕ over the surface as

$$\int \phi\,(\mathbf{r}(u,v))dS(u,v)$$

with (u, v) covering the appropriate range so that the entire surface S is covered. Sometimes, the equation of the surface is specified as $z = f\,(x, y)$. Then, (x, y) can be regarded as the independent parameters and the surface equation is given in parametric form by

$$\mathbf{r}\,(x,y) = x\hat{x} + y\hat{y} + f(x,y)\hat{z}$$

The displacement vectors on the surface in the x and y directions are respectively given by

$$\frac{\partial \mathbf{r}}{\partial x} = \hat{x} + \frac{\partial f}{\partial x}\hat{z}$$

$$\frac{\partial \mathbf{r}}{\partial y} = \hat{y} + \frac{\partial f}{\partial y}\hat{z}$$

This gives

$$\frac{\partial \mathbf{r}}{\partial x} \times \frac{\partial \mathbf{r}}{\partial y} = \hat{z} - \frac{\partial f}{\partial y}\hat{y} - \frac{\partial f}{\partial x}\hat{x}$$

so that the area element is given by

$$\left|\frac{\partial \mathbf{r}}{\partial x} \times \frac{\partial \mathbf{r}}{\partial y}\right| dxdy = \left(1 + \left(\frac{\partial f}{\partial x}\right)^2 + \left(\frac{\partial f}{\partial y}\right)^2\right)^{1/2} dxdy$$

The volume measure in Cartesian co-ordinates is given by $dxdydz$. To obtain the volume measure in cylindrical co-ordinates, we use the Jacobian rule:

$$x = \rho.\cos\phi,\; y = \rho.\sin\phi,\; z = z$$

so that

$$\frac{\partial(x,y,z)}{\partial(\rho,\phi,z)} = \det\begin{pmatrix} \cos\phi & \sin\phi & 0 \\ -\rho\sin\phi & \rho\cos\phi & 0 \\ 0 & 0 & 1 \end{pmatrix} = \rho$$

The volume element in the cylindrical system is therefore

$$dxdydz = \frac{\partial(x, y, z)}{\partial(\rho, \phi, z)} d\rho\, .d\phi\, .dz = \rho\, .d\rho\, .d\phi\, .dz$$

This can also be seen via the diagram of Fig. 1.9. The volume of the box shown in the figure is ρ.*d*ρ.*d*φ.*dz*. Likewise, the volume measure in spherical polar co-ordinates is

$$dxdydz = \frac{\partial(x, y, z)}{\partial(r, \theta, \phi)} dr\, .d\theta\, .d\phi$$

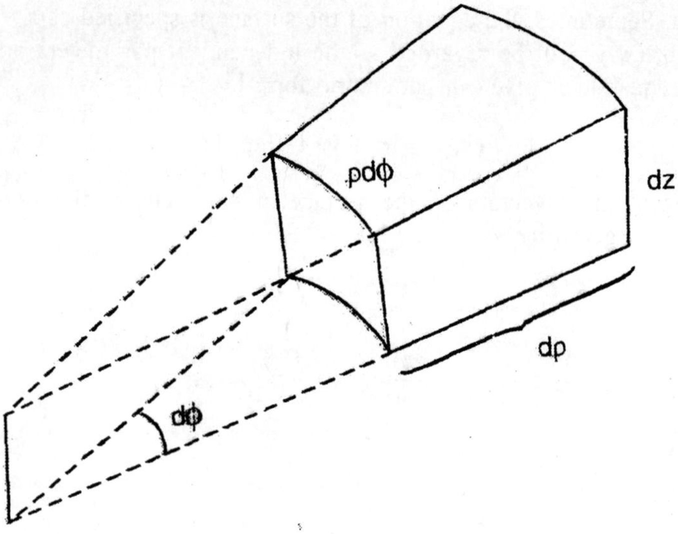

Fig. 1.9

with

$$x = r.\cos\phi.\sin\theta,\ y = r.\sin\phi.\sin\theta,\ z = r.\cos\theta$$

the Jacobian determinant is given by

$$\frac{\partial(x, y, z)}{\partial(r, \theta, \phi)}$$

$$= \det\begin{pmatrix} \cos\phi.\sin\theta & \sin\phi.\sin\theta & \cos\theta \\ r.\cos\phi.\cos\theta & r.\sin\phi.\cos\theta & -r.\sin\theta \\ -r.\sin\phi.\sin\theta & r.\cos\phi.\sin\theta & 0 \end{pmatrix}$$

$$= r^2.\sin\theta$$

This gives the volume element as $dV = r^2.\sin\theta.drd\theta.d\phi$. Finally, it is easy to see that the surface element on the walls of a cylinder of radius ρ_0 with axis coinciding with the $\hat{z}$ axis is $\rho_0 d\phi.dz$ and the surface element on the surface of a sphere of radius r_0 with origin as the centre is $r_0^2.\sin\theta.d\theta.d\phi$. These are depicted in Figs. 1.10 and 1.11.

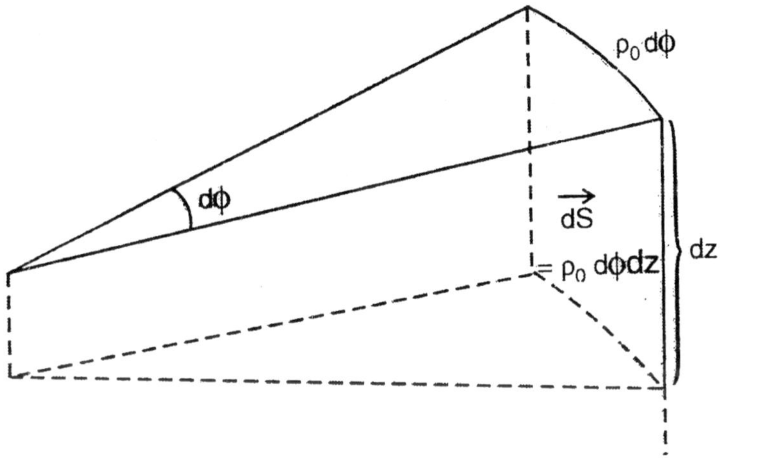

Fig. 1.10 Surface element on a Cylinder of radius r_0

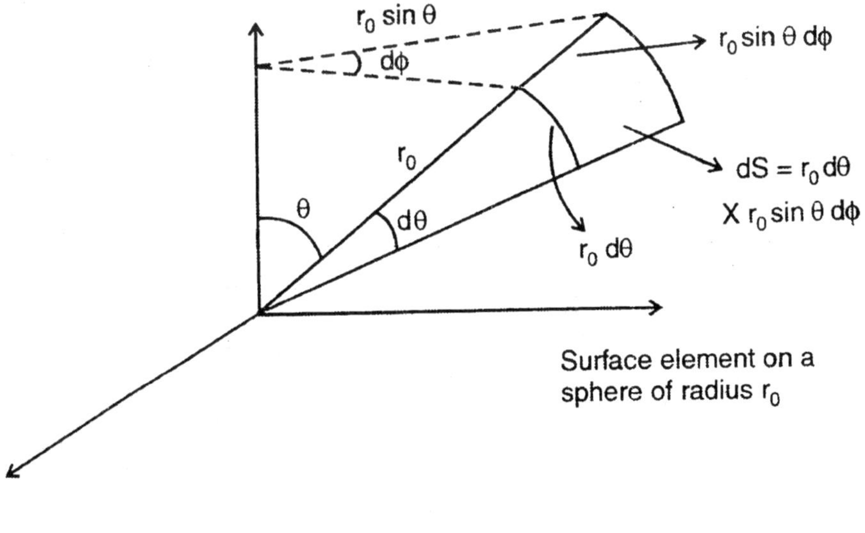

Surface element on a sphere of radius r_0

Fig. 1.11

7. Line integral and flux of a vector field

Given a vector field **F** (x, y, z) in terms of the Cartesian system, its line integral along a curve specified parametrically by $t \rightarrow \mathbf{r}(t) = x(t)\,\hat{x} + y(t)\,\hat{y} + z(t)\,\hat{z}$ from the point $\mathbf{r}_1 = \mathbf{r}(t_1)$ to the point $\mathbf{r}_2 = \mathbf{r}(t_2)$ is given by

$$\int_1^2 \mathbf{F.dr} = \int_{t_1}^{t_2} \mathbf{F(r}(t)).\frac{d\mathbf{r}(t)}{dt}\,dt$$

$$= \int_{t_1}^{t_2} \left(F_x(x(t), y(t), z(t))\frac{dx(t)}{dt} + F_y(x(t), y(t), z(t))\frac{dy(t)}{dt} + F_z(x(t), y(t), z(t))\frac{dz(t)}{dt}\right)dt$$

This follows from the fact that the line element in the Cartesian system is

$$d\mathbf{r} = \hat{x}.dx + \hat{y}.dy + \hat{z}.dz$$

In particular, the length along the curve can be calculated by integrating

$$\sqrt{dx^2 + dy^2 + dz^2} = \sqrt{\left(\frac{dx}{dt}\right)^2 + \left(\frac{dy}{dt}\right)^2 + \left(\frac{dz}{dt}\right)^2}\,dt$$

The position vector in cylindrical co-ordinates is $\mathbf{r} = \rho.\hat{\rho} + z.\hat{z}$. See Fig. 1.9. The line element is therefore,

$$d\mathbf{r} = d\rho.\hat{\rho} + \rho.d\hat{\rho} + dz.\hat{z}$$

since $d\hat{z} = 0$. Now,

$$d\hat{\rho} = d(\hat{x}.cos\phi + \hat{y}.sin\phi) = (-\hat{x}.sin\phi + \hat{y}.cos\phi) = \hat{\phi}.d\phi$$

Thus,

$$d\mathbf{r} = d\rho.\hat{\rho} + \rho.\hat{\phi}.d\phi + \hat{z}.dz$$

Given a vector field $\mathbf{F}(\rho, \phi, z)$ in cylindrical co-ordinates, its line integral along a curve specified parametrically as

$$\mathbf{r}(t) = \rho(t)\hat{\rho} + z(t).\hat{z}$$

between the points $\mathbf{r}(t_1)$ and $\mathbf{r}(t_2)$ is

$$\int_1^2 \mathbf{F.dr} = \int_{t_1}^{t_2} \left(F_\rho(\rho(t), \phi(t), z(t))\frac{d\rho(t)}{dt} + F_\varphi(\rho(t), \phi(t), z(t))\rho.\frac{d\phi(t)}{dt} + F_z(\rho(t), \phi(t), z(t))\frac{dz(t)}{dt}\right)dt$$

or equivalently,

$$\mathbf{F.dr} = F_\rho\, d\rho + F_\phi\, \rho.d\phi + F_z dz$$

Finally, the position vector in spherical polar co-ordinates is given by $\mathbf{r} = r\hat{r}$, where

$$\hat{r} = \hat{x}.\cos\phi.\sin\theta + \hat{y}.\sin\phi.\sin\theta + \hat{z}.\cos\theta$$

Thus the line element is

$$d\mathbf{r} = dr.\hat{r} + r.d\hat{r}$$

where

$$d\hat{r} = \frac{\partial\hat{r}}{\partial\theta}d\theta + \frac{\partial\hat{r}}{\partial\phi}d\phi$$

$$= (\hat{x}.\cos\phi.\cos\theta + \hat{y}.\sin\phi.\cos\theta - \hat{z}.\sin\theta)d\theta$$

$$+(-\hat{x}.\sin\phi.\sin\theta + \hat{y}.\cos\phi.\sin\theta)d\phi = \hat{\theta}.d\theta + \sin\theta.\hat{\phi}.d\phi$$

This gives

$$d\mathbf{r} = \hat{r}.dr + r\hat{\theta}.d\theta + r.\sin\theta.d\phi$$

The line integral of a vector field is obtained by setting

$$\mathbf{F.dr} = F_r.dr + F_\theta r d\theta + F_\phi r\sin\theta.d\phi$$

Given a surface S with unit normal $\mathbf{n}(\mathrm{r})$ at $\mathbf{r} \in S$, the flux of a vector field $\mathbf{F}(\mathbf{r})$ through this surface is defined as $\int \mathbf{F}.\mathbf{n}dS$. For the surface specified parametrically as $\mathbf{r}(u,v)$, the flux may be computed as

$$\int \mathbf{F}(\mathbf{r}(u,v)).\frac{\partial\mathbf{r}(u,v)}{\partial u} \times \frac{\partial\mathbf{r}(u,v)}{\partial v}dudv$$

which follows from the fact that

$$\mathbf{n} = \frac{\frac{\partial\mathbf{r}(u,v)}{\partial u} \times \frac{\partial\mathbf{r}(u,v)}{\partial v}}{\left|\frac{\partial\mathbf{r}(u,v)}{\partial u} \times \frac{\partial\mathbf{r}(u,v)}{\partial v}\right|}$$

is the unit normal at $\mathbf{r}(u,v)$ and $dS = dS = \left|\frac{\partial\mathbf{r}(u,v)}{\partial u} \times \frac{\partial\mathbf{r}(u,v)}{\partial v}\right|dudv$ is the surface element. In case the vector field is the velocity field of a fluid, its flux through a surface gives the volume of fluid flowing out through the surface per unit time.

8. Gradient, Divergence and Curl operations

Maxwell's equations are most conveniently expressed using these differential operators. Their significance will become clear in the next section when we discuss the Gauss' and Stokes' theorems. The gradient of a scalar function ϕ (x, y, z) is defined by

$$\nabla\phi = \hat{x}.\frac{\partial\phi}{\partial x} + \hat{y}\frac{\partial\phi}{\partial y} + \hat{x}\frac{\partial\phi}{\partial z}$$

$\nabla\phi$ is clearly a vector field. The gradient operator is defined to be

$$\nabla = \hat{x}.\frac{\partial}{\partial x} + \hat{y}.\frac{\partial}{\partial y} + \hat{z}.\frac{\partial}{\partial z}$$

If the surface euqation is $\phi\,(x,y,z) = 0$, then $\nabla\phi\,(x, y, z)$ determines the direction of the normal to the surface at the point (x, y, z) on it. This follows from the fact that if (x, y, z) and $(x + dx, y + dy, z + dz)$ are infinitely close points on the surface, then $\phi\,(x, y, z) = \phi\,(x + dx, y + dy, z + dz) = 0$. Taking the difference of these two equations, we get

$$\frac{\partial\phi}{\partial x}dx + \frac{\partial\phi}{\partial y}dy + \frac{\partial\phi}{\partial z}dz = 0$$

The vector $\hat{x}.dx + \hat{y}.dy + \hat{z}.dz$ is tangential to the surface at (x, y, z) and its direction can be chosen arbitrarily subject to this constraint. Since the scalar product of $\nabla\phi$ with this vector is zero, it follows that $\nabla\phi$ is orthogonal to the tangent plane to the surface at (x, y, z) and is hence directed along the normal to the surface at (x, y, z). The divergence of a vector field **F** is defined as $\nabla.\mathbf{F}$, where ∇ is regarded as a vector differential operator. In terms of components,

$$\nabla.\mathbf{F} = \left(\hat{x}\frac{\partial}{\partial x} + \hat{y}\frac{\partial}{\partial y} + \hat{z}\frac{\partial}{\partial z}\right).(\hat{x}.F_x + \hat{y}.F_y + \hat{z}.F_z)$$

$$= \frac{\partial F_x}{\partial x} + \frac{\partial F_y}{\partial y} + \frac{\partial F_z}{\partial z}$$

The curl of a vector field is $\nabla \times \mathbf{F}$, where once again, ∇ is a vector differential operator. Expressed in terms of components,

$$\nabla\times\mathbf{F} = \left(\frac{\hat{\partial}}{\partial x} + \hat{y}\frac{\partial}{\partial y} + \hat{z}\frac{\partial}{\partial z}\right)\times(\hat{x}.F_x + \hat{y}.F_y + \hat{z}.F_z)$$

$$= (\hat{x}\times y)\left(\frac{\partial F_y}{\partial x} - \frac{\partial F_x}{\partial y}\right) + (\hat{z}\times\hat{x})\left(\frac{\partial F_x}{\partial z} - \frac{\partial F_z}{\partial x}\right) + (\hat{y}\times\hat{z})\left(\frac{\partial F_z}{\partial y} - \frac{\partial F_y}{\partial z}\right)$$

$$= \left(\frac{\partial F_z}{\partial y} - \frac{\partial F_y}{\partial z}\right)\hat{x} + \left(\frac{\partial F_x}{\partial z} - \frac{\partial F_z}{\partial x}\right)\hat{y} + \left(\frac{\partial F_y}{\partial x} - \frac{\partial F_x}{\partial y}\right)\hat{z}$$

A convenient notation for this is using determinants:

$$\nabla \times \mathbf{F} = \det \begin{pmatrix} \hat{x} & \hat{y} & \hat{z} \\ \frac{\partial}{\partial x} & \frac{\partial}{\partial y} & \frac{\partial}{\partial z} \\ F_x & F_y & F_z \end{pmatrix}$$

We shall now compute the differential operators given above with respect to an arbitrary orthogonal co-ordinate system and then specialize to cylindrical and spherical polar co-ordinates. Suppose we introduce co-ordinates (q_1, q_2, q_3) to parametrize the points in space. Thus, a given point is defined by the vector

$$\mathbf{r}(q_1, q_2, q_3) = \sum_i x_i(q_1, q_2, q_3)\hat{e}_i$$

where for simplicity of notation, $x_1 = x$, $x_2 = y$, $x_3 = z$ are the Cartesian components and $\hat{e}_1$, $\hat{e}_2$, $\hat{e}_3$ are the unit vectors along the $\hat{x}$, $\hat{y}$, and $\hat{z}$ axes. The unit vectors in the directions q_1, q_2, q_3 are respectively

$$u_i = \frac{\frac{\partial \mathbf{r}}{\partial q_i}}{\left|\frac{\partial \mathbf{r}}{\partial q_i}\right|}, i = 1,2,3$$

We assume that the co-ordinate system introduced is orthogonal, which means that the unit vectors $\{u_1, u_2, u_3\}$ at each point in space are mutually orthogonal and form a right handed system at each point. Special examples of such systems are the cylindrical and spherical polar system. Owing to the orthogonality of the system, we first of all note that

$$u_i = \frac{\nabla q_i}{|\nabla q_i|}, i = 1,2,3$$

i.e., u_i is normal to the q_i = const. surface. To see this, we use the fact that the normal to q_1 = const. surface is $\frac{\partial \mathbf{r}}{\partial q_2} \times \frac{\partial \mathbf{r}}{\partial q_3}$, since both $\frac{\partial \mathbf{r}}{\partial q_2}$ and $\frac{\partial \mathbf{r}}{\partial q_3}$ are tangential to q_1 = const. surface at that point. Thus,

$$\frac{\nabla q_1}{|\nabla q_1|} = u_2 \times u_3 = u_1$$

in view of the orthogonality of $\{u_1, u_2, u_3\}$.

9. Gauss' and Stokes' Theorems

A. Gauss' Theorem: Consider a vector field $\mathbf{F}(\mathbf{r})$ and a closed surface S bounding a volume V. Let $\mathbf{n}(\mathbf{r})$ denote the unit normal to the surface at the point r on it. The flux of the vector field through the surface is given by

$$\Phi_F = \int \mathbf{F}.\mathbf{n}dS = \int \mathbf{F}(\mathbf{r}).\mathbf{n}(\mathbf{r})dS(\mathbf{r})$$

Gauss' theorem asserts that this flux also equals $\int_V \text{div}\mathbf{F}dV$, i.e. the volume integral of the divergence of the vector field. To see this, we divide the volume V into little cuboids. Take one such cuboid with corners at

$$(x,y,z),\ (x+dx,y,z),\ (x,y+dy,z),\ (x,y,z+dz),$$
$$(x+dx,y+dy,z),\ (x+dx,y,z+dz),\ (x,y+dy,z+dz),\ (x+dx,y+dy,z+dz)$$

The flux of the vector field through the two faces parallel to the yz plane, i.e. the top and bottom surfaces is approximately $F_z(x, y, z + dz)\, dxdy -$ $F_z\,(x, y, z)dxdy = \frac{\partial F_z}{\partial z} dxdydz$. Likewise, the flux of the field through the faces parallel to the xz planes is approximately $F_y(x, y + dy, z)dxdz -$ $Fy\,(x, y, z)\ dxdz = \frac{\partial F_y}{\partial y} dxdydz$. Finally, the flux through the surfaces parallel to the yz plane is approximately $F_x(x + dx, y, z)\ dydz - F_x(x, y, z)$ $dydz = \frac{\partial F_x}{\partial x} dxdydz$. Adding these three components, we get for the approximate flux through the infinitesimal cuboid, $\text{div}\mathbf{F}dV$. Now, the flux of the vector field through S, the surface bounding V equals the sum of the fluxes through the surfaces of the little cuboids inside V, the reason being that the flux contributions through the faces common to two cuboids cancels out, since the unit normals used in the computation of these contributions for the two volumes are equal and opposite in directions. Thus, while adding all the fluxes through the little cuboids, the contributions from the faces that fall inside V cancel out leaving only the contribution from the faces falling on the boundary of V, i.e. on S. This proves the divergence theorem:

$$\int_S \mathbf{F}.\mathbf{n}dS = \int_V \text{div}\mathbf{F}dV$$

We can furnish an alternate proof of the divergence theorem of Gauss. We take a closed surface S assuming it to cut the XY, YZ and ZX planes. We look at the volume integral

$$\int \text{div}\mathbf{F}dV = \int_V \left(\frac{\partial F_y}{\partial x} + \frac{\partial F_y}{\partial y} + \frac{\partial F_z}{\partial z} \right) dxdydz$$

For a fixed (y, z), x will vary from x_1 to x_2, where x_1, x_2 are respectively the minimum and maximum of the x co-ordinates of the points on the surface S intersected by the line parallel to the x axis and passing through the points (x', y, z), with x' varying and y, z fixed. See Fig. 1.12 Thus,

$$\int \frac{\partial F_x}{\partial x} dxdydz = \int (F_x(x_2, y, z) - F_x(x_1, y, z))dydz$$

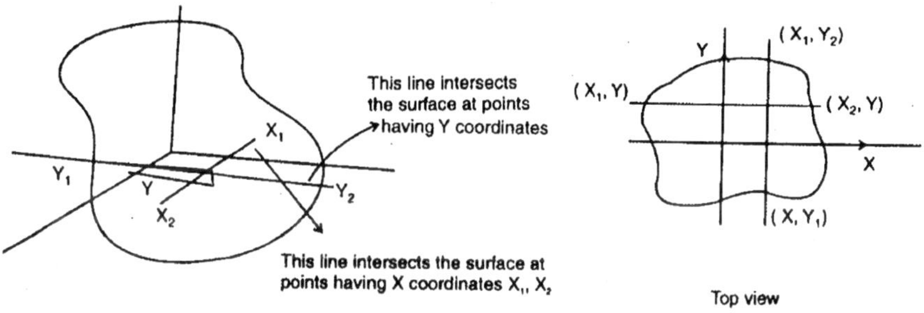

Fig. 1. 12

The latter integral is taken over the region of yz plane falling inside V. Note that x_1, x_2 are functions of y, z. These determine the equation of the surface: $x_1 = x_1\ (y, z)$, $x_2 = x_2\ (y, z)$. Likewise,

$$\int \frac{\partial F_y}{\partial y} dxdydz = \int (F_y(x, y_2, z) - F_y(x, y_1, z))dzdx$$

and

$$\int \frac{\partial F_z}{\partial z} dxdydz = \int (F_z(x, y, z_2) - F_z(x, y, z_1))dxdy$$

Summing up these contributions, and noting that the surface element with the unit normal is given by $dS\hat{n} = (dydz,\ dzdx,\ dxdy)$ we get the desired result. Note that while summing $\mathbf{F.n}dS = F_x dydz + Fydzdx + F_z dxdy$ we have three terms. The sum of the first term $F_x dydz$ is taken over the two parts of the surface into which S is divided by the yz plane, namely the part falling on the side $x > 0$ and the part falling on $x < 0$. For the first part, we sum $F_x(x_2,y,z)dydz$ and for the second part, $- F_x(x_1, y, z)dydz$, the minus sign corresponding to the fact that the x component of the unit normal on to the surface falling on the $x < 0$ side is negative. Likewise for the other components.

B. Stokes' Theorem: Consider a vector field **F** and a closed curve Γ that bounds a surface *S*. We can compute the line integral of the vector field **F** around this closed curve as $\int_\Gamma \mathbf{F}.d\mathbf{r}$. Stokes' theorem asserts that this line integral coincides with the surface integral $\int_S \nabla \times \mathbf{F}.\mathbf{n}dS$, where **n** is the unit normal to *S* such that looking in the direction opposite to **n**, the sense of Γ appears counterclockwise. To prove this, we make a simplifying assumption that the curve is planar, say falling in the *xy* plane so that $\mathbf{n} = \hat{z}$. We divide the region *S* into little pixels (rectangles) and consider the line integral around one such rectangle having vertices at say (x, y), $(x + dx, y)$, $(x, y + dy)$, $(x + dx, y + dy)$. The line integral of **F** around this rectangle is approximately

$$F_x(x,y,0)dx - F_x(x,y+dy,0)dx - F_y(x,y,0)dy + F_y(x+dx,y,0)dy$$

$$= \left(\frac{\partial F_y}{\partial x} - \frac{\partial F_x}{\partial y}\right)dxdy$$

If we sum up this little line integrals around the tiny rectangles, we get the line integral of **F** around Γ in the limit, the reason being that the contributions to the line integral around any little edge common to two adjacent rectangles cancel out, because for the two rectangles, they are taken in the opposite senses. Finally, after these cancellations, we are left with the contributions coming only from the edges of the boundary, that is the line integral around Γ. This shows that

$$\int_\Gamma \mathbf{F.dr} = \int_S \left(\frac{\partial F_y}{\partial x} - \frac{\partial F_x}{\partial y}\right)dxdy$$

which completes the proof of the theorem for planar curves. For nonplanar curves, it is an elementary generalization of this idea but we leave the details to the student. An alternate proof can also be given. Assume the loop to fall in the *xy* plane and the origin to fall inside the loop. We consider the line integral

$$\int_S \left(\frac{\partial F_y}{\partial x} - \frac{\partial F_x}{\partial y}\right)dxdy$$

Consider the first term. We draw a line parallel to the *x* axis passing through the points $(x', y, 0)$ with *y* fixed and x' varying from x_1 to x_2. The line intersects the loop at the points (x_1, y) and (x_2, y), with x_1 being the minimum abscissa falling in the region $x < 0$ and x_2 the maximum abscissa falling in the region $x > 0$. Thus,

$$\int_S \frac{\partial F_y}{\partial x} dxdy = \int (F_y(x_2, y) - F_y(x_1, y))dy$$

Here, y varies from y_{min} to y_{max} which are respectively the minimum and maximum values of the ordinates of the points on the curve. Similarly,

$$\int_S \frac{\partial F_x}{\partial y} dxdy = \int_{x_{min}}^{x_{max}} (F_x(x, y_2) - F_x(x, y_1))dx$$

Thus, the surface integral can be expressed as

$$\int (F_y(x_2, y) - F_y(x_1, y))dy + \int (F_x(x, y_1) - F_x(x, y_2))dx$$

This is precisely the line integral of **F** around the loop as Fig. 1.13 shows. Note that the signs of the various terms appearing here are appropriate: In the region $x > 0$, y increases along the loop, in the region $x < 0$, y decreases. In the region $y > 0$, x decreases and in the region $y < 0$ x increases. Note that x_1, x_2 are functions of y and y_1, y_2 are likewise functions of x.

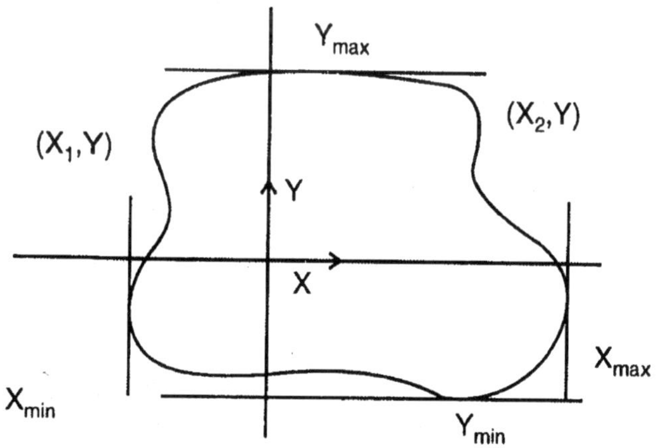

Fig. 1.13

C. Some simple Corollaries of the Gauss' and Stokes' theorems:
1. We shall evaluate the integral $\int_S \phi \mathbf{n} dS$ where ϕ is a scalar field and S a closed surface bounding a volume V. Take the dot product with a constant vector **a**. Then, application of Gauss' theorem gives

$$\mathbf{a}.\int_S \phi\mathbf{n}dS = \int \phi\mathbf{a}.\mathbf{n}dS = \int_V \nabla.(\phi\mathbf{a})\,dV = \int_V \nabla\phi.\mathbf{a}dV$$

From the arbitrariness of the vector **a**, it follows that

$$\int_S \phi\;\mathbf{n}dS = \int_V \nabla\phi\;dV$$

2. We take a vector field **F** and a closed surface bounding a volume V. We want to evaluate $\int_S \mathbf{n}\times\mathbf{F}dS$. Taking the dot product with a constant vector **a** gives

$$\int_S \mathbf{a}.\mathbf{n}\times\mathbf{F}dS = \int_S \mathbf{n}.\mathbf{F}\times\mathbf{a}dS = \int \nabla.(\mathbf{F}\times\mathbf{a})dV$$

Now from a vector identity and the constancy of **a**, it follows that

$$\nabla.(\mathbf{F}\times\mathbf{a}) = \nabla\times\mathbf{F}.\mathbf{a}$$

Thus, the above equals

$$\int \nabla\times\mathbf{F}.\mathbf{a}dV$$

and hence we derive the identity

$$\int_S \mathbf{n}\times\mathbf{F}ds = \int_V \nabla\times\mathbf{F}dV$$

3. We take a scalar field ϕ and a closed contour Γ bounding a surface S and we want to evaluate $\int_\Gamma \phi d\mathbf{r}$. Taking the dot product with a constant vector **a** gives

$$\int_\Gamma \phi\mathbf{a}.d\mathbf{r} = \int_S \nabla\times(\phi\mathbf{a}).\mathbf{n}dS = \int_S \nabla\phi\times\mathbf{a}.\mathbf{n}dS$$

$$= \mathbf{a}.\int \mathbf{n}\times\nabla\phi dS$$

This leads us to the identity

$$\int_\Gamma \phi d\mathbf{r} = \int_S \mathbf{n}\times\nabla\phi dS$$

4. We take a vector field **F** and a closed contour Γ bounding a surface S and we want to evaluate $\int_\Gamma \mathbf{F}\times d\mathbf{r}$. Taking the scalar product with a constant vector **a** gives

$$\int_\Gamma \mathbf{a}.\mathbf{F}\times d\mathbf{r} = \int_\Gamma \mathbf{a}\times\mathbf{F}.d\mathbf{r} = \int_S \nabla\times(\mathbf{a}\times\mathbf{F}).\mathbf{n}dS$$

Again using a vector identity and the constancy of **a**, we get

$$\nabla\times(\mathbf{a}\times\mathbf{F}) = \mathbf{a}\nabla.\mathbf{F} - \mathbf{a}.\nabla\mathbf{F}$$

Thus, the above evaluates to

$$\int_S (\mathbf{a}.\mathbf{n})(\nabla.\mathbf{F}) - \mathbf{n}.(\mathbf{a}.\nabla\mathbf{F}))dS$$

Now,

$$(\mathbf{n} \times \nabla) \times \mathbf{F} = = \nabla_F(\mathbf{n}.\mathbf{F}) - n\nabla.\mathbf{F}$$

and

$$\mathbf{a}.\nabla_F(\mathbf{n}.\mathbf{F}) = a_x\mathbf{n}.\frac{\partial \mathbf{F}}{\partial x} + a_y\mathbf{n}.\frac{\partial \mathbf{F}}{\partial y} + a_z\mathbf{n}.\frac{\partial \mathbf{F}}{\partial z}$$

$$= \mathbf{n}.(\mathbf{a}.\nabla)\mathbf{F}$$

Combining the previous two identities, we get

$$\mathbf{a}.(\mathbf{n} \times \nabla) \times \mathbf{F} = \mathbf{n}.(\mathbf{a}.\nabla)\mathbf{F} - (\text{a.n})\nabla.\mathbf{F}$$

This yields the following identity:

$$\int_\Gamma \mathbf{a}.\mathbf{F} \times d\mathbf{r} = -\int_S \mathbf{a}.(\mathbf{n} \times \nabla) \times \mathbf{F}dS$$

and since **a** is a constant vector, it follows that

$$\int_\Gamma \mathbf{F} \times d\mathbf{r} = -\int_S (\mathbf{n} \times \nabla) \times \mathbf{F}dS$$

10. Gradient, Divergence and Curl in Curvilinear Co-ordinates:

A. Gradient in curvilinear co-ordinates: Now, for any scalar field ϕ (q_1, q_2, q_3), we have from the chain rule for partial differentials

$$\nabla\phi = \sum_{i=1}^{3} \frac{\partial \phi}{\partial q_i} \nabla q_i = \sum_{i=1}^{3} \frac{1}{H_i} \frac{\partial \phi}{\partial q_i} u_i$$

where

$$H_i = \frac{1}{|\nabla q_i|}, \; i = 1, 2, 3$$

We shall show that

$$H_i = \left|\frac{\partial \mathbf{r}}{\partial q_i}\right|$$

as a consequence of the orthogonality of the curvilinear co-ordinate system (q_1, q_2, q_3). To see this, first note that

$$\sum_{j=1}^{3} \frac{\partial q_i}{\partial x_j} \frac{\partial x_j}{\partial q_k} = \delta_{ik}$$

This is equivalent to saying that the matrices $\left(\left(\frac{\partial x_i}{\partial q_j}\right)\right)$ and $\left(\left(\frac{\partial x_i}{\partial q_j}\right)\right)$ are inverses of each other. Note that

$$\left(\left(\frac{\partial q_i}{\partial x_j}\right)\right) = \begin{pmatrix} \frac{\partial q_1}{\partial x_1} & \frac{\partial q_1}{\partial x_2} & \frac{\partial q_1}{\partial x_3} \\ \frac{\partial q_2}{\partial x_1} & \frac{\partial q_2}{\partial x_2} & \frac{\partial q_2}{\partial x_3} \\ \frac{\partial q_3}{\partial x_1} & \frac{\partial q_3}{\partial x_2} & \frac{\partial q_3}{\partial x_3} \end{pmatrix}$$

$$= \begin{pmatrix} \nabla q_1^T \\ \nabla q_2^T \\ \nabla q_3^T \end{pmatrix}$$

where we are assuming ∇q_i to be a column vector. However the identities

$$\frac{\nabla q_i^T \nabla q_j}{|\nabla q_i||\nabla q_j|} = \delta_{ij}$$

when expressed in matrix notation, become

$$\begin{pmatrix} \frac{\nabla q_1^T}{|\nabla q_1|} \\ \frac{\nabla q_2^T}{|\nabla q_2|} \\ \frac{\nabla q_3^T}{|\nabla q_3|} \end{pmatrix} \begin{pmatrix} \frac{\nabla q_1}{|\nabla q_1|} & \frac{\nabla q_2}{|\nabla q_2|} & \frac{\nabla q_3}{\nabla q_3} \end{pmatrix} = \mathbf{I}$$

From this, we can easily derive the identity

$$\begin{pmatrix} \nabla q_1^T \\ \nabla q_2^T \\ \nabla q_3^T \end{pmatrix} = (\nabla q_1 \; \nabla q_2 \; \nabla q_3) \begin{pmatrix} |\nabla q_1|^{-2} & 0 & 0 \\ 0 & |\nabla q_2|^{-2} & 0 \\ 0 & 0 & |\nabla q_3|^{-2} \end{pmatrix}$$

From the earlier discussion, this must coincide with

$$\left(\left(\frac{\partial x_i}{\partial q_j}\right)\right) = \left(\frac{\partial \mathbf{r}}{\partial q_1} \; \frac{\partial \mathbf{r}}{\partial q_2} \; \frac{\partial \mathbf{r}}{\partial q_3}\right)$$

where $\mathbf{r} = [x_1, x_2, x_3]^T$ is a column vector. It follows that

$$\frac{\partial \mathbf{r}}{\partial q_i} = \frac{\nabla q_i}{|\nabla q_i|^2}$$

and taking magnitudes on both sides, we get the desired identity:

$$\left|\frac{\partial \mathbf{r}}{\partial q_i}\right| = \frac{\nabla q_1}{|\nabla q_i|}$$

B. Divergence in curvilinear co-ordinates: We use the formula just derived for the gradient operator:

$$\nabla = \sum_{i=1}^{3} \frac{u_i}{H_i} \frac{\partial}{\partial q_i}$$

Take any vector field $\mathbf{F} = \sum_{i=1}^{3} F_i u_i$ expressed in terms of the curvilinear system. The components F_i are functions of (q_1, q_2, q_3). Then,

$$\nabla.\mathbf{F} = \left(\sum \frac{u_i}{H_i} \frac{\partial}{\partial q_i}\right) . \left(\sum F_j u_j\right)$$

$$= \sum_{i,j} \left(\frac{u_i}{H_i}, \frac{\partial F_j u_j}{\partial q_i}\right)$$

$$= \sum_j \frac{1}{H_j} \frac{\partial F_j}{\partial q_j} + \sum_{i,j} \frac{1}{H_i} \left(u_i, \frac{\partial u_j}{\partial q_i}\right) F_j$$

Now, the coefficient of F_j in the second term is

$$\sum_i \frac{1}{H_i} \left(u_i, \frac{\partial u_j}{\partial q_i}\right)$$

It is easy to see that the $i = j$ term in the above expression vanishes since u_i is a unit vector. Let us look at this term for $j = 1$. It is given by

$$\frac{1}{H_2}\left(u_2, \frac{\partial u_1}{\partial q_2}\right) + \frac{1}{H_3}\left(u_3, \frac{\partial u_1}{\partial q_3}\right)$$

$$= \frac{\left(\frac{\partial \mathbf{r}}{\partial q_2}, \frac{\partial}{\partial q_2}\left(\frac{\partial \mathbf{r}}{\partial q_1} / |\frac{\partial \mathbf{r}}{\partial q_1}|\right)\right)}{|\frac{\partial \mathbf{r}}{\partial q_2}|^2} + \frac{\left(\frac{\partial \mathbf{r}}{\partial q_3}, \frac{\partial}{\partial q_3}\left(\frac{\partial \mathbf{r}}{\partial q_1} / |\frac{\partial \mathbf{r}}{\partial q_1}|\right)\right)}{|\frac{\partial \mathbf{r}}{\partial q_3}|^2}$$

$$= -\frac{\left(\frac{\partial \mathbf{r}}{\partial q_1}, \frac{\partial^2 \mathbf{r}}{\partial q_2^2}\right)}{|\frac{\partial \mathbf{r}}{\partial q_1}||\frac{\partial \mathbf{r}}{\partial q_2}|^2} - \frac{\left(\frac{\partial \mathbf{r}}{\partial q_1}, \frac{\partial^2 \mathbf{r}}{\partial q_3^2}\right)}{|\frac{\partial \mathbf{r}}{\partial q_1}||\frac{\partial \mathbf{r}}{\partial q_3}|^2}$$

We now note that

$$\frac{\partial}{\partial q_1}(H_2 H_3) = \frac{\partial}{\partial q_1}\left(|\frac{\partial \mathbf{r}}{\partial q_2}||\frac{\partial \mathbf{r}}{\partial q_3}|\right)$$

$$= \frac{\left(\frac{\partial^2 \mathbf{r}}{\partial q_1 \partial q_2}, \frac{\partial \mathbf{r}}{\partial q_2}\right)|\frac{\partial \mathbf{r}}{\partial q_3}|}{|\frac{\partial \mathbf{r}}{\partial q_2}|} + \frac{\left(\frac{\partial^2 \mathbf{r}}{\partial q_1 \partial q_3}, \frac{\partial \mathbf{r}}{\partial q_3}\right)|\frac{\partial \mathbf{r}}{\partial q_2}|}{|\frac{\partial \mathbf{r}}{\partial q_3}|}$$

This gives

$$\frac{1}{H_1 H_2 H_3}\frac{\partial}{\partial q_1}(H_2 H_3) = \frac{\left(\frac{\partial^2 \mathbf{r}}{\partial q_1 \partial q_2}, \frac{\partial \mathbf{r}}{\partial q_2}\right)}{|\frac{\partial \mathbf{r}}{\partial q_2}|^2|\frac{\partial \mathbf{r}}{\partial q_1}|} + \frac{\left(\frac{\partial^2 \mathbf{r}}{\partial q_1 \partial q_3}, \frac{\partial \mathbf{r}}{\partial q_3}\right)}{|\frac{\partial \mathbf{r}}{\partial q_3}|^2|\frac{\partial \mathbf{r}}{\partial q_1}|}$$

$$= -\frac{\left(\frac{\partial \mathbf{r}}{\partial q_1}, \frac{\partial^2 \mathbf{r}}{\partial q_2^2}\right)}{|\frac{\partial \mathbf{r}}{\partial q_1}||\frac{\partial \mathbf{r}}{\partial q_2}|^2} - \frac{\left(\frac{\partial \mathbf{r}}{\partial q_1}, \frac{\partial^2 \mathbf{r}}{\partial q_3^2}\right)}{|\frac{\partial \mathbf{r}}{\partial q_1}||\frac{\partial \mathbf{r}}{\partial q_3}|^2}$$

Comparing the two expressions above gives

$$\frac{1}{H_2}\left(u_2, \frac{\partial u_1}{\partial q_2}\right) + \frac{1}{H_3}\left(u_3, \frac{\partial u_1}{\partial q_3}\right) = \frac{1}{H_1 H_2 H_3}\frac{\partial}{\partial q_1}(H_2 H_3)$$

with similar expressions for the other terms. This leads us to the following expression for divergence in orthogonal curvilinear co-ordinates:

$$\nabla.\mathbf{F} = \frac{1}{H_1 H_2 H_3}\left(\frac{\partial}{\partial q_1}(F_1 H_2 H_3) + \frac{\partial}{\partial q_2}(F_2 H_3 H_1) + \frac{\partial}{\partial q_3}(F_3 H_1 H_2)\right)$$

C. Curl in curvilinear co-ordinates:

$$\nabla \times \mathbf{F} = \left(\sum \frac{u_i}{H_i}\frac{\partial}{\partial q_i}\right) \times \left(\sum_{i=1}^{3} F_i u_i\right)$$

$$= \sum_{i,j=1}^{3} \frac{1}{H_i} u_i \times \frac{\partial}{\partial q_i}(F_j u_j)$$

$$= \sum_{i,j=1}^{3} \frac{1}{H_i}\frac{\partial F_j}{\partial q_i} u_i \times u_j + \sum_{i,j=1}^{3} \frac{1}{H_i} F_j u_i \times \frac{\partial u_j}{\partial q_i}$$

$$= S_1 + S_2$$

say, where S_1 is the first sum and S_2 the second sum. The first sum expands to

$$S_1 = \left(\frac{1}{H_2}\frac{\partial F_3}{\partial q_2} - \frac{1}{H_3}\frac{\partial F_2}{\partial q_3}\right)u_1 + \left(\frac{1}{H_3}\frac{\partial F_1}{\partial q_3} - \frac{1}{H_1}\frac{\partial F_3}{\partial q_1}\right)u_2$$

$$+\left(\frac{1}{H_1}\frac{\partial F_2}{\partial q_1} - \frac{1}{H_2}\frac{\partial F_1}{\partial q_2}\right)u_3$$

For evaluating S_2, we need to evaluate the coefficients of F_1, F_2, F_3 separately. The coefficient of F_1 is

$$S_{21} = \sum_{i=1}^{3} \frac{1}{H_i} u_1 \times \frac{\partial u_1}{\partial q_i}$$

We note that since u_1 is a unit vector, all its partial derivatives $\frac{\partial u_1}{\partial q_i}$, $i = 1$, 2, 3 are orthogonal to u_1. Thus

$$\frac{\partial u_1}{\partial q_i} = \lambda_i u_2 + \mu_i u_3, i = 1, 2, 3$$

where

$$\lambda_i = \left(\frac{\partial u_1}{\partial q_i}, u_2 \right) \quad \mu_i = \left(\frac{\partial u_1}{\partial q_i}, u_3 \right)$$

This gives

$$S_{21} = \frac{1}{H_1} u_1 \times (\lambda_1 u_2 + \mu_1 u_3) + \frac{1}{H_2} u_2 \times (\lambda_2 u_2 + \mu_2 u_3) + \frac{1}{H_3} u_3 \times (\lambda_3 u_2 + \mu_3 u_3)$$

$$= \left(\frac{\mu_2}{H_2} - \frac{\lambda_3}{H_3} \right) u_1 - \frac{\mu_1}{H_1} u_2 + \frac{\lambda_1}{H_1} u_3$$

We evaluate the different coefficients appearing here.

$$\frac{\mu_2}{H_2} - \frac{\lambda_3}{H_3} = \frac{\left(\frac{\partial \mathbf{r}}{\partial q_3}, \frac{\partial}{\partial q_2} \left(\frac{\partial \mathbf{r}}{\partial q_1} / \left| \frac{\partial \mathbf{r}}{\partial q_1} \right| \right) \right)}{H_2 H_3} - \frac{\left(\frac{\partial \mathbf{r}}{\partial q_2}, \frac{\partial}{\partial q_3} \left(\frac{\partial \mathbf{r}}{\partial q_1} / \left| \frac{\partial \mathbf{r}}{\partial q_l} \right| \right) \right)}{H_2 H_3}$$

$$= -\frac{\left(\frac{\partial \mathbf{r}}{\partial q_1}, \frac{\partial^2 \mathbf{r}}{\partial q_2 q_3} \right)}{H_1 H_2 H_3} + \frac{\left(\frac{\partial \mathbf{r}}{\partial q_1}, \frac{\partial^2 \mathbf{r}}{\partial q_2 q_3} \right)}{H_1 H_2 H_3}$$

$$= 0$$

$$\mu_1 = \frac{\left(\frac{\partial \mathbf{r}}{\partial q_3}, \frac{\partial}{\partial q_1} \left(\frac{\partial \mathbf{r}}{\partial q_1} / H_1 \right) \right)}{H_3}$$

$$= -\frac{\left(\frac{\partial \mathbf{r}}{\partial q_1}, \frac{\partial^2 \mathbf{r}}{\partial q_1 \partial q_3} \right)}{H_1 H_3} = -\frac{1}{H_3} \frac{\partial H_1}{\partial q_3}$$

and likewise,

$$\lambda_1 = -\frac{1}{H_2} \frac{\partial H_1}{\partial q_2}$$

Thus,

$$S_{21} = -\frac{\mu_1}{H_1} u_2 + \frac{\lambda_1}{H_1} u_3$$

$$= \frac{1}{H_1 H_3} \frac{\partial H_1}{\partial q_3} u_2 - \frac{1}{H_1 H_2} \frac{\partial H_1}{\partial q_2} u_3$$

The term involving F_1 and its derivatives in the total sum $S_1 + S_2$ is thus given by

$$-\frac{1}{H_2} \frac{\partial F_1}{\partial q_2} u_3 + \frac{1}{H_3} \frac{\partial F_1}{\partial q_3} u_2 + F_1 \frac{1}{H_1 H_3} \frac{\partial H_1}{\partial q_3} u_2 - F_1 \frac{1}{H_1 H_2} \frac{\partial H_1}{\partial q_2} u_3$$

$$= -\frac{1}{H_1 H_2} \frac{\partial (F_1 H_1)}{\partial q_2} u_3 + \frac{1}{H_1 H_3} \frac{\partial (F_1 H_1)}{\partial q_3} u_2$$

We can do a similar thing for the terms involving F_2 and F_3. Combining all these three identities gives us the following neat expression for the curl of a vector field in curvilinear co-ordinates:

$$\nabla \times \mathbf{F} = \det \begin{pmatrix} \dfrac{u_1}{H_2 H_3} & \dfrac{u_2}{H_3 H_1} & \dfrac{u_3}{H_1 H_2} \\ \dfrac{\partial}{\partial q_1} & \dfrac{\partial}{\partial q_2} & \dfrac{\partial}{\partial q_3} \\ F_1 H_1 & F_2 H_2 & F_3 H_3 \end{pmatrix}$$

As a final exercise, we shall derive the expression for the Laplacian of a scalar function in curvilinear co-ordinates.

D. Laplacian in Curvilinear Co-ordinates: Let ϕ (q_1, q_2, q_3) be a scalar field specified as a function of curvilinear co-ordinates. Its Laplacian is

$$\nabla^2\phi = \nabla.(\nabla\phi)$$

Now,

$$\nabla\phi = \sum_{i=1}^{3} F_i \hat{u}_i = \mathbf{F}$$

where

$$F_i = \frac{1}{H_i} \frac{\partial \phi}{\partial q_i}, i = 1, 2, 3$$

Thus,

$$\nabla^2\phi = \nabla.\mathbf{F} = \frac{1}{H_1 H_2 H_3} \left(\frac{\partial}{\partial q_1} \left(F_1 H_2 H_3\right) + \frac{\partial}{\partial q_2} \left(F_2 H_3 H_1\right) + \frac{\partial}{\partial q_3} \left(F_3 H_1 H_2\right) \right)$$

$$= \frac{1}{H_1 H_2 H_3}\left(\frac{\partial}{\partial q_1}\left(\frac{H_2 H_3}{H_1}\frac{\partial \phi}{\partial q_1}\right) + \frac{\partial}{\partial q_2}\left(\frac{H_3 H_1}{H_2}\frac{\partial \phi}{\partial q_2}\right) + \frac{\partial}{\partial q_3}\left(\frac{H_1 H_2}{H_3}\frac{\partial \phi}{\partial q_3}\right)\right)$$

E. Gradient, Divergence, Curl and Laplacian in the Cylindrical and Spherical Polar Co-ordinate Systems:

(i) Cylindrical Co-ordinates: For the cylindrical co-ordinate system, we use $x = \rho.\cos\phi$, $y = \rho.\sin\phi$, $z = z$ and compute the coefficients

$$H_\rho = \left(\left(\frac{\partial x}{\partial \rho}\right)^2 + \left(\frac{\partial y}{\partial \rho}\right)^2 + \left(\frac{\partial z}{\partial \phi}\right)^2\right)^{1/2} = 1$$

$$H_\phi = \left(\left(\frac{\partial x}{\partial \phi}\right)^2 + \left(\frac{\partial y}{\partial \phi}\right)^2 + \left(\frac{\partial z}{\partial \phi}\right)^2\right)^{1/2} = \rho$$

$$H_z = \left(\left(\frac{\partial x}{\partial z}\right)^2 + \left(\frac{\partial y}{\partial z}\right)^2 + \left(\frac{\partial z}{\partial z}\right)^2\right)^{1/2} = 1$$

This gives for the gradient of a scalar field $f(\rho, \phi, z)$,

$$\nabla f = \hat{\rho}\frac{\partial f}{\partial \rho} + \frac{\hat{\phi}}{\rho}\frac{\partial f}{\partial \phi} + \hat{z}\frac{\partial f}{\partial z}$$

for the divergence of a vector field **F**

$$\text{div}\mathbf{F} = \left(\frac{1}{H_\rho H_\phi H_z}\left(\frac{\partial}{\partial \rho}\left(H_\phi H_z F_\rho\right) + \frac{\partial}{\partial \phi}\left(H_z H_\rho H_\phi\right) + \frac{\partial}{\partial z}\left(H_\rho H_\phi H_z\right)\right)\right)$$

$$= \frac{1}{\rho}\frac{\partial}{\partial \rho}(\rho.F_\rho) + \frac{1}{\rho}\frac{\partial}{\partial \phi}(F_\phi) + \frac{\partial}{\partial z}(F_z)$$

For the curl of the vector field **F**,

$$\nabla \times \mathbf{F} = \det\begin{pmatrix} \frac{\hat{\rho}}{\rho} & \hat{\phi} & \frac{\hat{z}}{\rho} \\ \frac{\partial}{\partial \rho} & \frac{\partial}{\partial \phi} & \frac{\partial}{\partial z} \\ F_\rho & \rho.F_\phi & F_z \end{pmatrix}$$

Finally, for the Laplacian of a scalar field f in cylindrical co-ordinates, we get

$$\nabla^2 f = \frac{1}{H_\rho H_\phi H_z}\left(\frac{\partial}{\partial\rho}\left(\frac{H_\phi H_z}{H_\rho}\frac{\partial f}{\partial_\rho}\right)+\frac{\partial}{\partial\phi}\left(\frac{H_z H_\rho}{H_\phi}\frac{\partial f}{\partial\phi}\right)+\frac{\partial}{\partial_z}\left(\frac{H_\rho H_\phi}{H_z}\frac{\partial f}{\partial z}\right)\right)$$

$$= \frac{1}{\rho}\frac{\partial}{\partial\rho}\rho\frac{\partial f}{\partial\rho}+\frac{1}{\rho^2}\frac{\partial^2 f}{\partial\phi^2}+\frac{\partial^2 f}{\partial z^2}$$

(ii) Spherical Polar Co-ordinates: Here, we can once again compute the coefficients from the defining relations

$$x = r.\cos\phi.\sin\theta,\ y = r.\sin\phi.\sin\theta,\ z = r.\cos\theta$$

We have

$$H_r = \left(\left(\frac{\partial x}{\partial r}\right)^2+\left(\frac{\partial y}{\partial r}\right)^2+\left(\frac{\partial z}{\partial r}\right)^2\right)^{1/2} = 1$$

$$H_\theta = \left(\left(\frac{\partial x}{\partial\theta}\right)^2+\left(\frac{\partial y}{\partial\theta}\right)^2+\left(\frac{\partial z}{\partial\theta}\right)^2\right)^{1/2} = r$$

$$H_\phi = \left(\left(\frac{\partial x}{\partial\phi}\right)^2+\left(\frac{\partial y}{\partial\phi}\right)^2+\left(\frac{\partial z}{\partial\phi}\right)^2\right)^{1/2} = r.\sin\theta$$

We leave it to you to show that

$$\nabla f = \hat{r}.\frac{\partial f}{\partial r}+\frac{\hat{\theta}}{r}.\frac{\partial f}{\partial\theta}+\frac{\hat{\phi}}{r.\sin\theta}\frac{\partial f}{\partial\phi}$$

$$\nabla.\mathbf{F} = \frac{1}{r^2}\frac{\partial}{\partial r}(r^2 F_r)+\frac{1}{r.\sin\theta}\frac{\partial}{\partial\theta}(\sin\theta.F_\theta)+\frac{1}{r.\sin\theta}\frac{\partial}{\partial\phi}F_\phi$$

$$\nabla^2 f = \frac{1}{r^2}\frac{\partial}{\partial r}\left(r^2\frac{\partial f}{\partial r}\right)+\frac{1}{r^2.\sin\theta}\frac{\partial}{\partial\theta}\left(\sin\theta\frac{\partial f}{\partial\theta}\right)+\frac{1}{r^2.\sin^2\theta}\frac{\partial^2 f}{\partial\phi^2}$$

and for the curl

$$\nabla \times \mathbf{F} = \det \begin{pmatrix} \dfrac{\hat{r}}{r^2.\sin\theta} & \dfrac{\hat{\phi}}{r.\sin\theta} & \dfrac{\hat{z}}{r} \\ \dfrac{\partial}{\partial r} & \dfrac{\partial}{\partial \theta} & \dfrac{\partial}{\partial \phi} \\ F_r & r.F_\theta & r.\sin\theta.F_\phi \end{pmatrix}$$

These expressions are useful in the solution of problems in electrodynamics when either the problem has cylindrical or radial symmetry.

11. Green's theorems:

Gauss' theorem can be used to derive two auxiliary results which are particularly useful in proving uniqueness of the solution to the Poisson's equation in electrostatics, obtaining the solution to the Poisson equation for electrostatics, and the solution to the Helmholtz and wave equation for the scalar and vector potential in electrodynamics. The theorems are that if S is a closed surface enclosing a volume V, and ϕ, ψ are two scalar fields, then (a)

$$\int_V (\phi\nabla^2\psi + (\nabla\phi, \nabla\psi))dV = \int_S \phi \frac{\partial\psi}{\partial \mathbf{n}} dS$$

and (b)

$$\int_V (\phi\nabla^2\psi - \psi\nabla^2\phi)dV = \int_S \left(\phi \frac{\partial\psi}{\partial \mathbf{n}} dS - \psi \frac{\partial\phi}{\partial \mathbf{n}}\right) dS$$

The first equation follows by applying Gauss' theorem to the vector field $\mathbf{F} = \phi\nabla\psi$ after nothing that

$$\Delta.\mathbf{F} = \phi\nabla^2\psi + (\nabla\phi, \nabla\psi)$$

and the secon equation follows by applying Gauss' theorem to the vector field $\mathbf{F} = \phi\nabla\psi - \phi\nabla\psi$ noting that

$$\nabla.\mathbf{F} = \phi\nabla^2\psi - \psi\nabla^2\phi$$

12. Review Questions and Solved Problems on Vector Analysis

[1] Give examples of scalars and vectors.

Scalar fields are the pressure in the atmosphere $p(\mathbf{r})$ as a function of the position $\mathbf{r}$, the temperature of the atmosphere $T(\mathbf{r})$ as a function of the position, the density of a fluid $\rho(\mathbf{r})$ as a function of position etc. Vector fields are the velocity field $\mathbf{v}(\mathbf{r})$ of a fluid as a function of the position, the gravitational field (force per unit mass) produced by a piece of matter in space, the electric field (electric force per unit charge) produced by a piece of charge in space etc. Another example of a vector field is the heat flow in a metal. At each point $\mathbf{h}(\mathbf{r})$ defines the amount of heat flowing per unit area with the direction of this vector being the direction along which the heat is flowing. It flows usually along the direction in which the temperature has maximum gradient.

[2] Can the reference vectors be a non-orthogonal set?

Yes. The only condition needed for the reference vectors $\mathbf{a}_1$, $\mathbf{a}_2$, $\mathbf{a}_3$ is that they be linearly independent, that is $c_1\mathbf{a}_1 + c_2\mathbf{a}_2 + c_3\mathbf{a}_3 = \mathbf{0}$ imply that $c_1 = c_2 = c_3 = 0$. If this holds, then given any vector $\mathbf{r}$, we can determine its components relative to this reference set:

$$\mathbf{r} = c_1\mathbf{a}_1 + c_2\mathbf{a}_2 = c_3\mathbf{a}_3$$

by inverting the equations $\mathbf{A}\mathbf{c} = \mathbf{y}$, where

$$\mathbf{y} = [(\mathbf{r},\mathbf{a}_1), (\mathbf{r}, \mathbf{a}_2), (\mathbf{r}, \mathbf{a}_3)]^T, \ \mathrm{c} = [c_1, c_2, c_3]^T,$$
$$\mathrm{A} = ((\mathbf{a}_i.\mathbf{a}_j))_{1\le i,j\le 3}$$

[3] Explain physically what you understand by the gradient, divergence and curl.

Ans. If $\phi(\mathrm{r})$ is a scalar field, then $|\nabla\phi|$ tells us the maximum magnitude of the change in the function ϕ at the point $\mathbf{r}$. The direction of $\nabla\phi(\mathbf{r})$ gives us the direction in which the maximum change occurs in the function at $\mathbf{r}$ on slight displacement. If $\hat{\mathbf{n}}$ is any unit vector, then $(\nabla\phi(\mathbf{r}), \hat{\mathbf{n}})$ tells us the change in the function at $\mathbf{r}$ along the direction $\hat{\mathbf{n}}$. Another interpretation of the gradient is that the direction $\nabla\phi(\mathbf{r})$ gives us the direction of the normal to the surface ϕ = const. at the point $\mathbf{r}$. A third intepretation of the gradient is the following: consider a curve specified parametrically as $\mathbf{r}(t)$. The rate of change of the quantity $\phi(\mathbf{r}(t))$ at the point $\mathbf{r}(t)$ can be found as

$$\frac{d\phi(\mathbf{r}(t))}{dt} = \left(\nabla\phi(\mathbf{r}(t)), \frac{d\mathbf{r}(t)}{dt}\right)$$

that is the gradient in the direction of the tangent of the curve at that point after an appropriate normalization. The interpretation of the divergence is obtained using Gauss' divergence theorem. The flux of a vector field $\mathbf{F}(\mathbf{r})$ through the surface bounding an infinitesimal volume element dV at the point $\mathbf{r}$ is given by $\nabla.\mathbf{F}dV$ and hence, $\nabla.\mathbf{F}$ is the flux out of a unit volume at $\mathbf{r}$. If $\mathbf{F}$ is the velocity field of a fluid, then $\Delta.\mathbf{F}$ is the volume of the fluid flowing out of a unit volume element located at

r. Finally, the interpretation of the curl of a vector field is obtained from Stokes' theorem. If Γ is a tiny closed curve bounding a surface dS at **r**, then the circulation of **F** around Γ (i.e. its line integral around Γ) is by Stokes' theorem, approximately $\nabla \times \mathbf{F}.\mathbf{n}dS$ where **n** is the normal in the direction of dS. The circulation per unit surface area at **r** is therefore, $\nabla \times \mathbf{F}.\mathbf{n}$. This quantity tells us how speedily the vector field circulates at a given point. If **F** is the velocity field of a fluid, then its circulation tells us how strong the vortices of the fluid are at a given point.

[3] Given four points (x_i, y_i, z_i), $i = 1, 2, 3, 4$ in Cartesian co-ordinates. Find the co-ordinates of the centre of the sphere passing through these four points.

Ans. Let (f, g, h) be the centre of the sphere and R the radius. Then,

$$(x_i - f)^2 + (y_i - g)^2 + (z_i - h)^2 = R^2, \; i = 1, 2, 3, 4$$

Taking the difference of the successive equations here gives

$$(x_i - x_{i+1})(x_i + x_{i+1} - 2f) + (y_i - y_{i+1})(y_i + y_{i+1} - 2g) + (z_i - z_{i+1})(z_i + z_{i+1} - 2h) = 0, \quad i = 1, 2, 3$$

This can be rearranged to give

$$(x_i - x_{i+1})f + (y_i - y_{i+1})g + (z_i - z_{i+1})h = (r_i^2 - r_{i+1}^2)/2, = i = 1,2,3$$

where

$$r_i^2 = x_i^2 + y_i^2 + z_i^2, i = 1,2,3,4$$

The solution to this simultaneous set of three linear equations is

$$\begin{pmatrix} f \\ g \\ h \end{pmatrix} = \begin{pmatrix} x_1 - x_2 & y_1 - y_2 & z_1 - z_2 \\ x_2 - x_3 & y_2 - y_3 & z_2 - z_3 \\ x_3 - x_4 & y_3 - y_4 & z_3 - z_4 \end{pmatrix}^{-1} \begin{pmatrix} r_1^2 - r_2^2 \\ r_2^2 - r_3^2 \\ r_3^2 - r_4^2 \end{pmatrix}$$

[4] What is a phasor?

Ans. Suppose we are given two sinusoidal real waveforms $x(t) = A.\cos(\omega t)$, $y(t) = B.\cos(\omega t + \phi)$, having the same frequency ω but possibly different amplitudes and phases. We can represent these two signals in the complex plane by means of the complex numbers $\mathbf{x} = A$, $\mathbf{y} = B.\exp(j\phi)$ respectively. When viewed as vectors, the first complex number is obtained by drawing a vector of length A along the x axis and the second complex number is obtained by drawing a vector of length B making an angle ϕ with the x axis. In general, given n signals $x_i(t) = A_i.\cos(\omega_i t + \phi_i)$, $i = 1, 2, ..., n$, all having the same frequency, we can represent these signals by the complex numbers $\mathbf{x}_i = A_i.\exp(j\phi_i)$, $i = 1, 2, ..., n$ respectively. $\mathbf{x}_i$ is called the phasor associated with the ith signal. The signals themselves can be computed from its phasor by the rule $x_i(t) = Re(\mathbf{x}_i.\exp(j\omega t))$. The advantage with using the phasor to represent sinusoidal signals can be seen from the following example. Suppose we want to obtain the steady state solution to the differential equation

$$\frac{d^2 x(t)}{dt^n} + c_1 \frac{d^{n-1} x(t)}{dt^{n-1}} + \ldots + c_n x(t) = \sum_{i=1}^{k} A_i.\cos(\omega t + \phi_i)$$

We let $x(t) = \mathrm{Re}(\mathbf{x}.\exp(j\omega t))$ and substituting this into the differential equation gives

$$\mathrm{Re}(\mathbf{x}.P(j\omega).\exp(j\omega t)) = \mathrm{Re}\left(\sum_{i=1}^{n} A_i \exp(j\phi_i).\exp(j\omega t)\right)$$

where $P(j\omega) = (j\omega)^n + c_1(j\omega)^{n-1} + \ldots + c_n$. Since this equation must hold for every t, its solution is

$$\mathbf{x} = \frac{1}{P(j\omega)} \sum_{i=1}^{n} A_i.\exp(j\phi_i)$$

We can now represent the different terms in the source input as well as the response using their respective phasors in the complex plane. The method of phasors is thus a convenient tool while solving linear circuits with sinusoidal excitations.

[5] Find the condition for a set of four vectors **A, B, C, D** to fall in a plane.

Ans. The condition is that if the pairs of points **A, B, A, C** and **A, D** are connected, then these three lines will define a parallelopiped with zero volume. This is the same as saying that the normal to the plane defined by the vectors **AB** and **AC** is perpendicular to the line **AD**. That is, $(\mathbf{B} - \mathbf{A}).\ (\mathbf{C} - \mathbf{A}) \times (\mathbf{D} - \mathbf{A}) = 0$.

[6] Show that $\mathbf{A} \times (\mathbf{B} \times \mathbf{C}) = (\mathbf{A.C})\mathbf{B} - (\mathbf{A.B})\mathbf{C}$.

Ans. We have

$(\mathbf{B} \times \mathbf{C})_1 = B_2C_3 - B_3C_2$, $(\mathbf{B} \times \mathbf{C})_2 = B_3C_1 - B_1C_3$, $(\mathbf{B} \times \mathbf{C})_3 = B_1C_2 - B_2C_1$

Thus,

$$(\mathbf{A} \times (\mathbf{B} \times \mathbf{C}))_1 = A_2(\mathbf{B} \times \mathbf{C})_3 - A_3(\mathbf{B} \times \mathbf{C})_2 = A_2(B_1C_2 - B_2C_1) - A_3(B_3C_1 - B_1C_3)$$

$$= (A_2C_2 + A_3C_3)\,B_1 - (A_2B_2 + A_3B_3)C_1 = (A_1C_1 + A_2C_2 + A_3C_3)\,B_1 - (A_1B_1 + A_2B_2 + A_3B_3)\,C_1 = (\mathbf{A.C})\,B_1 - (\mathbf{A.B})\,C_1$$

with similar expressions for the other components.

[7] Find the equation for the plane spanned by two vectors **A, B** originating from a common point $\mathbf{r}_1$.

Ans. If **r** be any point on the plane, then the vectors **A, B,** $\mathbf{r} - \mathbf{r}_1$ should be collinear. This is also a sufficient condition for **r** to fall on the plane. Thus, the equation of the plane is

$$\mathbf{A} \times \mathbf{B} \,.\, (\mathbf{r} - \mathbf{r}_1) = 0$$

[8] If **A, B, C** are three vectors, find the component of the vector joining **A** to **B** along the vector joining **A** to **C**.

Ans. The unit vector along the direction of the vector joining **A** to **C** is $\mathbf{n} = \frac{\mathrm{C-A}}{|\mathrm{C-A}|}$. Thus the component of the vector joining **A** to **B** along this direction is given by

$$(\mathbf{B}-\mathbf{A}).\mathbf{nn} = \frac{((\mathbf{B}-\mathbf{A}).(\mathbf{C}-\mathbf{A}))(\mathbf{C}-\mathbf{A})}{|\mathbf{C}-\mathbf{A}|^2}$$

[9] What is the significance of $\mathbf{A} \times (\mathbf{B} \times \mathbf{C}) = \mathbf{0}$?

Ans. $\mathbf{B} \times \mathbf{C}$ is directed along the normal to the plane spanned by the vectors **B** and **C**. So the significance is that for the above identity to hold, **A** should be directed along the normal to the plane spanned by the vectors **B** and **C**.

[10] Suppose we are given a curve that is the intersection of the surface $f(x, y) = 0$, $y = g(z)$. Explain how you would determine the differential length vector on the curve tangent to it at a given point.

Ans. We shall express the differential length in terms of dz. Note that $dy = g'(z)dz$, $f_x(x, y)dx + f_y(x, y)dy = 0$. where $f_x(x,y) = \frac{\partial f(x, y)}{\partial y}$, $f_y(x, y) = \frac{\partial f(x, y)}{\partial y}$. The second identity gives $dx = -\frac{f_y(x, y)}{f_x(x, y)}dy$.

Substituting for dy from the first identity, we get $dx = -\frac{f_y(x, y)}{f_x(x, y)}dy.$. This gives us the differential length vector along the curve as

$$d\mathbf{r} = dx.\hat{x} + dy.\hat{y} + dz.\hat{z} = \left(-\frac{f_y(x,y)g'(z)}{f_x(x,y)}\hat{x} - \frac{f_y(x,y)}{f_x(x,y)}\hat{y} + \hat{z} \right) dz$$

[11] Given a curve in parametric form $\mathbf{r}(t)$, find a vector normal to it at the point t and having no component along the direction $\mathbf{n}$.

Ans. Ahe vector tangent to the curve at t is $\mathrm{v} = \dfrac{d\mathbf{r}(t)}{dt}$. The vector $\mathbf{n} \times \mathbf{v}$ is normal to both $\mathbf{n}$ and $\mathbf{v}$ and hence, in particular, has no component along the $\mathbf{n}$ direction.

[12] For a scalar field V, show that $\nabla \times \nabla V = 0$, that is, the curl of the gradient of any scalar field vanishes.

Ans.

$$\nabla V = \hat{x}\frac{\partial V}{\partial x} + \hat{y}\frac{\partial V}{\partial y} + \hat{z}\frac{\partial V}{\partial z}$$

Thus,

$$\nabla \times \nabla V = \hat{x}\left(\frac{\partial}{\partial y}\frac{\partial V}{\partial z} - \frac{\partial}{\partial z}\frac{\partial V}{\partial y} \right) + \hat{y}\left(\frac{\partial}{\partial z}\frac{\partial V}{\partial x} - \frac{\partial}{\partial x}\frac{\partial V}{\partial z} \right) + \hat{z}\left(\frac{\partial}{\partial x}\frac{\partial V}{\partial y} - \frac{\partial}{\partial y}\frac{\partial V}{\partial x} \right) = \mathbf{0}$$

on using the commutativity of the partial derivatives:

$$\frac{\partial^2}{\partial x \partial y} = \frac{\partial^2}{\partial y \partial x} \qquad \frac{\partial^2}{\partial y \partial z} = \frac{\partial^2}{\partial z \partial y} \qquad \frac{\partial^2}{\partial z \partial x} = \frac{\partial^2}{\partial x \partial z}$$

An alternate method to prove this is to use Stoke's theorem: Let Γ be any closed contour bounding a surface S. Thus, Stoke's theorem gives

$$\int_S \nabla \times \nabla V.\mathbf{n}dS = \int_\Gamma \nabla V.d\mathbf{r}$$

But the left side equals the integral of the perfect differential dV once around Γ and this is clearly zero. Thus, the right side is zero. Since the surface S can be chosen arbitrarily, it follows that $\nabla \times \nabla V = \mathbf{0}$.

[13] Consider a vector field $\mathbf{F} = F_x(x, y, z)\hat{x} + F_y(x, y, z)\hat{y} + F_z(x, y, z)\hat{z}$. Consider a closed contour $ABCDA$ consisting of straight line segments AB, BC, CD, DA where AB is the segement $\mathbf{r} = (ta_x, ta_y, ta_z)$ with $0 < t < 1$, BC is the segment $\mathbf{r} = (a_x + tb_x, a_y + tb_y, a_z + tb_z)$ with $0 < t < 1$, CD is the segment $\mathbf{r} = (a_x + b_x + tc_x, a_y + b_y + tc_y, a_z + b_z + tc_z)$ with $0 < t < 1$ so that finally, DA will be the segment $\mathbf{r} = (a_x + b_x + c_x + td_x, a_y + b_y + c_y + td_y, a_z + b_z + c_z + td_z)$, $0 < t < 1$ where $d_x = -(a_x + b_x + c_x)$, $d_y = -(a_y + b_y + c_y)$, $d_z = -(a_z + b_z + c_z)$. Draw the contour assuming

appropriate values for the vectors **a**, **b**, **c** and express the line integral $\int_{ABCDA}$ **F**.d**r** in terms of ordinary integrals of the functions F_x, F_y, F_z.

Ans. Clearly

$$\int_{AB} \mathbf{F}.d\mathbf{r} = \int_0^1 (a_x F_x(ta_x, ta_y, ta_z) + a_y F_y(ta_x, ta_y, ta_z) + a_z F_z(ta_x, ta_y, ta_z))dt$$

since $d\mathbf{r} = (a_x, a_y, a_z)\, dt$ along AB. Likewise,

$$\int_{BC} \mathbf{F}.d\mathbf{r} = \int_0^1 (b_x F_x(a_x + tb_x, a_y + tb_y, a_z + tb_z) + b_y F_y(a_x + tb_x, a_y + tb_y, a_z + tb_z) + b_z F_z(a_x + tb_x, a_y + tb_y, a_z + tb_z))dt$$

since along the segment BC, $d\mathbf{r} = (b_x, b_y, b_z)\, dt$. Next,

$$\int_{CD} \mathbf{F}.d\mathbf{r} = \int_0^1 (c_x F_x(a_x + b_x + tc_x, a_y + b_y + tc_y, a_z + b_z + tc_z) + c_y F_y(a_x + b_x + tc_x, a_y + b_y + tc_y, a_z + b_z + tc_z + c_z F_z(a_x + b_x + tc_x, a_y + b_y + tc_y, a_z + b_z + tc_z))dt$$

Finally,

$$\int_{DA} \mathbf{F}.d\mathbf{r} = -\int_0^1 ((a_x + b_x + c_x) F_x((1-t)(a_x + b_x + c_x), (1-t)(a_y + b_y + c_y), (1-t)(a_z + b_z + c_z)) + (a_y + b_y + c_y) F_y((1-t)(a_x + b_x + c_x), (1-t)(a_y + b_y + c_y), (1-t)(a_z + b_z + c_z)) + (a_z + b_z + c_z) F_z((1-t)(a_x + b_x + c_x), (1-t)(a_y + b_y + c_y), (1-t)(a_z + b_z + c_z)))dt$$

[14] Consider a cuboid with faces parallel to the co-ordinate planes. Specifically, let the top and bottom faces parallel to the xy plane be P_z, Q_z, the faces parallel to the yz plane be P_x, Q_x, with P_x being the plane having larger value of x co-ordinate and Q_x the plane having smaller x co-ordinate, and the faces parallel to the zx plane be P_y, Q_y, with P_y having larger y co-ordinate and Q_y smaller y co-ordinate. Then determine expressions for the following: (a) d**r** along the edge common to the faces P_z, P_x, (b) d**r** along the edge common to the faces P_x, Q_y, (c) d**r** along the edge common to the faces P_z, Q_y, (d) d**S** for the surface P_x, (e) d**S** for the surface Q_y, (f) d**S** for the surface Q_z, (g) d**S** for the surface P_z.

Ans. (a) The edge common to the faces P_z, P_x, is parallel to $\hat{y}$. Thus, d**r** for this edge is $\pm\, \hat{y}dy$ with the plus sign being taken if we go along the direction of increasing y and the minus sign being taken if we go along the direction of decreasing y. (b) The edge common to the faces P_x, P_y is parallel to $\hat{z}$. Thus, d**r** for this edge is $\pm\, \hat{z}dz$ with the plus sign being taken if we go along the direction of increasing z and the minus sign if we go along the direction of decreasing z. (c) The edge common to the faces P_z, Q_y is parallel to $\hat{x}$. For this edge, we have $d\mathbf{r} = \pm\hat{x}dx$ with the plus or minus sign chosen appropriately. (d) $d\mathbf{S}$ for the surface P_x is $dydz\hat{x}$. (e) $d\mathbf{S}$ for the surface P_z is $dxdy\hat{z}$.

[15] If $\mathbf{F} = F_x(x, y, z)\,\hat{x} + F_y(x, y, z)\,\hat{y} + F_z(x, y, z)\,\hat{z}$ is a given vector field and S the surface $\rho = R$, $a < z < b$, $0 < \phi < 2\pi$, then calculate $\int_S \mathbf{F}.\mathbf{n}d\mathbf{S}$.

Ans. A point on the surface is parametrized in cylindrical co-ordinates by (R, ϕ, z) with ϕ, z varying over the appropriate range. The unit normal to the surface at this point in terms of Cartesian components is $\mathbf{n} = (\cos\phi, \sin\phi, 0)$ and hence

$$\mathbf{F}\,.\,\mathbf{n} = F_x(R\cos\phi, R\sin\phi, z)\cos\phi + F_y(R\cos\phi, R\sin\phi, z)\sin\phi$$

dS for this surface is $Rd\phi.dz$. Thus, the required surface integral can be expressed as

$$\int_S \mathbf{F}.\mathbf{n}dS = \int_0^{2\pi}\int_a^b (F_x(R\cos\phi, R\sin\phi, z)\cos\phi + F_y(R\cos\phi, R\sin\phi, z)\sin\phi)R.d\phi.dz$$

[16] Sketch the isosceles triangle ABC in the xy plane where AB is the segment $\mathbf{r} = (x, 0)$, $0 < x < a$, BC is the segment $\mathbf{r} = (a - t.\cos\theta, t.\sin\theta)$, $0 < t < L$ so that CA is the segment $\mathbf{r} = ((a - L.\cos\theta)(1 - t), L(\sin\theta)(1 - t))$, $0 < t < 1$. Here, $a = 2L.\cos\theta$ and evaluate the line integral of $\mathbf{F} = F_x(x,y)\,\hat{x} + F_y(x, y)\,\hat{y}$ around it directly and using Stokes' theorem. Verify that the two expressions tally.

Ans. The given triangle is isosceles with base of length a along the x axis from $x = 0$ up to $x = a$ and the remaining two sides of length L. The region A enclosed by the triangle can be expressed in two different ways. The first expression is

$$A = \{0 < y < L.\sin\theta,\ y.\cot\theta < x < a - y.\cot\theta\}$$

and the second expression is

$$A = \{0 < x < a/2,\ 0 < y < x.\tan\theta\} \cup \{a/2 < x < a,\ 0 < y < (a - x)\tan\theta\}$$

Thus, the line integral evaluated by Stokes' theorem is

$$\int_A \left(\frac{\partial F_y}{\partial x} - \frac{\partial F_x}{\partial y}\right)dxdy$$

$$= \int_0^{L.\sin\theta}\int_{y.\cot\theta}^{a-y.\cot\theta}\frac{\partial F_y}{\partial x}dydx$$

$$-\int_0^{a/2}\int_0^{x.\tan\theta}\frac{\partial F_x}{\partial y}dxdy - \int_{a/2}^{a}\int_0^{(a-x)\tan\theta}\frac{\partial F_x}{\partial y}dxdy$$

$$= \int_0^{L.\sin\theta} (F_y(a - y.\cot\theta, y) - F_y(y.\cot\theta, y))dy - \int_0^{a/2} (F_x(x, x.\tan\theta) - F_x(x,0))dx$$

$$- \int_{a/2}^{a} (F_x(x,(a-x)\tan\theta) - F_x(x,0))dx$$

Let us now compute the line integral using the direct method. We have

$$\int_{AB} \mathbf{F}, d\mathbf{r} = \int_0^a F_x(x,0)\, dx$$

$$\int_{BC} \mathbf{F}.d\mathbf{r} = \int_0^L F_x(a - t.\cos\theta, t.\sin\theta)(-\cos\theta)dt + F_y(a - t.\cos\theta, t.\sin\theta)(\sin\theta)dt$$

and finally,

$$\int_{CA} \mathbf{F}.d\mathbf{r} = -\int_0^1 F_x((a - L\cos\theta)(1-t), (L\sin\theta)(1-t))(a - L\cos\theta)dt$$

$$+ F_y((a - L\cos\theta)(1-t), L(\sin\theta)(1-t))(L\sin\theta)dt$$

We leave it to the student to show by using an appropriate change of variables that the sum of these three integrals equals the expression derived using Stokes' theorem.

[17] A vector field is given in terms of mixed Cartesian and cylindrical variables as

$$\mathbf{F} = F_x(\rho, x, \phi)\hat{x} + F_y(y, z, \rho)\hat{\phi} + F_z(x, \rho)\hat{z}$$

Express **F** completely in the spherical system.

Ans. We have

$$\rho = \sqrt{x^2 + y^2} = \sqrt{r^2\cos^2\phi.\sin^2\theta + r^2\sin^2\phi.\sin^2\theta} = r.\sin\theta$$

$$\hat{x} = (\hat{x}.\hat{r})\hat{r} + (\hat{x}.\hat{\theta})\hat{\theta} + (\hat{x}.\hat{\phi})\hat{\phi}$$

$$= \hat{r}.\cos\phi.\sin\theta + \hat{\theta}.\cos\phi\cos\theta - \hat{\phi}.\sin\phi$$

$$\hat{z} = (\hat{z}.\hat{r})\hat{r} + (\hat{z}.\hat{\theta})\hat{\theta} + (\hat{z}.\hat{\phi})\hat{\phi}$$

$$= \hat{r}.\cos\theta - \hat{\theta}.\sin\theta$$

Plugging these into the given expression for the vector field gives

$$\mathbf{F} = F_x(r.\sin\theta, r.\cos\phi.\sin\theta, \phi)(\hat{r}.\cos\phi.\sin\theta + \hat{\theta}.\cos\phi\cos\theta - \hat{\phi}.\sin\phi)$$

$$+ F_y(r.\sin\phi.\sin\theta, r.\cos\theta, r.\sin\theta)\hat{\phi} + F_z(r.\cos\phi.\sin\theta, r.\sin\theta)(\hat{r}.\cos\theta - \hat{\theta}.\sin\theta)$$

We leave the job of simplifying this to the student.

[18] Let $\mathbf{F} = F_\rho(\rho, \phi, z)\hat{\rho} + F_z(\rho, \phi, z)\hat{z}$ be a vector field specified in the cylindrical system. Consider the region V defined by $0 < \rho < R$, $a < z < b$, $0 < \phi < 2\pi$. Let S denote the surface bounding V. Evaluate $\int_S \mathbf{F}.\mathbf{n}dS$ and $\int_V \nabla.\mathbf{F}dV$ and verify their equality.

Ans.

$$\nabla.\mathbf{F} = \frac{1}{\rho}\frac{\partial \rho F_\rho}{\partial \rho} + \frac{\partial F_z}{\partial z}$$

Thus,

$$\int_V \nabla.\mathbf{F}dV = \int_{0<\rho<R, a<z<b, 0<\phi<2\pi} \nabla.\mathbf{F}\rho.d\rho.d\phi.dz$$

$$= \int_{0<\rho<R, a<z<b, 0<\phi<2\pi} \frac{\partial \rho F_\rho}{\partial \rho} d\rho d\phi dz$$

$$+ \int_{0<\rho<R, a<z<b, 0<\phi<2\pi} \frac{\partial F_z}{\partial z} \rho d\rho d\phi dz$$

$$= \int_{0<\phi<2\pi\ <a<z<b} F_\rho(R, \phi, z) R d\phi dz + \int_{0<\rho<R, 0<\phi<2\pi} (F_z(\rho, \phi, b)$$
$$- F_z(\rho, \phi, a))\rho d\rho d\phi$$

The first integral is the flux of **F** through the side of the cylinder and the second integral is the flux through the top and bottom surfaces of the cylinder. This proves the validity of Gauss' theorem.

[19] Given a surface $g(\rho, \phi, z) = 0$ in cylindrical co-ordinates and a vector $\mathbf{F} = F_\rho\hat{\rho} + F_\phi\hat{\phi} + F_z\hat{z}$, determine the component of **F** tangential to the surface at the point (ρ_0, ϕ_0, z_0) on it.

Ans. The normal to the surface at the given point is given by

$$\mathbf{a} = \nabla g = \hat{\rho}\frac{\partial g}{\partial \rho} + \frac{\hat{\phi}}{\rho}\frac{\partial g}{\partial \phi} + \hat{z}\frac{\partial g}{\partial z}$$
$$= a_\rho\hat{\rho} + a_\phi\hat{\phi} + a_z\hat{z} \text{ say,}$$

where the partial derivatives are evaluated at (ρ_0, ϕ_0, z_0). The normal component of **F** to the surface at this point is given by $(\mathbf{F}.\mathbf{a})\mathbf{a}/|\mathbf{a}|^2$ and the tangential component by

$$\mathbf{F}_t = \mathbf{F} - (\mathbf{F}.\mathbf{a})\mathbf{a}/|\mathbf{a}|^2 = F_{t,\rho}\hat{\rho} + F_{t,\phi}\hat{\phi} + F_{t,z}\hat{z}$$

where

$$F_{t,\rho} = F_\rho - C\left(F_\rho \frac{\partial g}{\partial \rho} + \frac{F_\phi}{\rho}\frac{\partial g}{\partial \phi} + F_z \frac{\partial g}{\partial z} \right)\frac{\partial g}{\partial \rho}$$

$$F_{t,\phi} = F_\phi - C\left(F_\rho \frac{\partial g}{\partial \rho} + \frac{F_\phi}{\rho}\frac{\partial g}{\partial \phi} + F_z \frac{\partial g}{\partial z} \right)\frac{1}{\rho}\frac{\partial g}{\partial \phi}$$

$$F_{t,z} = F_z - C\left(F_\rho \frac{\partial g}{\partial \rho} + \frac{F_\phi}{\rho}\frac{\partial g}{\partial \phi} + F_z \frac{\partial g}{\partial z} \right)\frac{\partial g}{\partial z}$$

where

$$C = \left((\frac{\partial g}{\partial \rho})^2 + \frac{1}{\rho^2}(\frac{\partial g}{\partial \phi})^2 + \left(\frac{\partial g}{\partial z}\right)^2 \right)^{-1}$$

[20] Let

$$\mathbf{A} = A_\rho(\rho,\phi,z)\hat{\rho} + A_\phi(\rho,\phi,z)\hat{\phi} + A_z(\rho,\phi,z)\hat{z}$$

$$\mathbf{B} = B_r(r,\theta,\phi)\hat{r} + B_\theta(r,\theta,\phi)\hat{\theta} + B_\phi(r,\theta,\phi)\hat{\phi}$$

Calculate the component of **A** along **B** at the point (*x, y, z*) specified in Cartesian co-ordinates. We determine first the cylindrical and spherical polar co-ordinates of the point (*x, y, z*). These are respectively given by

$$\rho = \sqrt{x^2 + y^2}, \phi = \tan^{-1}(y/x), z = z$$

$$r = \sqrt{x^2 + y^2 + z^2},\ \theta = \cos^{-1}(z/r), \phi = \tan^{-1}(y/x)$$

We evaluate A_ρ, A_ϕ, A_z, B_r, B_θ, B_ϕ at these values. Then, the component of **A** along **B** is given by $(\mathbf{A}.\mathbf{B})/|\mathbf{B}|$, where

$$|\mathbf{B}| = \sqrt{B_r^2 + B_\theta^2 + B_\phi^2}$$

and

$$\mathbf{A.B} = A_\rho B_r \hat{\rho}.\hat{r} + A_\rho B_\theta \hat{\rho}.\hat{\theta} + A_\rho B_\varphi \hat{\rho}.\hat{\phi}$$

$$+ A_\phi B_r \hat{\phi}.\hat{r} + A_\phi B_\theta \hat{\phi}.\hat{\theta} + A_\phi B_\phi$$

$$+ A_z B_r \hat{z}.\hat{r} + A_z B_\theta \hat{z}.\hat{\theta} + A_z B_\phi \hat{z}.\hat{\phi}$$

We can evaluate the various scalar products

$$\hat{\rho}.\hat{r} = (x\hat{x} + y\hat{y}).(x\hat{x} + y\hat{y} + z\hat{z}) / \rho.r = (x^2 + y^2) / \rho.r = \rho / r = \sin\theta$$

$$\hat{\rho}.\hat{\theta} = (\hat{x}\cos\phi + \hat{y}\sin\phi).(\hat{x}.\cos\phi.\cos\theta + \hat{y}.\sin\phi.\cos\theta - \hat{z}.\sin\theta)$$

$$= \cos^2\phi\cos\theta + \sin^2\phi\cos\theta = \cos\theta$$

$$\hat{\rho}.\hat{\phi} = (\hat{x}\cos\phi + \hat{y}\sin\phi).(-\hat{x}\sin\phi + \hat{y}\cos\phi) = 0$$

$$\hat{\phi}.\hat{r} = (-\hat{x}.\sin\phi + \hat{y}.\cos\phi).(\hat{x}.\cos\phi.\sin\theta + \hat{y}.\sin\phi.\sin\theta + \hat{z}.\cos\theta)$$

$$= -\sin\phi.\cos\phi.\sin\theta + \cos\phi.\sin\phi.\sin\theta = 0$$

$$\hat{\phi}.\hat{\theta} = (-\hat{x}.\sin\phi + \hat{y}.\cos\phi).(\hat{x}.\cos\phi.\cos\theta + \hat{y}.\sin\phi.\cos\theta - \hat{z}.\sin\theta)$$

$$= -\sin\phi.\cos\phi.\cos\theta + \cos\phi.\sin\phi.\cos\theta = 0$$

$$\hat{z}.\hat{r} = \cos\theta, \hat{z}.\hat{\theta} = -\sin\theta, \hat{z}.\hat{\phi} = 0$$

Substituting these values of the scalar products we finally get

$$\mathbf{A.B} = \sin\theta\, A_\rho B_r + \cos\theta\, A_\rho B_\theta + \cos\theta\, A_z B_r + A_\phi B_\phi$$

[21] Verify the divergence theorem for a vector field specified in Cartesian co-ordinates where the surface is a cuboid of side lengths *a, b, c* along the three co-ordinate axes.

Ans.

$$\int_V \nabla.\mathbf{F} dV = \int_0^a \int_0^b \int_0^c \left(\frac{\partial F_x}{\partial x} + \frac{\partial F_y}{\partial y} + \frac{\partial F_z}{\partial z} \right) dxdydz$$

The first integral is

$$\int_0^a \int_0^b \int_0^c \frac{\partial F_x}{\partial x} dxdydz = \int_0^b \int_0^c (F_x(a,y,z) - F_x(0,y,z))\, dydz$$

The second integral is

$$\int_0^a \int_0^b \int_0^c \frac{\partial F_y}{\partial y} dxdydz = \int_0^a \int_0^c (F_y(x,b,z) - F_y(x,0,z))\, dxdz$$

and finally, the third integral is

$$\int_0^a \int_0^b \int_0^c \frac{\partial F_z}{\partial z} dxdydz = \int_0^a \int_0^b (F_z(x,y,c) - F_z(x,y,0))\, dxdy$$

The integral of the divergence over the volume equals the sum of these three expressions and it is easily seen that this equals the flux of **F** through the surface, i.e. the surface integral of **F.n** over the surface of the cuboid.

[22] Given a surface in the shape of an ice-cream built out of the union of the conical surface $S_1 : \theta = \alpha,\ 0 < r < R$ and the section of the hemisphere above it S_2 having its centre as the origin and radius R. First of all, sketch the surface. Second, if $\mathbf{F} = F_x(x, y, z)\,\hat{x} + F_y(x, y, z)\,\hat{y}$, then evaluate (a) $\int_L \mathbf{F}.\mathbf{dr}$, where L is the circular edge $\rho = R.\sin\alpha$, $z = R.\cos\alpha$, (b) $\int_{S_1} \nabla \times \mathbf{F}.\mathbf{n}dS$, (c) $\int_{S_2} \nabla \times \mathbf{F}.\mathbf{n}dS$, (d) $\int_{S_1} \mathbf{F}.\mathbf{n}dS$, (e) $\int_{S_2} \mathbf{F}.\mathbf{n}dS$, (f) $\int_V \nabla.\mathbf{F}dV$. Also deduce the relations between these quantities.

Ans. (a,b): First observe that in terms of Cartesian components,

$$\nabla \times \mathbf{F} = \left(-\frac{\partial F_y}{\partial z}, \frac{\partial F_x}{\partial z}, \frac{\partial F_y}{\partial x} - \frac{\partial F_x}{\partial y}\right)$$

We assume that F_x, F_y have been expressed as functions of r, θ, ϕ. Then,

$$\frac{\partial F_y}{\partial z} = \frac{\partial r}{\partial z}\frac{\partial F_y}{\partial r} + \frac{\partial \theta}{\partial z}\frac{\partial F_y}{\partial \theta} + \frac{\partial \phi}{\partial z}\frac{\partial F_y}{\partial \phi}$$

$$= \cos\theta\frac{\partial F_y}{\partial r} - \frac{\sin\theta}{r}\frac{\partial F_y}{\partial \theta}$$

since

$$\frac{\partial r}{\partial z} = z/r = \cos\theta$$

$$\frac{\partial \theta}{\partial z} = \frac{\partial}{\partial z}\cos^{-1}(z/r) = -\sqrt{r^2 - z^2}\,/r = -(\sin\theta)/r$$

Likewise,

$$\frac{\partial F_x}{\partial z} = \cos\theta\frac{\partial F_x}{\partial r} - \frac{\sin\theta}{r}\frac{\partial F_x}{\partial \theta}$$

$$\frac{\partial F_y}{\partial x} = \frac{\partial r}{\partial x}\frac{\partial F_y}{\partial r} + \frac{\partial \theta}{\partial x}\frac{\partial F_y}{\partial \theta} + \frac{\partial \phi}{\partial x}\frac{\partial F_y}{\partial \phi}$$

$$= \cos\phi\sin\theta\frac{\partial F_y}{\partial r} + \cos\phi\cos\theta\frac{\partial F_y}{\partial \theta} - \frac{\sin\phi}{r\sin\theta}\frac{\partial F_y}{\partial \phi}$$

$$\frac{\partial F_x}{\partial y} = \frac{\partial r}{\partial y}\frac{\partial F_x}{\partial r} + \frac{\partial \theta}{\partial y}\frac{\partial F_x}{\partial \theta} + \frac{\partial \phi}{\partial y}\frac{\partial F_x}{\partial \phi}$$

$$= \sin\phi\sin\theta\frac{\partial F_x}{\partial r} + \sin\phi\cos\theta\frac{\partial F_x}{\partial \theta} + \frac{\cos\phi}{r\sin\theta}\frac{\partial F_x}{\partial \phi}$$

On the conical surface S_1, we have $\theta = \alpha$ and

$$\mathbf{n} = (\cos\phi\cos\alpha, \sin\phi\cos\alpha, -\sin\alpha),\quad dS = dr.r.\sin\alpha.d\phi$$

Plugging these values, we get

$$\int_{S_1} \nabla \times \mathbf{F}.\mathbf{n}dS = \int_{0<r<R,0<\phi<2\pi} \left(-\frac{\partial F_y}{\partial z}\cos\phi\cos\alpha + \frac{\partial F_x}{\partial z}\sin\phi\cos\alpha \right.$$

$$\left. -\left(\frac{\partial F_y}{\partial x} - \frac{\partial F_x}{\partial y}\right)\sin\alpha \right) r\sin\alpha dr d\phi$$

$$= \int_0^R \int_0^{2\pi} \left\{ \left(-\cos\alpha\frac{\partial F_y}{\partial r} + \frac{\sin\alpha}{r}\frac{\partial F_y}{\partial \theta} \right)\cos\phi\cos\alpha \right.$$

$$+ (\cos\alpha\frac{\partial F_x}{\partial r} - \frac{\sin\alpha}{r}\frac{\partial F_x}{\partial \theta})\sin\phi\cos\alpha$$

$$+ (\sin\phi\sin\alpha\frac{\partial F_x}{\partial r} + \frac{\sin\phi\cos\alpha}{r}\frac{\partial F_x}{\partial \theta}$$

$$+ \frac{\cos\phi}{r\sin\alpha}\frac{\partial F_x}{\partial \phi} - \cos\phi\sin\alpha\frac{\partial F_y}{\partial r} - \frac{\cos\phi\cos\alpha}{r}\frac{\partial F_y}{\partial \theta}$$

$$\left. + \frac{\sin\phi}{r\sin\alpha}\frac{\partial F_y}{\partial \phi} \right)\sin\alpha \Bigg\} r\sin\alpha dr d\phi$$

$$= \int_0^R \int_0^{2\pi} \left\{ -\cos\phi\frac{\partial F_y}{\partial r} + \sin\phi\frac{\partial F_x}{\partial r} + \frac{\cos\phi}{r}\frac{\partial F_x}{\partial \phi} + \frac{\sin\phi}{r}\frac{\partial F_y}{\partial \phi} \right\} r\sin\alpha dr d\phi$$

$$= T_1 + T_2$$

The integrand is evaluated at (r, α, ϕ). Here,

$$T_1 = \int_0^R \int_0^{2\pi} \left(-\cos\phi\frac{\partial F_y}{\partial r} + \sin\phi\frac{\partial F_x}{\partial r} \right) r\sin\alpha dr d\phi$$

$$= \sin\alpha \int_0^R \int_0^{2\pi} \frac{\partial}{\partial r}(-r\cos\phi\, F_y + r\sin\phi\, F_x) dr d\phi$$

$$+ \sin\alpha \int_0^R \int_0^{2\pi} (\cos\phi\, F_y - \sin\phi\, F_x) dr d\phi$$

$$= \sin\alpha \int_0^{2\pi} (\sin\phi\, F_x(R,\alpha,\phi) - \cos\phi\, F_y(R,\alpha,\phi)) R d\phi$$

$$+ \sin\alpha \int_0^R \int_0^{2\pi} (\cos\phi\, F_y - \sin\phi\, F_x) dr d\phi$$

where we have used integration by parts. Likewise,

$$T_1 = \int_0^R \int_0^{2\pi} (\cos\phi \frac{\partial F_x}{\partial \phi} + \sin\phi \frac{\partial F_y}{\partial \phi}) \sin\alpha dr d\phi$$

$$= \sin\alpha \int_0^R \int_0^{2\pi} \frac{\partial}{\partial \phi}(F_x \cos\phi + F_y \sin\phi) dr d\phi$$

$$+ \sin\alpha \int_0^R \int_0^{2\pi} (F_x \sin\phi - F_y \cos\phi) dr d\phi$$

$$= +\sin\alpha \int_0^R \int_0^{2\pi} (F_x \sin\phi - F_y \cos\phi) dr d\phi$$

Adding these two expressions, we get

$$\int_{S_1} \nabla \times \mathbf{F}.\mathbf{n} dS = T_1 + T_2 = \int_0^{2\pi} (\sin\phi\, F_x(R,\alpha,\phi) - \cos\phi\, F_y(R,\alpha,\phi)) R\sin\alpha d\phi$$

This is precisely the line integral $-\int_L \mathbf{F}.d\mathbf{r}$. Thus Stokes' theorem is verified. Note the negative sign appearing in the line integral which is due to the fact that the surface S_1 extends below L rather than above it.

(c). We next evaluate $\int_{S_2} \nabla \times \mathbf{F}.\mathbf{n} dS$. To do this, we first note that on this surface has Cartesian components as calculated above:

$$\nabla \times \mathbf{F} = -\left(\cos\theta \frac{\partial F_y}{\partial r} - \frac{\sin\theta}{r} \frac{\partial F_y}{\partial \theta} \right),$$

$$\cos\theta \frac{\partial F_x}{\partial r} - \frac{\sin\theta}{r} \frac{\partial F_x}{\partial \theta},$$

$$\cos\phi \sin\theta \frac{\partial F_y}{\partial r} + \cos\phi \cos\theta \frac{\partial F_y}{\partial \theta} - \frac{\sin\phi}{r\sin\theta} \frac{\partial F_y}{\partial \phi}$$

$$- \sin\phi \sin\theta \frac{\partial F_x}{\partial r} - \sin\phi \cos\theta \frac{\partial F_x}{\partial \theta} - \frac{\cos\phi}{r\sin\theta} \frac{\partial F_x}{\partial \phi}$$

evaluated at (R, θ, ϕ). The scalar product of this with the unit normal on S_2 having Cartesian components

$$\mathbf{n} = (\cos\phi\sin\theta, \sin\phi\sin\theta, \cos\theta)$$

equals

$$\nabla \times \mathbf{F}.\mathbf{n} =$$

$$-\cos\phi\sin\theta\left(\cos\theta\frac{\partial F_y}{\partial r}-\frac{\sin\theta}{r}\frac{\partial F_y}{\partial\theta}\right)$$

$$+\sin\phi\sin\theta\left(\cos\theta\frac{\partial F_x}{\partial r}-\frac{\sin\theta}{r}\frac{\partial F_x}{\partial\theta}\right)$$

$$+\cos\theta\left(\cos\phi\sin\theta\frac{\partial F_y}{\partial r}+\frac{\cos\phi\cos\theta}{r}\frac{\partial F_y}{\partial\theta}-\frac{\sin\phi}{r\sin\theta}\frac{\partial F_y}{\partial\phi}\right.$$

$$\left.-\sin\phi\sin\theta\frac{\partial F_x}{\partial r}-\frac{\sin\phi\cos\theta}{r}\frac{\partial F_x}{\partial\theta}-\frac{\cos\phi}{r\sin\theta}\frac{\partial F_x}{\partial\phi}\right)$$

$$=\frac{\cos\phi}{r}\frac{\partial F_y}{\partial\theta}-\frac{\sin\phi}{r}\frac{\partial F_x}{\partial\theta}-\frac{\cos\theta}{r\sin\theta}\left(\cos\phi\frac{\partial F_x}{\partial\phi}+\sin\phi\frac{\partial F_y}{\partial\phi}\right)$$

Also, on S_2, $dS = R^2\sin\theta d\theta d\phi$. Thus,

$$\int_{S_2}\nabla\times\mathbf{F}.\mathbf{n}dS=$$

$$\int_0^{\alpha}\int_0^{2\pi}\left(R.\sin\theta\cos\phi\frac{\partial F_y}{\partial\theta}-R.\sin\theta\sin\phi\frac{\partial F_x}{\partial\theta}\right)d\theta d\phi$$

$$-\int_0^{\alpha}\int_0^{2\pi}\left(R\cos\theta\cos\phi\frac{\partial F_x}{\partial\phi}+R\cos\theta\sin\phi\frac{\partial F_y}{\partial\phi}\right)d\theta d\phi$$

$$=\int_0^{\alpha}\int_0^{2\pi}\frac{\partial}{\partial\theta}(R\sin\theta\cos\phi\, F_y-R\sin\theta\sin\phi\, F_x)d\theta d\phi$$

$$-\int_0^{\alpha}\int_0^{2\pi}(R\cos\theta\cos\phi\, F_y-R\cos\theta\sin\phi\, F_x)d\theta d\phi$$

$$-\int_0^{\alpha}\int_0^{2\pi}\frac{\partial}{\partial\phi}(R\cos\theta\cos\phi\, F_x+R\cos\theta\sin\phi\, F_y)d\theta d\phi$$

$$-\int_0^{\alpha}\int_0^{2\pi}(R\cos\theta\sin\phi\, F_x-R\cos\theta\cos\phi\, F_y)d\theta d\phi$$

$$=\int_0^{2\pi}(-F_x(R,\alpha,\phi)\sin\phi+F_y(R,\alpha,\phi)\cos\phi)R\sin\alpha d\phi$$

which is precisely the line integral $\int_L \mathbf{F}.d\mathbf{r}$

Exercises

[1] Consider a radial vector $\mathbf{A} = a(r)\mathbf{r}$, where the scalar function $a(r)$ depends only on the radial distance r. Evaluate $\nabla.\mathbf{A}$ and $\nabla \times \mathbf{A}$.

[2] If $f(r)$ is a radial scalar function, then evaluate ∇f.

[3] Prove Euler's theorem: If $\phi(x_1,...,x_n)$ is a homogeneous function of n variables, i.e. $\phi(cx_1,...,cx_n) = c^r\,\phi(x_1,...,x_n)$ where r is a constant and c, $x_1,...,x_n$ are arbitrary, then

$$\sum_{i=1}^{n} x_i \frac{\partial \phi}{\partial x_i} = r\phi$$

[4] Evaluate $\nabla^2 V$ where $V = \rho z^2 \sin(2m\phi)$ using the Cartesian co-ordinates as well as cylindrical coordinates and prove the equality of these two expressions. Assume m to be a positive integer.

[5] Evaluate $\nabla^2 V$ where $V = r^2(A + B\cos\phi.\sin\theta + C.\sin\phi.\sin\theta + D.\cos\theta)$ with A, B, C, D are constants. Do the evaluation in both the Cartesian and spherical system and prove their equalities.

[6] Newton's equations of motion are given by

$$m\frac{d^2\mathbf{r}}{dt^2} = \mathbf{F}$$

where

$$\mathbf{r} = x\hat{x} + y\hat{y} + z\hat{z}$$

is the position vector and

$$\mathbf{F} = F_x\hat{x} + F_y\hat{y} + F_z\hat{z}$$

is the force vector. Express these equations in terms of cylindrical and spherical polar co-ordinates. To do this express the position vector $\mathbf{r} = \rho\hat{\rho} + z\hat{z} = r\hat{r}$ and calculate the acceleration in cylindrical co-ordinates in terms of $\frac{d\rho}{dt}, \frac{\phi}{dt}, \frac{dz}{dt}$ and in the spherical polar co-ordinates in terms of $\frac{dr}{dt}, \frac{d\theta}{dt}$ and $\frac{d\phi}{dt}$. For example, you will use

$$\frac{d\hat{\rho}}{dt} = \frac{\partial\hat{\rho}}{\partial\phi}\frac{d\phi}{dt} = \hat{\phi}\frac{d\phi}{dt}$$

and

$$\frac{d\hat{r}}{dt} = \frac{\partial\hat{r}}{\partial\theta}\frac{d\theta}{dt} + \frac{\partial\hat{r}}{\partial\phi}\frac{d\phi}{dt}$$

$$= r\frac{d\theta}{dt}\hat{\theta} + r\sin\theta\frac{d\phi}{dt}\hat{\phi}$$

and so on.

[7] The cross product of two vectors

$$\mathbf{A} = A_x\hat{x} + A_y\hat{y} + A_z\hat{z}$$

and

$$\mathbf{B} = B_x\hat{x} + B_y\hat{y} + B_z\hat{z}$$

is given by

$$\mathbf{A} \times \mathbf{B} = (A_y B_z - A_z B_y)\,\hat{x} + (A_z B_x - A_x B_z)\,\hat{y} + (A_x B_y - A_y B_x)\,\hat{z}$$

Given two vectors with one expressed in Cartesian and the other in cylindrical, or one in cylindrical and the other in spherical or one in spherical and the other in Cartesian co-ordinates, find their cross product in each of the systems.

[8] Let

$$\mathbf{F} = Ax^n\hat{x} + By^n\hat{y} + Cz^n\hat{z}$$

be a vector field and Γ the contour obtained by joining the points (a_1, b_1, c_1), (a_2, b_2, c_2), (a_3, b_3, c_3), (a_4, b_4, c_4), (a_1, b_1, c_1) in that order. These are the Cartesian components of the points. Determine the line integral $\int_\Gamma \mathbf{F}.d\mathbf{r}$. Also determine the line integral for the contour Γ given by the intersection of the plane $z = d$ with the sphere $x^2 + y^2 + z^2 = R^2$ where $0 < d < R$. Assume that n is an integer.

[9] Given a vector field $\mathbf{F} = f(r)\hat{r}$ in the spherical system, determine its divergence and curl. Given a vector field $\mathbf{F} = f(\rho)\hat{\rho}$, determine its divergence and curl.

[10] Derive an expression for the volume enclosed by a surface of revolution $\rho = f(z)$ between the planes $z = a$ and $z = b$. Derive also an expression for the surface area of this surface of revolution. Finally, derive an expression for the unit normal to this surface at any given point on it.

[11] Given a vector field $\mathbf{F} = f(\rho)\hat{\rho} + g(z)\,\hat{z}$, evaluate both sides of the divergence theorem for the volume enclosed by the surfaces $\rho = a$, $z = 0$, $z = d$ and prove their equality.

[12] Given $\mathbf{F} = f(\rho)\hat{\rho} + g(z)\,\hat{z}$, evaluate both sides of the divergence theorem for the volume enclosed by the surfaces $\rho = a$, $\rho = b$, $z = 0$, $z = d$.

[13] The potential of a dipole is given by the expression

$$V(\mathbf{r}) = \frac{K\mathbf{p.r}}{r^3}$$

where $\mathbf{p}$ is a constant vector. Express this potential in Cartesian and spherical polar co-ordinates assuming that $\mathbf{p}$ is along the $\hat{z}$ axis. Evaluate the electric field in both the systems. It is defined by $\mathbf{E} = -\nabla V$. Show that $\nabla.\mathbf{E} = 0$ at all points except the origin using the spherical system.

[14] Given the vector field

$$\mathbf{F} = (f_1(x) + g_1(y) + h_1(z))\,\hat{x} + (f_2(x) + g_2(y) + h_2(z))\,\hat{y} + (f_3(x) + g_3(y) + h_3(z))\,\hat{z}$$

evaluate its divergence in the Cartesian system. Also using the divergence theorem, evaluate its flux out of a cuboid with faces $x = 0$, $x = a$, $y = 0$, $y = b$, $z = 0$, $z = c$. Evaluate the flux directly and verify the Gauss' divergence theorem.

[15] The force field in which a body moves is radial. It is given by $\mathbf{F} = f(r)\hat{r}$. Assuming that the body moves in a plane, show that the Newtonian equations of motion of the body are given in plane polar co-ordinates (r, ϕ), $(\theta = \pi/2)$ by

$$\frac{d^2 r}{dt^2} - r\left(\frac{d\theta}{dt}\right)^2 = f(r)$$

$$r\frac{d^2\theta}{dt^2} + 2\frac{dr}{dt}\frac{d\theta}{dt} = 0$$

Integrate the second equation to show that $r^2\dfrac{d\theta}{dt}$ = const. Hence derive an ordinary differential equation for the trajectory, i.e. r as a function of θ.

[15] Consider a vector field

$$\mathbf{F} = F_r(r, \theta, \phi)\,\hat{r} + F_\theta(r, \theta, \phi)\hat{\theta} + F_\phi(r, \theta, \phi)\hat{\phi}$$

Derive an expression for its flux out of a sphere of radius R. Derive an expression for the volume integral of $\nabla.\mathbf{F}$ inside this sphere and prove using basic integration results that both are equal.

[16] Derive an expression for the volume of the region $d < z < R$, $x^2 + y^2 + z^2 \leq R^2$ using triple integration. Derive also an expression for the surface area of the surface bounding this region.

[17] Given a curve on a sphere of radius R specified parametrically by $(\theta(t), \phi(t))$, $0 < t < 1$, and a vector field

$$\mathbf{F} = F_r(r, \theta, \phi)\,\hat{r} + F_\theta(r, \theta, \phi)\hat{\theta} + F_\phi(r, \theta, \phi)\hat{\phi}$$

derive an expression for the line integral of this vector field along the given curve.

[18] A particle moves on the inner surface of the paraboloid $z = a\rho^2$. Determine an expression for the velocity of the particle as well as its acceleration. Assuming that there are two forces acting on the particle, one a normal reaction from the surface of the paraboloid and two the downward force of gravity, obtain two differential equations that describe the motion of the particle, i.e. for the functions $\rho(t)$, $\phi(t)$.

[19] Assume that the charge density in space is given in terms of

spherical co-ordinates by the formula $\rho(r, \phi) = f(r)g(\phi)$. Determine an expression for the charge enclosed in the region between two concentric spherical shells of radii a and b respectively.

[20] Given a radial force field $\mathbf{F} = f(r)\hat{r}$, evaluate its line integral along the contour $ABCA$, where AB is the segment $0 < r < a$, $\theta = \pi/2$, $\phi = 0$, BC the arc $r = a$, $\theta = \pi/2$, $0 < \phi < \phi_0$ and CA the segment $-a < -r < 0$, $\theta = \pi/2$, $\phi = \phi_0$. Check that it tallies with the result obtained by applying Stokes' theorem.

[21] The electric field produced by an infinitesimal volume element of charge $\rho(\mathbf{r})dV$ at $\mathbf{R}$ is given by $\dfrac{C\rho(\mathbf{r})dV(\mathbf{R}-\mathbf{r})}{|\mathbf{R}-\mathbf{r}|^3}$. The total integral of this over a volume gives the electric field produced by the entire volume distribution of charge. If $\mathbf{F}(\mathbf{r}) = F_x(\mathbf{r})\hat{x} + F_y(\mathbf{r})\hat{y} + F_z(\mathbf{r})\hat{z}$ is a vector, then its volume integral over V is defined by

$$\int_V F(\mathbf{r})dV = \hat{x}\int_V F_x(\mathbf{r})dV + \hat{y}\int_V F_y(\mathbf{r})dV + \hat{z}\int_V F_z(\mathbf{r})dV$$

Show by integration that the electric field produced by a sphere of charge having a uniform charge distribution ρ outside it is the same as that produced by a point charge of the same total charge as that of the sphere placed at the origin.

[22] Consider a parallelopiped subtended by the vectors **a, b, c**. Sketch this parallelopiped and show that the volume enclosed by it is given by $\mathbf{a} \times \mathbf{b.c}$.

[23] A simple pendulum consists of a bob of mass m attached to a string from a point on the wall. The oscillations take place in a plane. Describe the position, velocity and acceleration vector of the pendulum in terms of the angle. θ, its rate of change $r^2\dfrac{d\theta}{dt}$ and its rate of change $\dfrac{d^2\theta}{dt^2}$ The force on the pendulum consists of two components, one a gravitational force downward and two the tension force of the string. Write down the expression for the total force vector acting on the bob and the equation of motion of the pendulum. Eliminate the tension component.

[24] A surface is parametrically described by the equation $\mathbf{r} = \mathbf{r}(u, v)$. A curve on the surface is defined by specifying (u,v) as functions of a parameter t, i.e. $(u(t), v(t))$. Show that the length of the curve between the points specified by the values t_1, t_2 of the parameter t is given by

$$L(t_1,t_2) = \int_{t_1}^{t_2} \left(\left| \frac{\partial \mathbf{r}(u(t),v(t))}{\partial u} \right|^2 + \left| \frac{\partial \mathbf{r}(u(t),v(t))}{\partial v} \right|^2 + 2\left(\frac{\partial \mathbf{r}(u(t),v(t))}{\partial u} \cdot \frac{\partial \mathbf{r}(u(t),v(t))}{\partial v} \right) \right)^{1/2} dt$$

[25] Write down the equation of the curve corresponding to a general longitude and latitude on a sphere of radius R and centre at the origin in (a) the Cartesian system (b) the cylindrical system (c) the spherical system. Assume that the north pole is at (0, 0, R).

[26] The surface of an ellipsoid is described by the equation $x^2/a^2 + y^2/b^2 + z^2/c^2 = 1$. By using the change of variables

$$x = a\cos\phi\sin\theta,\ y = b\sin\phi\sin\theta,\ z = c\cos\theta$$

calculate the volume enclosed by the elipsoid.

[26] Show that in plane polar co-ordinates, the length of a curve in the plane specified by the equation $r = r(\phi)$ is given by $L = \int \sqrt{dr^2 + r^2 d\phi^2}$.

[27] Let $r = r(\phi)$, $0 < \phi < 2\pi$ describe the equation of a closed curve in the plane. Show that the area enclosed by this curve is given by $A = \frac{1}{2}\int_0^{2\pi} r^2(\phi)d\phi$.

[28] A curve in the plane is given by the equation $r = r(\phi)$. A vector field in the plane is given by $\mathbf{F} = F_r(r, \phi)\,\hat{r} + F_z(r,\phi)\hat{\phi}$. Show that the line integral of this vector field around the curve can be expressed as

$$\int_\Gamma \mathbf{F}.d\mathbf{r} = \int_0^{2\pi} F_r(r,\phi)dr + F_\phi(r,\phi)rd\phi$$

Write down the expression for $\mathbf{A} = \nabla \times \mathbf{F}$ using the expression for curl in cylindrical co-ordinates. Hence write down the expression for $\int_S \mathbf{A}.\mathbf{n}dS$ where S is the area enclosed by the curve. To do this, use $dS = rdrd\phi$ with the limits $0 < \phi < 2\pi$, $0 < r < r(\phi)$. Evaluate the latter integral and hence prove Stokes' theorem for planar curves and planar vector fields using plane polar co-ordinates.

[29] Let $\mathbf{A} = A_x\hat{x} + A_y\hat{y} + A_z\hat{z}$ and $\mathbf{T} = T_x\hat{x} + T_y\hat{y} + T_z\hat{z}$ be two vector fields. Write down the expressions for

(a) $(\nabla.\mathbf{A})\,\mathbf{T}$, (b) $(\mathbf{A}, \nabla)\,\mathbf{T} + (\mathbf{T}, \nabla)\,\mathbf{A}$, (c) $\nabla\,(\mathbf{A}, \mathbf{T})$

(d) Show that

$$\nabla \times (\mathbf{A} \times \mathbf{T}) = (\nabla, \mathbf{T})\,\mathbf{A} - (\mathbf{A}, \nabla)\,\mathbf{T} + (\mathbf{T}, \nabla)\,\mathbf{A} - (\nabla, \mathbf{A})\,\mathbf{T}$$

(e) Show that

$$\nabla.\mathbf{A} \times \mathbf{T} = (\nabla \times \mathbf{A}).\mathbf{T} - \mathbf{A}.\nabla \times \mathbf{T}$$

2

CHAPTER

ELECTROSTATICS, COULOMB'S LAW AND GAUSS' LAW

Contents

INTRODUCTION

The subject of electrostatics deals with the computation of electric fields and forces on charges produced by static charges, i.e. by charges at rest. Coulomb experimentally discovered that the force between two charges at rest is directly proportional to the product of the charges and inversely to the square of the distance between them. The force is attractive when the charges are of the same sign and repulsive when they are of the opposite sign. In nature, negative charges occur with electrons as the basic unit and positive charges with protons as the basic unit. Of course, it is known that there exists a positive charge called the positron having the same mass as the electron and exactly the opposite charge. So we may equivalently take the positron as our unit. However, its discovery came much later than that of the proton and to give respect to history we take the proton as our basic unit. The idea of the electric field permeating space is fundamental to physics and engineering. The electric field does not require any medium for it to exist, it can exist even in vacuum. However, its strength in a given medium depends upon the dielectric constant of the medium, a notion that is explained later on in this chapter. The electric field follows the principle of superposition, i.e. the total field at any point due to several systems of charges is obtained by superposing the electric field vectors at that point due to each system. This fact combined with Coulomb's law is used to derive a general expression for the electric field due to a distribution of charges in space which may have either a discrete component, a continuous component or both. The field produced by a static charge distribution is irrotational or conservative. This means that its line integral around any closed path is zero. This fact is a simple consquence of the Coulomb law according to which the field due to a point charge is directed radially outward from the charge and its magnitude depends upon only the distance of the point in question from the charge producing the field. The conservative property also means that the work done by the field in transporting a test charge from one point to another is independent of the path chosen. This fact gives rise to the notion of the electrostatic potential function, a scalar function that at each point equals the work done by the electric field in transporting a unit positive charge from ∞ to the point in question. The potential depends only on the point and not on the path chosen to transport the charge from ∞ to the point the reason being that the electric field of a static charge distribution is conservative. The potential difference between two points is the work done by the field in transporting a unit positive charge from one point to another and again for a conservative field, i.e. that produced by a static distribution of charges, this is path independent. This fact shows that the electric field can be derived as the negative gradient of the potential function. It is a very useful idea since computation of the potential due to a charge distribution involves just a single integration while computation of the electric field involves three integrations. The idea of the potential function can be exploited to obtain the potential energy stored in a system of charges, i.e. the energy involved in bringing the charges from infinite separation to their present configuration. Using Coulomb's law and superposition, Karl Friederich Gauss, the great German mathematician, derived a law that states that the flux of the electric field out of a closed surface equals the net charge enclosed by the surface. This law goes by his name. Although since this derivation is based on Coulomb's law, it is also valid for moving charges and non static situations as experiment confirms. One can, using spherical symmetry equivalently, arrive at Coulomb's law starting with Gauss' law. This chapter explains and illustrates Gauss' law in considerable detail. Using the relation between the electric field and the potential function and Gauss' law, we arrive at the Poisson's equation for the electrostatic field. The Poisson's equation plays a fundamental role in applied mathematics and electrostatics is one important application. All of electrostatics is contained in the Poisson's equation. We can derive the Coulomb's law and superposition of electric fields as the solution of the Poisson's equation. This solution corresponds to a Green's function $\frac{1}{|\mathbf{r}-\mathbf{r}'|}$. In free space, the Poisson equation reduces to Laplace's equation, the solution to which we discuss under various kinds of boundary conditions corresponding to rectangular, cylindrical or spherical symmetry. Using the Green's theorems, it is shown that the Poisson's equation with the Dirichlet or Neumann boundary conditions has a unique solution. The Dirichlet boundary condition corresponds to specifying the potential on a closed surface while the Neumann boundary condition corresponds to specifying the normal component to the electric field on the surface. Various examples involving the computation of the potential function and electric field corresponding to different kinds of charge distributions are discussed in this chapter. They illustrate a variety of methods that can be used to compute the electric fields. Some of the applications are taken from the physics literature and some others have useful engineering applications. The field produced by a dipole is of great importance in applications. A dipole consists of two equal and opposite charges separated by a finite distance. For an infinitesimal dipole, the distance between the opposite charges is infinitesimal. The dipole field can be used to discuss electric properties of dielectric materials. A dielectric material consists of atoms or molecules which can be polarized by application of an external electric field. This means that the centres of positive and negative charges within the dielectric body can be separated by the action of an external field. Thus inside a dielectric that has been polarized, an additional electric field is produced by polarization charges that opposes the applied field and hence the field magnitude within the dielectric body is reduced. The properties of a dielectric are most conveniently discussed by means of a polarization vector which is the dipole strength per unit volume. Gauss' law for a dielectric can be formulated by defining the electric displacement vector involving both the electric field and the polarization vector. This provides a formulation of the law involving only external charges and the dielectric constant or permittivity of the dielectric medium. The potential due to a charge distribution at distances large compared to the dimensions of the distribution has a convenient expansion called the multipole expansion. This involves expanding the potential in powers of $1/r$ where r is the distance of the point in question from a point which we can call as the centre of the distribution. The coefficients are integrals of the charge density multiplied by certain functions of the polar angles where the distribution is to be determined. The cofficients of the first few leading terms in this expansion turn out to be the total charge, the total dipole moment, the total quadrupole moment of the distribution etc. In summary, the entire subject of electrostatics deals with integration of the Poisson's equation to determine the electric fields in terms of the charge distribution. Another aspect however deals with the motions of charged particles in an electrostatic field. However, we shall discuss this in the chapter on Maxwell equations.

1. Coulomb's Law and Superposition of the Electrostatic Force:

Coulomb discovered that the force between two charges Q_1 and Q_2 at rest is proportional to the product of the two charges and inversely proportional to the square of the distance between the two charges. Mathematically, this law can be expressed as $F = \dfrac{KQ_1Q_2}{r^2}$, where r is the distance between the two charges. The force is directed along the line joining the two charges, is repulsive if the two charges are of the same sign and attractive otherwise. The constant of proportionality K is set equal to $\dfrac{1}{4\pi\varepsilon_0}$ when the two charges are placed in vacuum. Inside a dielectric (which we shall discuss later), owing to the introduction of polarization charges, the force decreases by a factor of k, where k is a constant called the dielectric constant. If charge Q_1 is placed at $\mathbf{r}_1$ and charge Q_2 at $\mathbf{r}_2$, then the force on Q_2 due to Q_1 can be expressed in vector notation as

$$\mathbf{F} = \frac{Q_1Q_2(\mathbf{r}_2 - \mathbf{r}_1)}{4\pi\varepsilon_0 \mid \mathbf{r}_2 - \mathbf{r}_1 \mid^3}$$

This is called the electrostatic force, corresponding to the fact that it arises from charges at rest. For charges in motion, as we shall see later, Coulomb's law breaks down. The electric field at a point $\mathbf{r}_2$ in space produced by a charge Q_1 at rest and located at $\mathbf{r}_1$ is given by $\mathbf{E}(\mathbf{r}) = \mathbf{F}/Q_2$, where $\mathbf{F}$ is the force on charge Q_2 located at $\mathbf{r}_2$ due to Q_1. More generally, if we have a system of charges, then the electric field at a point in space is obtained as the force exerted on a unit positive charge located at that point. For a single charge Q located at $\mathbf{r}_0$, the electric field at $\mathbf{r}$ is thus

$$\mathbf{E}(\mathbf{r}) = \frac{Q(\mathbf{r} - \mathbf{r}_0)}{4\pi\varepsilon_0 \,|\, \mathbf{r} - \mathbf{r}_0 \,|^3}$$

The superposition principle holds for the electrostatic field. This is mathematically formulated along the following lines. Let charges $Q_i, i = 1, 2, ..., n$ be located at points $\mathbf{r}_i, i = 1, 2, ..., n$. Then, the electrostatic force on a charge Q located at $\mathbf{r}$ due to Q_i is given by $\mathbf{F}_i = \dfrac{QQ_i(\mathbf{r} - \mathbf{r}_i)}{4\pi\varepsilon_0 |r - r_i|^3}$. The electrostatic force on Q due to all the charges $\{Q_i\}$ is given by the sum

$$\mathbf{F} = \sum_{i=1}^{n} \mathbf{F}_i$$

Taking $Q = 1$ gives

$$\mathbf{E}(\mathbf{r}) = \sum_{i=1}^{n} \mathbf{E}_i = \sum_{i=1}^{n} \frac{Q_i(\mathbf{r} - \mathbf{r}_i)}{4\pi\varepsilon_0 |\mathbf{r} - \mathbf{r}_i|^3}$$

for the total electric field at $\mathbf{r}$ due to the system $\{Q_i\}$ of charges. Here, $\mathbf{E}_i$ is the electric field at $\mathbf{r}$ solely due to Q_i.

Problem 1: Express the electric field due to a charge Q at $\mathbf{r}_0$ at the point $\mathbf{r}$ in (a) Cartesian co-ordinates, (b) cylindrical co-ordinates, (c) spherical polar co-ordinates.

Solution: In Cartesian co-ordinates,

$$\mathbf{E} = E_x\hat{x} + E_y\hat{y} + E_z\hat{z}$$

where

$$E_x = \frac{Q(x - x_0)}{4\pi\varepsilon_0((x - x_0)^2 + (y - y_0)^2 + (z - z_0)^2)^{3/2}}$$

$$E_y = \frac{Q(y - y_0)}{4\pi\varepsilon_0((x - x_0)^2 + (y - y_0)^2 + (z - z_0)^2)^{3/2}}$$

$$E_z = \frac{Q(z - z_0)}{4\pi\varepsilon_0((x - x_0)^2 + (y - y_0)^2 + (z - z_0)^2)^{3/2}}$$

To express it in cylindrical co-ordinates, we note that the unit vectors along the co-ordinate directions are given by

$$\hat{\rho} = \hat{x}.\cos\phi + \hat{y}.\sin\phi, \qquad \hat{\phi} = -\hat{x}.\sin\phi + \hat{y}.\cos\phi$$

Thus,

$$E_\rho = \mathbf{E}.\hat{\rho} = Q\frac{(x-x_0)\cos\phi + (y-y_0).\sin\phi}{4\pi\varepsilon_0((x-x_0)^2+(y-y_0)^2+(z-z_0)^2)^{3/2}}$$

$$= Q\frac{(\rho.\cos\phi - \rho_0\cos\phi_0)\cos\phi + (\rho.\sin\phi - \rho_0.\sin\phi_0).\sin\phi}{4\pi\varepsilon_0(\rho^2+\rho_0^2+(z-z_0)^2 - 2\rho.\rho_0.\cos(\phi-\phi_0))^{3/2}}$$

$$= Q\frac{\rho - \rho_0\cos(\phi-\phi_0)}{4\pi\varepsilon_0(\rho^2+\rho_0^2+(z-z_0)^2 - 2\rho.\rho_0.\cos(\phi-\phi_0))^{3/2}}$$

Here, we are using the cylindrical co-ordinates (ρ, ϕ, z) and (ρ_0, ϕ_0, z_0) of **r** and $\mathbf{r}_0$ respectively. We note that $x = \rho.\cos\phi$, $y = \rho.\sin\phi$ and likewise for $\mathbf{r}_0$. The distance between **r** and $\mathbf{r}_0$ has been evaluated as

$$|\mathbf{r}-\mathbf{r}_0|^2 = (x-x_0)^2 + (y-y_0)^2 + (z-z_0)^2 = (\rho.\cos\phi - \rho_0.\cos\phi_0)^2 + (\rho.sin\phi - \rho_0.\sin\phi_0)^2 + (z-z_0)^2$$

$$= \rho^2 + \rho_0^2 + (z-z_0)^2 - 2\rho.\rho_0.\cos(\phi-\phi_0)$$

Likewise,

$$E_\phi = \mathbf{E}.\hat{\phi} = -E_x\sin\phi + E_y\cos\phi = Q\frac{-(x-x_0)\sin\phi + (y-y_0)\cos\phi}{4\pi\varepsilon_0((x-x_0)^2+(y-y_0)^2+(z-z_0)^2)^{3/2}}$$

$$= Q\frac{-(\rho.\cos\phi - \rho_0\cos\phi_0)\sin\phi + (\rho.\sin\phi - \rho_0.\sin\phi_0)\cos\phi}{4\pi\varepsilon_0(\rho^2+\rho_0^2+(z-z_0)^2 - 2\rho.\rho_0.cos(\phi-\phi_0))^{3/2}}$$

$$= Q\frac{\rho_0.sin(\phi-\phi_0)}{4\pi\varepsilon_0(\rho^2+\rho_0^2+(z-z_0)^2 - 2\rho.\rho_0.\cos(\phi-\phi_0))^{3/2}}$$

Finally,

$$E_z = \frac{Q(z-z_0)}{4\pi\varepsilon_0(\rho^2+\rho_0^2+(z-z_0)^2 - 2\rho.\rho_0.\cos(\phi-\phi_0))^{3/2}}$$

For the spherical co-ordinate system, we have the point **r** defined by (r, θ, ϕ) and the point $\mathbf{r}_0$ by (r_0, θ_0, ϕ_0), where

$$x = r.\cos\phi.\sin\theta,\ y = r.\sin\phi.\sin\theta,\ z = r.\cos\theta$$

and similar relations for $\mathbf{r}_0$. The unit vectors along the co-ordinate directions are

$$\hat{r} = \hat{x}.\cos\phi.\sin\theta + \hat{y}.\sin\phi.\sin\theta + \hat{z}.\cos\theta$$

$$\hat{\theta} = \hat{x}.\cos\phi.\cos\theta + \hat{y}.\sin\phi.\cos\theta - \hat{z}.\sin\theta$$
$$\hat{\phi} = -\hat{x}.\sin\phi + \hat{y}.\cos\phi$$

These three unit vectors are mutually orthogonal. We have

$$\mathbf{E}_r = \mathbf{E}.\hat{r} = \frac{Q(\mathbf{r}-\mathbf{r}_0).\hat{r}}{4\pi\varepsilon_0|\mathbf{r}-\mathbf{r}_0|^3}$$

$$\mathbf{E}_\theta = \mathbf{E}.\hat{\theta} = \frac{Q(\mathbf{r}-\mathbf{r}_0).\hat{\theta}}{4\pi\varepsilon_0|\mathbf{r}-\mathbf{r}_0|^3}$$

$$\mathbf{E}_\phi = \mathbf{E}.\hat{\phi} = \frac{Q(\mathbf{r}-\mathbf{r}_0).\hat{\phi}}{4\pi\varepsilon_0|\mathbf{r}-\mathbf{r}_0|^3}$$

Now,

$$|\mathbf{r}-\mathbf{r}_0|^2 = (r.\cos\phi.\sin\theta - r_0.\cos\phi_0.\sin\theta_0)^2 + (r.\sin\phi.\sin\theta - r_0.\sin\phi_0.\sin\theta_0)^2$$
$$+ (r.\cos\theta - r_0.\cos\theta_0)^2 = r^2 + r_0^2 - 2rr_0(\cos\theta.\cos\theta_0 + \sin\theta.\sin\theta_0.\cos(\phi-\phi_0))$$

Also,

$$(\mathbf{r}-\mathbf{r}_0).\hat{r} = r - \mathbf{r}_0.\hat{r} = r - r_0(\cos\theta.\cos\theta_0 + \sin\theta.\sin\theta_0.\cos(\phi-\phi_0))$$
$$(\mathbf{r}-\mathbf{r}_0).\hat{\theta} = -\mathbf{r}_0.\hat{\theta} = r_0(\sin\theta.\cos\theta_0 - \sin\theta_0.\cos\theta.\cos(\phi-\phi_0))$$
$$(\mathbf{r}-\mathbf{r}_0).\hat{\phi} = -\mathbf{r}_0.\hat{\phi} = r_0.\sin\theta_0.\sin(\phi-\phi_0)$$

Substituting these into the expression for the electric field, we get

$$E_r = \frac{Q}{4\pi\varepsilon_0}\frac{r - r_0(\cos\theta.\cos\theta_0 + \sin\theta.\sin\theta_0.\cos(\phi-\phi_0))}{(r^2 + r_0^2 - 2rr_0(\cos\theta.\cos\theta_0 + \sin\theta.\sin\theta_0.\cos(\phi-\phi_0)))^{3/2}}$$

$$E_\theta = \frac{Q}{4\pi\varepsilon_0}\frac{r_0(\sin\theta.\cos\theta_0 - \sin\theta_0.\cos\theta.\cos(\phi-\phi_0))}{(r^2 + r_0^2 - 2rr_0(\cos\theta.\cos\theta_0 + \sin\theta.\sin\theta_0.\cos(\phi-\phi_0)))^{3/2}}$$

$$E_\phi = \frac{Q}{4\pi\varepsilon_0}\frac{r_0.\sin\theta_0.\sin(\phi-\phi_0)}{(r^2 + r_0^2 - 2rr_0(\cos\theta.\cos\theta_0 + \sin\theta.\sin\theta_0.\cos(\phi-\phi_0)))^{3/2}}$$

2. Electric Field produced by a Distribution of Charges

When there is a continuous distribution of charge over a volume, we can characterize the distribution by means of a charge density, i.e. by specifying the quantity of charge per unit volume at each point in space. We denote this quantity by $\rho(\mathbf{r})$. Thus, the total amount of charge inside a volume V is given by the integral $Q(V) = \int_V \rho(\mathbf{r})d^3\mathbf{r}$. For a discrete distribution of point charges $\{Q_i\}$ at the respective points $\{\mathbf{r}_i\}$, the charge density is given by the generalized function $\rho(\mathbf{r}) = \Sigma_i Q_i\delta(\mathbf{r}-\mathbf{r}_i)$. The superposition principle implies that the electric field produced by such a static distribution of charge specified by the density $\rho(\mathbf{r})$ is given by

$$\mathbf{E}(\mathbf{r}) = \frac{1}{4\pi\varepsilon_0} \int \frac{\rho(\mathbf{R})(\mathbf{r}-\mathbf{R})}{|\mathbf{r}-\mathbf{R}|^3} d^3\mathbf{R}$$

In terms of rectangular components,

$$E_x(x,y,z) = \frac{1}{4\pi\varepsilon_0} \int \frac{\rho(X,Y,Z)(x-X)}{((x-X)^2+(y-Y)^2+(z-Z)^2)^{3/2}} dXdYdZ$$

$$E_y(x,y,z) = \frac{1}{4\pi\varepsilon_0} \int \frac{\rho(X,Y,Z)(y-Y)}{((x-X)^2+(y-Y)^2+(z-Z)^2)^{3/2}} dXdYdZ$$

$$E_z(x,y,z) = \frac{1}{4\pi\varepsilon_0} \int \frac{\rho(X,Y,Z)(z-Z)}{((x-X)^2+(y-Y)^2+(z-Z)^2)^{3/2}} dXdYdZ$$

If the charge density σ (ρ, ϕ, z) is specified in cylindrical co-ordinates, we can calculate the cylindrical components of the electric field by integrating the equations derived earlier:

$$E_\rho(\rho,\phi,z) = \int \frac{\sigma(\rho_0,\phi_0,z_0)(\rho-\rho_0\cos(\phi-\phi_0))}{4\pi\varepsilon_0(\rho^2+\rho_0^2+(z-z_0)^2-2\rho.\rho_0.\cos(\phi-\phi))^{3/2}} \rho_0 d\rho_0 d\phi_0 dz$$

$$E_\phi(\rho,\phi,z) = \int \frac{\sigma(\rho_0,\phi_0,z_0)\rho_0.\sin(\phi-\phi_0)}{4\pi\varepsilon_0(\rho^2+\rho_0^2+(z-z_0)^2-2\rho.\rho_0.\cos(\phi-\phi_0))^{3/2}} \rho_0 d\rho_0 d\phi_0 dz$$

$$E_z(\rho,\phi,z) = \int \frac{\sigma(\rho_0,\phi_0,z_0)(z-z_0)}{4\pi\varepsilon_0(\rho^2+\rho_0^2+(z-z_0)^2-2\rho.\rho_0.\cos(\phi-\phi_0))^{3/2}} \rho_0 d\rho_0 d\phi_0 dz_0$$

If however, the charge density is specified in spherical polar co-ordinates $\sigma(r,\theta,\phi)$, then the various components of the field can be expressed by integrating the corresponding expressions derived above:

$$E_r(r,\theta,\phi) = \frac{1}{4\pi\varepsilon_0} \int \frac{\sigma(r_0,\theta_0,\phi_0)(r-r_0(\cos\theta.\cos\theta_0+\sin\theta.\sin\theta_0.\cos(\phi-\phi_0))}{(r^2+r_0^2-2rr_0(\cos\theta.\cos\theta_0+\sin\theta.\sin\theta_0.\cos(\phi-\phi_0)))^{3/2}} r_0^2\sin\theta_0 dr_0 d\theta_0 d\phi_0$$

$$E_\theta(r,\theta,\phi) = \frac{1}{4\pi\varepsilon_0} \int \frac{\sigma(r_0,\theta_0,\phi_0)r_0(\sin\theta.\cos\theta_0-\sin\theta_0.\sin\theta.\cos(\phi-\phi_0))}{(r^2+r_0^2-2rr_0(\cos\theta.\cos\theta_0+\sin\theta.\sin\theta_0.\cos(\phi-\phi_0)))^{3/2}} r_0^2\sin\theta_0 dr_0 d\theta_0 d\phi_0$$

$$E_\phi(r,\theta,\phi) = \frac{1}{4\pi\varepsilon_0}\int \frac{\sigma(r_0,\theta_0,\phi_0)r_0.\sin\theta_0.\sin(\phi-\phi_0)}{(r^2+r_0^2-2rr_0(\cos\theta.\cos\theta_0+\sin\theta.\sin\theta_0.\cos(\phi-\phi_0)))^{3/2}}$$
$$r_0^2\sin\theta_0 dr_0 d\theta_0 d\phi_0$$

Problem 2: (a) Compute the electric field produced in space by a thin straight wire of length L carrying a charge σ (z) per unit length where z is the distance along the wire measured from one end. (b) What happens to the field as the length of the wire extends to ∞ on both sides and the linear charge density is a constant along the wire.

Solution: (a) We assume that the wire is placed along the z axis extending from $z = 0$ up to $z = L$. Coulomb's law and the principle of superposition imply that in Cartesian co-ordinates,

$$\mathbf{E}(x,y,z) = \frac{1}{4\pi\varepsilon_0}\int_0^L \frac{\sigma(Z)(x,y,z-Z)}{(x^2+y^2+(z-Z)^2)^{3/2}}dZ$$

(b)

$$\mathbf{E}(x,y,z) = \frac{\sigma}{4\pi\varepsilon_0}\int_{-\infty}^{\infty} \frac{(x,y,z-Z)}{(x^2+y^2+(z-Z)^2)^{3/2}}dZ$$

We use the following values of the integrals:

$$\int_{-\infty}^{\infty} \frac{1}{(\rho^2+z^2)^{3/2}}dz = \frac{2}{\rho^2}$$

$$\int_{-\infty}^{\infty} \frac{z}{(\rho^2+z^2)^{3/2}}dz = 0$$

to get

$$\mathbf{E}(x,y,z) = \frac{(x,y,0)}{2\pi\varepsilon_0\rho^2}$$

or in terms of cylindrical co-ordinates,

$$E_\rho = \frac{\sigma}{2\pi\varepsilon_0\rho},\ E_\phi = E_z = 0$$

Problem 3: Find the electric field produced by a thin circular loop of wire carrying a charge density of σ Coulombs per unit length.

Solution: First we assume that the loop is placed on the $X - Y$ plane with its centre coinciding with the origin of the co-ordinate system. With the angle ϕ measured from the $X - Z$ plane, we assume that the linear charge density is σ (ϕ), $0 < \phi < 2\pi$. Then, application of Coulombs's law shows that

$$\mathbf{E}(x,y,z) = \int_0^{2\pi} \frac{\sigma(\phi_0)Rd\phi_0(x - R.\cos\phi_0, y - R.\sin\phi_0, z)}{4\pi\varepsilon_0((x - R.\cos\phi_0)^2 + (y - R.\sin\phi_0)^2 + z^2)^{3/2}}$$

Letting $x = \rho.\cos\phi$, $y = \rho.\sin\phi$, from this expression, we get

$$E_\rho = \int \frac{\sigma(\phi_0)Rd\phi_0((x - R.\cos\phi_0).\cos\phi + (y - R.\sin\phi_0).\sin\phi)}{4\pi\varepsilon_0(\rho^2 + z^2 + R^2 - 2R.\rho.\cos(\phi - \phi_0))^{3/2}}$$

$$= \int \frac{\sigma(\phi_0)Rd\phi_0(\rho - R.\cos(\phi - \phi_0))}{4\pi\varepsilon_0(\rho^2 + z^2 + R^2 - 2R.\rho.\cos(\phi - \phi_0))^{3/2}}$$

$$E_\phi = \int \frac{\sigma(\phi_0)Rd\phi_0(-(x - R.\cos\phi_0).\sin\phi + (y - R.\sin\phi_0).\cos\phi)}{4\pi\varepsilon_0(\rho^2 + R^2 + z^2 - 2\rho.R.\cos(\phi - \phi_0))^{3/2}}$$

$$= \int \frac{\sigma(\phi_0).R^2.\sin(\phi - \phi_0)d\phi_0}{4\pi\varepsilon_0(\rho^2 + R^2 + z^2 - 2\rho.R.\cos(\phi - \phi_0))^{3/2}} = 0$$

The integral here is from $\phi_0 = 0$ to $\phi_0 = 2\pi$ and it can be transformed into an integral with respect to $x = \cos(\phi - \phi_0)$ from $x = \cos(\phi)$ to $x = \cos(\phi)$. It therefore, evaluates to zero. Finally,

$$E_z = \int \frac{\sigma(\phi_0)R.d\phi_0.z}{4\pi\varepsilon_0(\rho^2 + R^2 + z^2 - 2\rho.R.\cos(\phi - \phi_0))^{3/2}}$$

When σ is a constant, these integrals can be expressed in terms of elliptic integrals.

Problem 4: A charge is located at the origin. However, it tends to move about randomly to a point δ**r**, where δ**r** is random and small compared to R, the distance of the point from the origin where the electric field is being evaluated. Assume that the average value of δ**r** is zero and its second order moments are given by $< \delta x_a\ \delta x_b > = R_{ab}$. Find the approximate electric field correlation functions at **R**.

Solution:

$$E_a(\mathbf{R}) = \frac{Q}{4\pi\varepsilon_0} \frac{X_a - \delta x_a}{|\mathbf{R} - \delta\mathbf{r}|^3}$$

We make a Taylor series expansion of the field in powers of δr, retaining terms only up to first order.

$$|\mathbf{R} - \delta\mathbf{r}|^{-3} \approx (R^2 - 2\mathbf{R}.\delta\mathbf{r})^{-3/2} \approx R^{-3}(1 + 3\frac{\mathbf{R}.\delta\mathbf{r}}{R^2})$$

This gives

$$E_a(\mathbf{R}) \approx \frac{Q}{4\pi\varepsilon_0 R^3}\left(X_a - \delta x_a + \frac{3}{R^2}\sum_{b=1}^{3} X_b X_a \delta x_b\right)$$

Hence,

$$< E_a(\mathbf{R}) > \approx \frac{QX_a}{4\pi\varepsilon_0 R^3}$$

and

$$\delta E_a(\mathbf{R}) = \sum_{b=1}^{3} F_{ba}\delta x_b$$

where

$$F_{ba} = \frac{Q}{4\pi\varepsilon_0 R^3}\left(\frac{3}{R^2}X_b X_a - \delta_{ba}\right)$$

This gives the field correlations as

$$< \delta E_a . \delta E_b > = \sum_{i,j=1}^{3} F_{ai}F_{bj}R_{ij}$$

3. Electrostatic Potential

Consider the electric field produced by a point charge Q located at the origin. It is given by

$$\mathbf{E}(\mathbf{r}) = \frac{Q\mathbf{r}}{4\pi\varepsilon_0 r^3} = -\frac{Q}{4\pi\varepsilon_0}\nabla\frac{1}{r}$$

since direct differentiation shows that $\nabla\frac{1}{r} = -\frac{\mathbf{r}}{r^3}$. In terms of components,

$$\frac{\partial}{\partial x}(x^2 + y^2 + z^2)^{-1/2} = -\frac{x}{(x^2 + y^2 + z^2)^{3/2}}$$

$$\frac{\partial}{\partial y}(x^2 + y^2 + z^2)^{-1/2} = -\frac{y}{(x^2 + y^2 + z^2)^{3/2}}$$

$$\frac{\partial}{\partial z}(x^2 + y^2 + z^2)^{-1/2} = -\frac{z}{(x^2 + y^2 + z^2)^{3/2}}$$

It follows from translation invariance and linearity of the gradient operator that the electric field of a system of charges $\{Q_i\}$ located respectively at $\{\mathbf{r}_i\}$ is given by

$$\mathbf{E}(\mathbf{r}) = -\nabla V(\mathbf{r})$$

where

$$V(\mathbf{r}) = \sum_i \frac{Q_i}{4\pi\varepsilon_0|\mathbf{r}-\mathbf{r}_i|}$$

The work done by the field in taking a unit positive charge from $\mathbf{r}_a$ to $\mathbf{r}_b$ along any path is then given by

$$W_{ab} = \int_{\mathbf{r}_a}^{\mathbf{r}_b} E.dr = -\int_{\mathbf{r}_a}^{\mathbf{r}_b} \nabla V.d\mathbf{r} = V(\mathbf{r}_a) - V(\mathbf{r}_b)$$

This is called the potential difference between $\mathbf{r}_a$ and $\mathbf{r}_b$ due to the system of charges. Note that $\mathbf{E}$ is a conservative field, i.e. $\nabla \times \mathbf{E} = \mathbf{0}$. This follows from the fact that the curl of the gradient of any scalar field is identically zero (Established in the previous chapter). More generally, when we have a distribution of charges in space specified by the density $\rho(\mathbf{r})$, the electric field due to this is given by

$$\mathbf{E}(\mathbf{r}) = \frac{1}{4\pi\varepsilon_0}\int \frac{\rho(\mathbf{R})(\mathbf{r}-\mathbf{R})}{|\mathbf{r}-\mathbf{R}|^3} d^3\mathbf{R}$$

$$= -\frac{1}{4\pi\varepsilon_0}\int \rho(\mathbf{R})\nabla\frac{1}{|\mathbf{r}-\mathbf{R}|} d^3\mathbf{R} = -\nabla V(\mathbf{r})$$

where

$$V(\mathbf{r}) = \frac{1}{4\pi\varepsilon_0}\int \frac{\rho(\mathbf{R})}{|\mathbf{r}-\mathbf{R}|} d^3\mathbf{R}$$

V is called the potential of the charge distribution. Equivalently, it can be defined as the work done by an external field in bringing a unit positive charge from ∞ to the point $\mathbf{r}$:

$$\int_{\mathbf{r}}^{\infty} \mathbf{E}.dr = -\int_{\mathbf{r}}^{\infty} \nabla V.d\mathbf{r} = V(\mathbf{r}) - V(\infty) = V(\mathbf{r})$$

taking the reference $V(\infty) = 0$. The potential is easier to compute than the electric field since only one integral needs to be evaluated in contrast to the latter where we need to evaluate three integrals. Once the potential of a static charge distribution has been obtained, we can obtain the electric field by taking its gradient.

Problem 5: Consider a solid cylinder of radius a and length L carrying a uniform volume charge of ρ Coulombs. Find the electrostatic potential at a point on its axis and use this to calculate the electric field on the axis.

Solution: Assume that the base of the cylinder is on the X-Y plane with its centre coinciding with the origin and that the top face of the cylinder is on the plane $z = L$. The electric field at (0, 0, z) has only the z component and it is a function of only z. Hence

$$E_z(0,0,z) = -\frac{\partial V(0,0,z)}{\partial z}$$

Now integration shows that

$$V(0,0,z) = \frac{\rho}{4\pi\varepsilon_0}\int \frac{rdrd\phi d\xi}{\sqrt{r^2+(z-\xi)^2}}$$

the region of integration being $0 < r < a$, $0 < \xi < L$, $0 < \phi < 2\pi$. This gives

$$V(0,0,z) = \frac{\rho}{2\varepsilon_0}\int_0^a \int_0^a \frac{rdrd\xi}{\sqrt{r^2+(\xi-z)^2}}$$

$$= \frac{\rho}{2\varepsilon_0}\int_0^L (\sqrt{a^2+(\xi-z)^2} - |\xi - z|)d\xi$$

The integral $I_1 = \int_0^L (\sqrt{a^2+(\xi-z)^2}d\xi$ is evaluated by setting $\xi - z = a.\sinh\theta$. We get

$$I_1 = a^2\int_{\theta_1}^{\theta_2} \cosh^2\theta.d\theta$$

where $\theta_1 = -\sinh^{-1}(z/a)$ and $\theta_2 = \sinh^{-1}((L-z)/a)$. Replacing $\cosh^2\theta$ by $(1 + \cosh(2\theta))/2$ and performing the integration gives

$$I_1 = \frac{a^2}{2}(\theta_2 - \theta_1 + \sinh\theta_2\cosh\theta_2 - \sinh\theta_1\cosh\theta_1)$$

$$= \frac{a^2}{2}(\sinh^{-1}((L-z)/a) + \sinh^{-1}(z/a) + \frac{(L-z)}{a^2}\sqrt{a^2+(L-z)^2} + \frac{z}{a^2}\sqrt{a^2+z^2})$$

We also find

$$I_2 = \int_0^L |\xi - z|d\xi = \frac{1}{2}(z^2 + (L-z)^2)$$

The expression for the potential on the axis is finally

$$V(0, 0, z) = \frac{\rho}{2\varepsilon_0}(I_1 + I_2)$$

Problem 6: Calculate the potential for the same solid cylinder of charge at the point $(x, 0, z)$ where $|x| << a$. Make appropriate approximations.

Problem 7: Consider an infinitely long straight wire carrying a charge of σ Coulombs per unit length. The wire is parallel to the z axis and can suffer small displacements $\delta\mathbf{r}$ in the $X-Y$ plane where the displacements $\delta\mathbf{r} = (\delta x, \delta y)$ are random. Find the correlations in the electric field vectors at two points $(x_1, y_1, 0)$ and $(x_2, y_2, 0)$ on the X-Y plane. Make the approximation of small displacements.

Solution: The x and y components of the electric field at $(x_1, y_1, 0)$ and $(x_2, y_2, 0)$ are respectively given by the formulas

$$(E_{1x}, E_{1y}) = \frac{\sigma(x_1 - \delta x, y_1 - \delta y)}{2\pi((x_1 - \delta x)^2 + (y_1 - \delta y)^2)}$$

$$(E_{2x}, E_{2y}) = \frac{\sigma(x_2 - \delta x, y_2 - \delta y)}{2\pi((x_2 - \delta x)^2 + (y_2 - \delta y)^2)}$$

Let $\rho_1^2 = x_1^2 + y_1^2$ so that up to linear orders in δx, δy

$$\frac{(x_1 - \delta x; y_1 - \delta y)}{(x_1 - \delta x)^2 + (y_1 - \delta y)^2} = \frac{(x_1 - \delta x, y_1 - \delta y)}{\rho_1^2 - 2(x_1\delta x + y_1\delta y)}$$

$$= \frac{1}{\rho_1^2}(x_1 - \delta x, y_1 - \delta y)\left(1 + \frac{2}{\rho_1^2}(x_1\delta x + y_1\delta y)\right) =$$

$$\frac{1}{\rho_1^2}(x_1, y_1) + \frac{1}{\rho_1^4}((2x_1^2 - \rho_1^2)\delta x + 2x_1y_1\delta y, (2y_1^2 - \rho_1^2)\delta y + 2x_1y_1\delta x)$$

The fluctuation term in $\mathbf{E}_1$ is thus given by

$$(\delta E_{1x}, \delta E_{1y}) = \frac{\sigma}{2\pi\rho_1^4}(A_1\delta x + B_1\delta y, C_1\delta x + D_1\delta y)$$

where

$$A_1 = 2x_1^2 - \rho_1^2 = x_1^2 - y_1^2, B_1 = 2x_1y_1, D_1 = 2y_1^2 - \rho_1^2 = y_1^2 - x_1^2, C_1 = 2x_1y_1$$

Likewise, the fluctuation in $\mathbf{E}_2$ is given by

$$(\delta E_{2x}, \delta E_{2y}) = \frac{\sigma}{2\pi\rho_2^4}(A_2\delta x + B_2\delta y, C_2\delta x + D_2\delta y)$$

Letting

$$< \delta x^2 > = R_{xx}, < \delta x.\delta y > = R_{xy}, < \delta y^2 > = R_{yy}$$

we then get the following field correlations:

$$< \delta E_{1x}\delta E_{2x} >= \frac{\sigma^2}{4\pi^2(\rho_1\rho_2)^4}(A_1A_2R_{xx} + B_1B_2R_{yy} + (A_1B_2 + A_2B_1)R_{xy})$$

$$< \delta E_{1x}\delta E_{2y} >= \frac{\sigma^2}{4\pi^2(\rho_1\rho_2)^4}(A_1C_2R_{xx} + B_1D_2R_{yy} + (A_1D_2 + B_1C_2)R_{xy})$$

$$< \delta E_{1y}\delta E_{2x} >= \frac{\sigma^2}{4\pi^2(\rho_1\rho_2)^4}(C_1A_2R_{xx} + D_1B_2R_{yy} + (C_1B_2 + D_1A_2)R_{xy})$$

$$< \delta E_{1y}\delta E_{2y} >= \frac{\sigma^2}{4\pi^2(\rho_1\rho_2)^4}(C_1C_2R_{xx} + D_1D_2R_{yy} + (C_1D_2 + D_1C_2)R_{xy})$$

4. Energy stored in a Distribution of Charges

First assume that the charge field consists of isolated point charges $Q_1,\ldots,Q_n$ at $\mathbf{r}_1,\ldots, \mathbf{r}_n$ respectively. We want to determine the potential energy in this distribution. First assume that Q_1 alone is present at $\mathbf{r}_1$. The work done in bringing Q_2 from ∞ to $\mathbf{r}_2$ in the field generated by Q_1 is given by $W_{21} = Q_2\int_{\mathbf{r}_2}^{\infty}\mathbf{E}_1.d\mathbf{r} = \frac{Q_2Q_1}{4\pi\varepsilon_0|\mathbf{r}_2 - \mathbf{r}_1|}$. The work done in bringing Q_3 to $\mathbf{r}_3$ in the field generated by Q_1 and Q_2 is

$$Q_3\int_{\mathbf{r}}^{\infty}(\mathbf{E}_1 + \mathbf{E}_2).d\mathbf{r} = W_{31} + W_{32}$$

Here, $\mathbf{E}_1$ and $\mathbf{E}_2$ are respectively the fields generated by Q_1 and Q_2. The integrals evaluate to $W_{31} = \frac{Q_3Q_1}{4\pi\varepsilon_0|\mathbf{r}_3 - \mathbf{r}_1|}$, $W_{31} = \frac{Q_3Q_2}{4\pi\varepsilon_0|\mathbf{r}_3 - \mathbf{r}_2|}$. In this way, suppose that $Q_1,\ldots, Q_k$ have been brought to their respective positions $\mathbf{r}_1,\ldots,\mathbf{r}_k$. Then the work done in bringing Q_{k+1} to $\mathbf{r}_{k+1}$ from ∞ in the field generated by the k charges is given by $\sum_{i=1}^{k} W_{k+1,i}$. Adding all these contributions, we find that the total work done in assembing the n charges is given by

$$W = \sum_{k=2}^{n} \sum_{i=1}^{k-1} W_{ki} = \sum_{k=2}^{n} \sum_{i=1}^{k-1} Q_k \int_{\mathbf{r}_k}^{\infty} \mathbf{E}_i . d\mathbf{r} = \sum_{1 \le i < k \le n} \frac{Q_k Q_i}{4\pi\varepsilon_0 |\mathbf{r}_k - \mathbf{r}_i|}$$

This can be expressed in an alternate form

$$W = \sum_{k \neq i} \frac{Q_k Q_i}{8\pi\varepsilon_0 |\mathbf{r}_k - \mathbf{r}_i|}$$

Note that the potential field due to this system of charges at **r** is given by

$$V(\mathbf{r}) = \sum_{i=1}^{n} \frac{Q_i}{4\pi\varepsilon_0 |\mathbf{r} - \mathbf{r}_i|}$$

If we want to determine the potential at $\mathbf{r}_j$, then the contribution from the jth charge will be absent. Thus we obtain the following alternate form for the energy of the charge distribution:

$$W = \frac{1}{2} \sum_{j=1}^{n} Q_j V(\mathbf{r}_j)$$

This expression can be adapted to obtain the energy of a continuous distribution of charge specified by the density $\rho(\mathbf{r})$. If $V(\mathbf{r})$ is the potential of this distribution, then the energy is given by $W = \frac{1}{2} \int \rho(\mathbf{r}) V(\mathbf{r}) d^3\mathbf{r}$

5. The Poisson equation for the potential of a static charge distribution

Consider now the function $\frac{1}{r}$, where $r = \sqrt{x^2 + y^2 + z^2}$. We shall first show that $\nabla^2 \frac{1}{r} = -4\pi\delta(\mathbf{r})$. The spatial Fourier transform of $\nabla^2 \frac{1}{r}$ is $-k^2 \int \exp(-i\mathbf{k}.\mathbf{r}) d^3\mathbf{r}/r$. Taking as the pole of our polar co-ordinates the unit vector in the direction of **k**, we have

$$\int \exp(-i\mathbf{k}.\mathbf{r}) d^3\mathbf{r} / r = 2\pi \int_0^{\infty} \int_0^{\pi} \exp(-ikr.\cos\theta) r.\sin\theta.dr.d\theta$$

$$\frac{2\pi i}{k} \int_0^{\infty} (\exp(-ikr) - \exp(ikr))dr$$

We evaluate this integral as the limit of

$$\int_0^{\infty} \exp(-\varepsilon r)(\exp(-ikr) - \exp(ikr))dr = \frac{1}{\varepsilon + ik} - \frac{1}{\varepsilon - ik} = -\frac{2ik}{\varepsilon^2 + k^2}$$

as $\varepsilon \to 0$. Thus, the Fourier transform of $\nabla^2 \frac{1}{r}$ is the limit of $-\frac{4\pi k^2}{\varepsilon^2 + k^2}$ with $\varepsilon \to 0$. This limit is -4π and hence $\nabla^2 \frac{1}{r} = -4\pi\delta(\mathbf{r})$. Since the gradient operator is invariant under translations, it also follows that $\nabla^2 \frac{1}{|\mathbf{r}-\mathbf{r}_0|} = -4\pi\delta(\mathbf{r}-\mathbf{r}_0)$. Now consider the potential due to a continuous charge distribution specified by the density $\rho(\mathbf{r})$. The potential is

$$V(\mathbf{r}) = \frac{1}{4\pi\varepsilon_0} \int \frac{\rho(\mathbf{R})}{|\mathbf{r}-\mathbf{R}|} d^3\mathbf{R}$$

Taking the Laplacian of this equation gives

$$\nabla^2 V(\mathbf{r}) = \frac{1}{4\pi\varepsilon_0} \int \rho(\mathbf{R}) \nabla^2 \frac{1}{|\mathbf{r}-\mathbf{R}|} d^3\mathbf{R}$$

$$= -\frac{1}{\varepsilon_0} \int \rho(\mathbf{R})\delta(\mathbf{r}-\mathbf{R}) d^3\mathbf{R} = -\frac{\rho(\mathbf{r})}{\varepsilon_0}$$

6. Gauss' law for electrostatics

Noting from the earlier discussion that the electric field is given by $\mathbf{E} = -\nabla V$, where V satisfies Poisson's equation, it follows that $\nabla.\mathbf{E} = \rho/\varepsilon_0$. This is the differential form of Gauss' law. Its integral form can be obtained from Gauss' theorem relating volume integral of the divergence of a vector field to its surface integral over a closed surface. It is give by $\int_S \mathbf{E}.\mathbf{n}dS = \frac{Q}{\varepsilon_0}$, where $Q = \int_V \rho(\mathbf{r})dV$ is the total charge contained inside the volume V enclosed by the surface S. We can derive this law in another way starting directly from Coulomb's law for the field produced by a point charge. Assume that the point charge Q is located at the origin and let S be a closed surface surrounding the charge. If $\mathbf{r}$ is a point on the surface, the electric field at this point due to the charge at the origin is given by

$$\int \mathbf{E}(\mathbf{r}).\mathbf{n}dS(\mathbf{r}) = \frac{Q}{4\pi\varepsilon_0} \int_S \frac{dS.\cos\theta}{r^2}$$

where θ is the angle between the unit normal $\mathbf{n}$ to S at $\mathbf{r}$ and the radial vector $\hat{r}$. If we draw a cone from the origin to the surface covering the infinitesimal element dS and with the cone angle 2θ, then it is clear that $dS.\cos\theta$ is the projection of the surface element dS onto the plane perpendicular to the cone axis at **r**. Thus, $\frac{dS.\cos\theta}{r^2} = d\Omega(\mathbf{r})$, the solid angle subtended by the cone, and its integral over the sphere is 4π. This shows that for a point charge Q inside the surface, the field generated **E**(**r**) satisfies the equation

$$\int_S \mathbf{E}(\mathbf{r}).\mathbf{n}dS = \frac{Q}{\varepsilon_0}$$

More generally, assume that charges $Q_1,...,Q_n$ are inside the volume V bounded by the surface S. Let $\mathbf{E}_i(\mathbf{r})$ denote the electric field produced by Q_i. Then, the total electric field produced by all the charges $\{Q_i\}$ is by the superposition principle $\mathbf{E}(\mathbf{r}) = \sum_{i=1}^{n} \mathbf{E}_i(\mathbf{r})$. The flux of this field through S satisfies

$$\int \mathbf{E}.\mathbf{n}dS = \sum_{i=1}^{n} \int_S \mathbf{E}_i(\mathbf{r}).\mathbf{n}dS = \sum_{i=1}^{n} \frac{Q_i}{\varepsilon_0}$$

Assume now that a charge Q is located outside the volume V bounded by S. Assume that Q is located at the origin, which means that the origin falls outside V. Let **E**(**r**) denote the electric field produced by Q. Draw a ray from Q. This ray will meet the surface S at two points making angles θ_1, θ_2 with the outer normal to the surface at the two points. The solid angles subtended by a cone with apex at Q at these two points and axis along the ray are $d\Omega_1 = dS_1\cos\theta_1/r^2$ and $d\Omega_2 = dS_2.\cos\theta_2/r_2^2$. A simple picture shows that $d\Omega_2 = -d\Omega_1$ and hence by pairwise cancellation, we get $\int d\Omega = 0$. Thus, for Q outside the surface, $\int \mathbf{E}.\mathbf{n}dS = 0$. Now assume that charges $Q_1,...,Q_n$ are located inside the volume and $Q'_1,...,Q'_m$ are located outside the volume. Let $\mathbf{E}_i$ denote the electric field due to Q_i and $\mathbf{E}'_i$ the electric field due to Q'_j. By the above arguments, $\int_S \mathbf{E}_i.\mathbf{n}dS = Q_i/\varepsilon_0$ and $\int \mathbf{E}'_j.\mathbf{n}dS = 0$ for $i = 1, 2,..., n, j = 1, 2,...m$. Hence the total field due to all the charges inside and outside V, $\mathbf{E}(\mathbf{r}) = \Sigma_i \mathbf{E}_i + \Sigma_i \mathbf{E}'_i$ satisfies

$$\int \mathbf{E}.\mathbf{n}dS = \sum_i \int \mathbf{E}i.\mathbf{n}dS + \sum_i \int \mathbf{E}'_i.\mathbf{n}dS = \sum Q_i / \varepsilon_0$$

More generally if we have any charge distribution continuous or discrete or a combination of both, then the electric field due to all the charges satisfies

$$\int_S \mathbf{E}.\mathbf{n}dS = Q_{\text{encl}} / \varepsilon_0$$

where Q_{encl} is the charge inside V, i.e. the charge enclosed by the surface S. If $\rho(\mathbf{r})$ denotes the charge density in space, then we get from Gauss' law just derived

$$\int \mathbf{E}.\mathbf{n}dS = \int_V \rho.dV / \varepsilon_0$$

This is called the integral form of Gauss' law for electrostatics. The left hand side evaluates to $\int_V \nabla.\mathbf{E}dV$ by Gauss' theorem discussed in Chapter I. Thus, we end up with the equation

$$\int_V (\nabla.\mathbf{E} - \rho / \varepsilon_0)dV = 0$$

Since the volume V can be arbitrarily chosen, it follows that $\nabla.\mathbf{E} = \rho/\varepsilon_0$. This is the local differential form of Gauss' law for electrostatics. Written down in terms of the Cartesian system, this expands to

$$\frac{\partial E_x}{\partial x} + \frac{\partial E_y}{\partial y} + \frac{\partial E_z}{\partial z} = \frac{\rho}{\varepsilon_0}$$

7. Alternate derivation of the Poisson's equation

Consider a static system of charges. Let **E** denote the electric field produced by it. By the above discussion, **E** must be a conservative field, i.e. $\nabla \times \mathbf{E} = 0$. Combining this with Gauss' law, we obtain the complete set of equations for electrostatics:

$$\nabla.\mathbf{E} = \rho/\varepsilon_0, \; \nabla \times \mathbf{E} = \mathbf{0}$$

The latter equation implies that $\mathbf{E} = -\nabla V$, where V is a scalar function which we have termed as the electrostatic potential. Substituting this into the former equation gives

$$\nabla.\nabla V = \nabla^2 V = -\rho/\varepsilon_0$$

8. Uniqueness of the solution to Poisson's equation for a given boundary condition

In a region υ bounded by a surface S and having charge density $\rho(\mathbf{r})$, assume that there are two functions $V_1(\mathbf{r})$ and $V_2(\mathbf{r})$ satisfying $\nabla^2 V_1 = \nabla^2 V_2 = -\rho/\varepsilon_0$ inside the volume υ, with $V_1 = V_2 = \phi$ on the surface S. Letting $V = V_1 - V_2$, we get $\nabla^2 V = 0$ in the region υ and $V = 0$ on S. Application of Green's theorem then gives

$$\int_S (V\nabla^2 V + |\nabla V|^2)dv = \int_S V\frac{\partial V}{\partial \mathbf{n}}dS = 0$$

since $V = 0$ on S. Since $\nabla^2 V = 0$ inside υ, it follows that $\int_V |\nabla V|^2 dv = 0$

and hence V = const. inside V. $V = 0$ on S implies that the constant = 0 and therefore, $V = 0$, i.e. $V_1 = V_2$ inside υ. This proves uniqueness of the solution to Poisson's equation under the Dirichlet boundary condition. More generally, if the boundary condition is such that at each point on S either the potential or its normal derivative is specified, then we have with $V = V_1 - V_2$, the fact that the product $V\frac{\partial V}{\partial \mathbf{n}} = 0$ on the entire surface S. This again gives $\int_V |\nabla V|^2 dv = 0$ and hence V is a constant inside the volume υ. The potential is determined uniquely up to an additive constant in this case.

9. Solution to Poisson's equation in infinite space

We have already derived the Poisson's equation for the electrostatic potential starting from Coulomb's law and superposition. It is also possible to derive Coulomb's law starting from the Poisson's equation. This derivation is an elementary application of Green's theorem. Consider a spherical surface of very large radius R with origin as its centre. Noting that $\nabla^2 1/r = -4\pi\delta(\mathbf{r})$, it follows from the equation $\nabla^2 V = -\rho/\varepsilon_0$ that

$$\int_V (V\nabla^2 1/r - (1/r)\nabla^2 V)dv = -4\pi V(0) + \int_V \rho.dv/\varepsilon_0 r$$

On the other hand, by Green's theorem, the left side is $\int_S \left(-V/R^2 - (1/R)\left.\frac{\partial V}{\partial r}\right|_{r=R}\right) dS$ which tends to zero as $R \to \infty$ provided we assume that $V \to 0$ and $R\frac{\partial V}{\partial r} \to 0$ as $R \to \infty$. This proves that $V(0) = \frac{1}{4\pi\varepsilon_0}\int_V \rho.dv/r$, which is the Coulomb formula.

10. Green's theorem and boundary value problems of electrostatics

We consider the problem of determining the potential in the region V bounded by a surface S. The region V is filled with charge of density $\rho(\mathbf{r})$. The potential on the surface is specified. We shall now show that the determination of the potential in V is equivalent to the solution of the partial differential equation

$$\nabla^2 G(\mathbf{r}|\mathbf{r}') = \delta(\mathbf{r} - \mathbf{r}')$$

for $\mathbf{r}, \mathbf{r}' \in V$, with the boundary condition that $G(\mathbf{r}|\mathbf{r}') = 0$ for all $\mathbf{r}$ on the surface and all $\mathbf{r}' \in V$. The function $G(\mathbf{r}|\mathbf{r}')$ is called the Green's function for the problem. To see how the potential is determined, we assume that the potential in V is given by $V(\mathbf{r}')$. Application of Green's theorem shows that

$$\int_V (G(\mathbf{r}\,|\,\mathbf{r}')\nabla^2 V(\mathbf{r}) - V(\mathbf{r})\nabla^2 G(\mathbf{r}\,|\,\mathbf{r}'))d^3\mathbf{r}$$

$$= \int_S \left(G(\mathbf{r}\,|\,\mathbf{r}')\frac{\partial V(\mathbf{r})}{\partial \mathbf{n}} - V(\mathbf{r})\frac{\partial G(\mathbf{r}\,|\,\mathbf{r}')}{\partial \mathbf{n}} \right) dS(\mathbf{r})$$

The first term in the surface integrand on the right vanishes because $G(\mathbf{r}|\mathbf{r}') = 0$ for $\mathbf{r} \in S$. Moreover, the first term in the volume integrand on the left evaluates to $-\int_V G(\mathbf{r}|\mathbf{r}')\,\rho(\mathbf{r})d^3\mathbf{r}/\varepsilon_0$. The integral of the second term on the left side integrand evaluates to $-V(\mathbf{r}')$, since $\nabla^2 G(\mathbf{r}|\mathbf{r}') = \delta\,(\mathbf{r} - \mathbf{r}')$. This means that the above identity reduces to

$$V(\mathbf{r}') = -\int \rho(\mathbf{r})G(\mathbf{r}\,|\,\mathbf{r}')d^3\mathbf{r}/\varepsilon_0 + \int_S V(\mathbf{r})\frac{\partial G(\mathbf{r}\,|\,\mathbf{r}')}{\partial \mathbf{n}} dS(\mathbf{r})$$

The expression on the right here involves all known quantities: The charge density inside V and the potential on the surface S of V. In the special case when the volume is the entire $\mathbb{R}^3$, the surface may be taken as a sphere of infinite radius centred at $\mathbf{r}'$ and the appropriate Green's function is $\frac{-1}{4\pi|\mathbf{r}-\mathbf{r}'|}$. We have already noted that this satisfies the equation $\nabla^2 \frac{1}{4\pi|\mathbf{r}-\mathbf{r}'|} = \delta(\mathbf{r} - \mathbf{r}')$ and obviously, it vanishes when $r \to \infty$. Plugging this into the above equation assuming that the potential V satisfies the vanishing boundary condition at ∞ yields the Coulomb solution

$$V(\mathbf{r}') = \int_{\mathbb{R}^3} \frac{\rho(\mathbf{r})}{4\pi\varepsilon_0|\mathbf{r}-\mathbf{r}'|} d^3\mathbf{r}$$

Problem 7: Evaluate the Green's function for a one dimensional potential problem with the boundary being $x = a$ and $x = b$, where $a < b$.

Solution: We've got to solve $\frac{d^2G(x\,|\,x')}{dx^2} = \delta(x - x')$, with the condition $G(a|x') = G(b|x') = 0$, where $a < x' < b$. The solution in the region $a < x < x'$ is $G(x|x') = A_1x + B_1$ and the solution in the region $x' < x < b$ is $G(x|x') = A_2x+B_2$. Integration of the equation from $x' - 0$ to $x' + 0$ gives $\frac{dG(x'+0\,|\,x')}{dx} - \frac{dG(x'-0\,|\,x')}{dx} = 1$. This implies $A_2 - A_1 = 1$. Integration of the equation from $x = a$ to x gives $\frac{dG(x\,|\,x')}{dx} = u\,(x - x')$. Integration of this equation from $x' - 0$ to $x' + 0$ gives $G(x'+0|x') - G(x'-0|x') = 0$ and hence $A_1x' + B_1 = A_2x' + B_2$. Finally vanishing of $G(x|x')$ at $x = a$ and x

$= b$ leads to the equations $A_1 a + B_1 = A_2 b + B_2 = 0$. We have thus obtained four equations for A_1, B_1, A_2, B_2, the solution to which is

$$A_1 = \frac{x'-b}{b-a}, \quad A_2 = \frac{x'-a}{b-a}, \quad B_1 = \frac{a(b-x')}{b-a}, \quad B_2 = \frac{b(a-x')}{b-a}$$

and hence

$$G(x\,|\,x') = \frac{(x-a)(x'-b)}{b-a}, a < x < x', \quad G(x\,|\,x') = \frac{(x'-a)(x-b)}{b-a}, x' < x < b$$

Letting $x_> = \max(x, x')$ and $x_< = \min(x, x')$, it follows that the Green's function can be expressed as

$$G(x\,|\,x') = \frac{(x_< - a)(x_> - b)}{b-a}$$

Problem 8: Evaluate the two dimensional Green's function for a square of sidelengths a and b.

Solution: Assume that the region in which the Green's function is to be evaluated is $\{(x, y) : 0 < x < a, 0 < y < b\}$. The equation to be solved is

$$\frac{\partial^2 G(x,y\,|\,x',y')}{\partial x^2} + \frac{\partial^2 G(x,y\,|\,x',y')}{\partial y^2} = \delta(x-x')\delta(y-y')$$

where $0 < x, x' < a$, $0 < y, y' < b$. To solve this, we perform the half wave Fourier series expansion of $\delta(y - y')$:

$$\delta(y-y') = \frac{2}{b}\sum_{n=1}^{\infty} \sin(n\pi y/b).\sin(n\pi y'/b)$$

We likewise expand the Green's function as a function of y in a half wave Fourier series:

$$G(x,y\,|\,x',y') = \sum_{n=1}^{\infty} g_n(x,x',y')\sin(n\pi y/b)$$

Substituting these two expressions into the differential equation gives

$$\frac{d^2 g_n}{dx^2} - (n\pi/b)^2 g_n = \frac{2}{b}\delta(x-x').\sin(n\pi y'/b) \qquad (\alpha)$$

For $x < x'$ and $x > x'$, the right side is zero and we get the solution

$$g_n(x, x', y') = A_1.\cosh(n\pi x/b) + B_1.\sinh(n\pi x/b),\ 0 < x < x'$$

$g_n(x, x', y') = A_2.\cosh(n\pi(a - x)/b + B_2.\sinh(n\pi(a - x)/b), x' < x < a$

The boundary conditions on the Green's function imply that $g_n(0, x', y') = 0 = g_n(a, x', y')$ and hence $A_1 = 0 = A_2$. Further integrating (α) first from $x = 0$ to $x = x$ and then again from $x = x' - 0$ to $x = x' + 0$ with respect to x gives $g_n(x' - 0, x', y') = gn(x' + 0, x', y')$. This translates to

$$B_1.\sinh(n\pi x'/b) = B_2.\sinh(n\pi(a - x')/b)$$

Again integrating (α) from $x' - 0$ to $x' + 0$ with respect to x gives

$$\frac{dg_n(x'+0,x',y')}{dx} - \frac{dg_n(x'-0,x',y')}{dx} = \frac{2}{b}\sin(n\pi y'/b)$$

which translates to

$$-(n\pi B_2/b).\cosh(n\pi(a-x')/b - (n\pi B_1/b).\cosh(n\pi x'/b) = \frac{2}{b}\sin(n\pi y'/b)$$

Solving for B_1, B_2 gives

$$B_1 = -\frac{2}{n\pi}\frac{\sin(n\pi y'/b).\sinh(n\pi(a-x')/b)}{\sinh(n\pi a/b)}$$

$$B_2 = -\frac{2}{n\pi}\frac{\sin(n\pi y'/b).\sinh(n\pi x'/b)}{\sinh(n\pi a/b)}$$

This results in the following series expansion for the Green's function

$$G(x,y\,|\,x',y') = -\sum_{n=1}^{\infty}\frac{2}{n\pi}.$$

$$\frac{\sin(n\pi y/b).\sin(n\pi y'/b).\sinh(n\pi x/b).\sinh(n\pi(a-x')/b)}{\sinh(n\pi a/b)}, 0 < x < x'$$

$$G(x,y\,|\,x',y') = -\sum_{n=1}^{\infty}\frac{2}{n\pi}.$$

$$\frac{\sin(n\pi y/b).\sin(n\pi y'/b).\sinh(n\pi x'/b).\sinh(n\pi(a-x)/b)}{\sinh(n\pi a/b)}, x' < x < a$$

Letting $x_> = \max(x, x')$ and $x_< = \min(x, x')$, the expression for the Green's function can be compressed into the single equation

$$G(x,y\,|\,x',y')$$

$$= -\frac{2}{\pi}\sum_{n=1}^{\infty}\frac{\sin(n\pi y/b)\sin(n\pi y'/b)\sinh(n\pi x_</b).\sinh(n\pi(a-x_>)/b)}{\sinh(n\pi a/b)}$$

11. Field produced by a Dipole and the Multipole Expansion

Consider now a charge $+Q$ located at **d**/2 and a charge $-Q$ at $-$**d**/2. Here, **d** is an arbitrary vector. We say that this pair of charges comprises an electric dipole of strength $\mathbf{p} = Q\mathbf{d}$. The electric potential due this pair of charges is given by

$$V(\mathbf{r}) = \frac{Q}{4\pi\varepsilon_0|\mathbf{r}-\mathbf{d}/2|} - \frac{Q}{4\pi\varepsilon_0|\mathbf{r}+\mathbf{d}/2|}$$

If we now assume that the separation $d = |\mathbf{d}|$ between the two charges is very small, then with neglect of $O(d^2)$ terms, the electric potential at **r** due to this pair is given by

$$V(\mathbf{r}) \approx -\frac{Q}{4\pi\varepsilon_0}\mathbf{d}.\nabla(1/r)$$

This follows from the Taylor expansion: $f(\mathbf{r} + \mathbf{h}) = f(\mathbf{r}) + \mathbf{h},\nabla f(\mathbf{r}) + O(h^2)$. Noting that $\nabla\ (1/r) = -\mathbf{r}/r^3$, we obtain the the approximate dipole potential as

$$V(\mathbf{r}) = \frac{Q\mathbf{d.r}}{4\pi\varepsilon_0 r^3} = \frac{\mathbf{p.r}}{4\pi\varepsilon_0 r^3}$$

In particular, if $\mathbf{d} = d\hat{z}$ (this can be assumed without any loss of generality by assuming the $\hat{z}$ axis to be pointed along the d direction), then in polar co-ordinates,

$$V(r,\theta) = \frac{p\cos\theta}{4\pi\varepsilon_0 r^2}$$

where $p = Qd$ is the dipole strength. The above formula is called the potential due to an infinitesimal dipole. We can calculate the components of the electric field by differentiation.

$$\mathbf{E}(\mathbf{r}) = \frac{1}{4\pi\varepsilon_0}\nabla(\frac{\mathbf{p.r}}{r^3}) = \frac{1}{4\pi\varepsilon_0}\frac{r^3\nabla(\mathbf{p.r}) - (\mathbf{p.r})\nabla r^3}{r^6}$$

$$= \frac{\mathbf{p} - 3(\mathbf{p}.\hat{\mathbf{r}})\hat{r}}{4\pi\varepsilon_0 r^3}$$

We now consider a charge distribution in space specified by the density $\rho(\mathbf{r})$. The potential due to this distribution is given by

$$V(\mathbf{r}) = \frac{1}{4\pi\varepsilon_0}\int\frac{\rho(\mathbf{R})}{|\mathbf{r}-\mathbf{R}|}d^3\mathbf{R}$$

We make an expansion of this potential by assuming $R < r$, i.e. that the point where the potential is to be computed is further in distance from the origin than any point where the charge density is nonzero. Let θ denote the angle between the vectors **R** and **r**. Then,

$$|\mathbf{r}-\mathbf{R}|^{-1} = (r^2 + R^2 - 2\mathbf{r}.\mathbf{R})^{-1/2} = (r^2 + R^2 - 2rR.\cos\theta)^{-1/2} =$$

$$= r^{-1}(1+(R/r)^2 - 2(R/r).\cos\theta)^{-1/2} = r^{-1}\sum_{n=0}^{\infty} P_n(\cos\theta)(R/r)^n$$

where $P_n(\xi)$ are the Legendre polynomials. They are defined by their generating function

$$\sum_{n=0}^{\infty} t^n P_n(\xi) = (1+t^2 - 2t\xi)^{-1/2}$$

Noting that $\cos\theta = \frac{\mathbf{r}.\mathbf{R}}{rR}$, it follows that

$$V(\mathbf{r}) = \frac{1}{4\pi\varepsilon_0}\sum_{n=0}^{\infty} r^{-n-1} \int \rho(\mathbf{R})P_n\left(\frac{\mathbf{r}.\mathbf{R}}{rR}\right)R^n d^3\mathbf{R}$$

This is called the multipole expansion of the potential. It is a useful expansion for the computation of the far field potential due to a charge distribution. We may compute the first few terms in this expansion

$$P_0(\xi) = 1,\ P_1(\xi) = \xi,\ P_2(\xi) = (3\xi^2-1)/2$$

The corresponding truncated potential expansion is given by

$$V(\mathbf{r}) = \frac{1}{4\pi\varepsilon_0}\left(\frac{\mu_0}{r} + \frac{\mathbf{p}.\mathbf{r}}{r^3} + \frac{\sum Q_{ij}x_i x_j}{r^5} + \ldots\right)$$

where $\mu_0 = \int \rho(\mathbf{R})dV$ is the total charge of the distribution and **p** is the dipole moment for the charge distribution defined by the integral

$$\mathbf{p} = \int \rho(\mathbf{R})\mathbf{R}d^3\mathbf{R}$$

and Q_{ij}, $i, j = 1,2,3$ are the components of the quadrupole moment of the charge distribution. They are defined by the integrals

$$Q_{ij} = \frac{1}{2}\int \rho(\mathbf{R})(3X_i X_j - R^2\delta_{ij})d^3\mathbf{R}$$

Here, (X_1, X_2, X_3) and (x_1, x_2, x_3) are respectively the Cartesian components of the vectors **r** and **R**.

Problem 9: Assume that a charge $+Q$ is located at $(0, 0, d)$, a charge $+Q$ at $(0, 0, -d)$ and finally a charge $-2Q$ at $(0, 0, 0)$. Evaluate the approximate potential at large distances retaining up to quadrupole terms. The exact potential at (x, y, z) is given by

$$V(x,y,z) = \frac{Q}{4\pi\varepsilon_0}\left(\frac{1}{x^2+y^2+(z-d)^2)^{1/2}} + \frac{1}{(x^2+y^2+(z+d)^2)^{1/2}} - \frac{2}{(x^2+y^2+z^2)^{1/2}}\right)$$

We now note that

$$(x^2+y^2+(z-d)^2)^{-1/2} = (r^2 - 2dz + d^2)^{-1/2} = r^{-1}(1 + dz/r^2 - d^2/2r^2 + 3(d^2-2dz)^2/8r^4) + O(d^3/r^4)$$

$$= 1/r + dz/r^3 - d^2/2r^3 + 3d^2z^2/2r^5 + O(d^3/r^4)$$

Changing d to $-d$ in the above expression gives

$$(x^2+y^2+(z+d)^2)^{-1/2} = 1/r - dz/r^3 - d^2/2r^3 + 3d^2z^2/2r^5 + O(d^3/r^4)$$

Thus,

$$V(x,y,z) = \frac{Q}{4\pi\varepsilon_0}(-d^2/r^3 + 3d^2z^2/r^5) + O(d^3/r^4) =$$

$$= \frac{Qd^2(3\cos^2\theta - 1)}{4\pi\varepsilon_0 r^3} + O(d^3/r^4)$$

12. Polarization and Dielectrics

A dielectric is a medium in which the atoms can be polarized easily by applying an external electric field. By polarization we mean the separation of positive and negative charges. The atoms in a dielectric are initially neutral with the centres of positive and negative charges coinciding. An electric field when applied to this medium causes the centres of positive and negative charges to separate out thus causing an overall charge density to prevail at each point inside the dielectric despite the fact that the entire dielectric is overall neutral. The polarization of a dielectric is best described by means of the polarization vector **P(r)**. At the point **r** within the dielectric, **P(r)** equals the dipole moment per unit volume, i.e. if δV is a small volume element centred at **R**, then $\delta\mathbf{p}(\mathbf{R}) = \mathbf{P}(\mathbf{R})\,\delta V$ is the net dipole moment of the atoms inside this volume element. We have already computed the electrostatic potential due to such a dipole moment. Its value at **r** is given by

$$\delta V(\mathbf{r}) = \frac{\delta\mathbf{p}(\mathbf{R}).(\mathbf{r}-\mathbf{R})}{4\pi\varepsilon_0|\mathbf{r}-\mathbf{R}|^3} = \frac{\mathbf{P}(\mathbf{R}).(\mathbf{r}-\mathbf{R})}{4\pi\varepsilon_0|\mathbf{r}-\mathbf{R}|^3}\delta V$$

The electrostatic potential $V(\mathbf{r})$ due to all the little dipoles that fill the dielectric is obtained according to the superposition principle by integrating the above expression over **R**. In place of δV, we use $d^3\mathbf{R}$ indicating that the volume element is at **R**:

$$V(\mathbf{r}) = \frac{1}{4\pi\varepsilon_0}\int \frac{\mathbf{P}(\mathbf{R}).(\mathbf{r}-\mathbf{R})}{|\mathbf{r}-\mathbf{R}|^3}d^3\mathbf{R}$$

$$= \frac{1}{4\pi\varepsilon_0}\int \mathbf{P}(\mathbf{R}).\nabla_R\frac{1}{|\mathbf{r}-\mathbf{R}|}d^3\mathbf{R} =$$

$$\frac{1}{4\pi\varepsilon_0}\int_V \nabla_R.\frac{\mathbf{P}(\mathbf{R})}{|\mathbf{r}-\mathbf{R}|}d^3\mathbf{R} - \frac{1}{4\pi\varepsilon_0}\int_V \frac{\nabla_R.\mathbf{P}(\mathbf{R})}{|\mathbf{r}-\mathbf{R}|}d^3\mathbf{R}$$

$$= \frac{1}{4\pi\varepsilon_0}\int_S \frac{\mathbf{P}(\mathbf{R}).\mathbf{n}}{|\mathbf{r}-\mathbf{R}|}dS(\mathbf{R}) - \frac{1}{4\pi\varepsilon_0}\int_V \frac{\nabla_R.\mathbf{P}(\mathbf{R})}{|\mathbf{r}-\mathbf{R}|}d^3\mathbf{R}$$

where S is the surface bounding the volume V of the dielectric. Here, Gauss' theorem has been used to convert the volume integral of the divergence of a vector field to a surface integral of the vector field. We now know that from Coulomb's law and the principle of superposition for the electrostatic potential that if a surface S bounds a volume V and the volume V is filled with charge of density $\rho(\mathbf{R})$ and the surface is covered with charges of surface charge density $\sigma(\mathbf{R})$ (charge per unit area), then the electrostatic potential at **r** due to the volume and surface charges is given by

$$V(\mathbf{r}) = \frac{1}{4\pi\varepsilon_0}\int_S \frac{\sigma(\mathbf{R})}{|\mathbf{r}-\mathbf{R}|}dS(\mathbf{R}) + \frac{1}{4\pi\varepsilon_0}\int_V \frac{\rho(\mathbf{R})}{|\mathbf{r}-\mathbf{R}|}d^3\mathbf{R}$$

Comparing this expression with that above, we can conclude that the volume density of polarization charges is given by $\rho_{pol} = -\nabla.\mathbf{P}$ and the surface density of polarization charges is given by $\sigma_{pol} = \mathbf{P}.\mathbf{n}$. Another way to look at the phenomenon of polarization (Jackson) is to regard the dielectric body as being composed of a sequence of pairs of parallel layers, each pair being very closely separated. Consider a given pair of layers. The two layers in this pair are oppositely charged due to polarization. Assume that the surface charge density on the bottom layer is $-\sigma(\mathbf{R})$ and that on the top layer is $\sigma(\mathbf{R})$ Let **d** denote the vector that determines the separation between the two layers. The electrostatic potential due to this pair of layers is by Coulomb's law

$$V(\mathbf{r}) = -\frac{1}{4\pi\varepsilon_0}\int \frac{\sigma(\mathbf{R})}{|\mathbf{r}-\mathbf{R}|}dS(\mathbf{R}) + \frac{1}{4\pi\varepsilon_0}\int \frac{\sigma(\mathbf{R})}{|\mathbf{r}-\mathbf{R}-\mathbf{d}|}dS(\mathbf{R})$$

Assuming **d** to be small and making a Taylor expansion in the second integrand, we have

$$V(\mathbf{r}) = \frac{1}{4\pi\varepsilon_0}\int \sigma(\mathbf{R})\mathbf{d}.\nabla_R \frac{1}{|\mathbf{r}-\mathbf{R}|}dS(\mathbf{R})$$

Now the dipole moment of the pair of layers due to charges in the cylinder bounded by the surface element $dS(\mathbf{R})$ is given by $\sigma(\mathbf{R})dS(\mathbf{R})\mathbf{d}$. If we replace this by the expression $\mathbf{P}(\mathbf{R})d^3\mathbf{R}$, we then arrive at the starting point for the above formula. Now assume that apart from the polarization charges, we also have external charges inside the dielectric body. Let $\rho_{ext}(\mathbf{r})$ denote the volume density of external charges and $\rho_{pol}(\mathbf{r})$ the volume density of polarization charges. Then the total charge density is $\rho_{ext}(\mathbf{r})+\rho_{pol}(\mathbf{r})$. Gauss' law for the dielectric can then be expressed as

$$\varepsilon_0\nabla.\mathbf{E} = \rho_{ext} + \rho_{pol} = \rho_{ext} - \nabla.\mathbf{P}$$

or

$$\nabla.\mathbf{D} = \rho_{ext}$$

where $\mathbf{D} = \varepsilon_0\mathbf{E} + \mathbf{P}$. The above form of Gauss' law involves only the external charges and in place of the electric field vector **E**, the electric displacement vector **D** has appeared. It usually turns out that the induced dipole moment in a dielectric due to an applied external electric field is proportional to the electric field. This approximation is good provided the electric field amplitude is not too large. The reason for this is that the bonding of the electrons of an atom to its nucleus can to a good degree of approximation be regarded as a spring obeying Hooke's law. In fact, if the potential energy of the bonding is described by the function $U(x)$ where x is the distance of the electron from the nucleus, then we can expand U about the equilibrium distance x_0 to get $U(x) \approx U(x_0) + \frac{1}{2}U''(x_0) \times (x - x_0)^2$. The distance $x - x_0$ is taken as our variable y and then the force of attraction of the nucleus on the electron is approximately ky, where $k = U''(x_0)$. At equilibrium this force must compensate the force on the electron due to the external electric field E. Thus, $y = eE/k$ showing that the displacement from equilibrium is proportional to the external field. The induced dipole moment is however proportional to y and hence also proportional to E. This leads us to the empirical relation connecting the polarization vector with the total electric field: $\mathbf{P}(\mathbf{r}) = \varepsilon_0\chi.\mathbf{E}(\mathbf{r})$. Here, χ is a constant of proportionality called the susceptibility of the dielectric and it varies from medium to medium. It can also vary from point to point when the medium is non-uniform. For large electric field amplitudes, χ would generally depend upon the electric field itself and the polarization would not be proportional to the field. Such models are used in non-linear optics and in the analysis of the propagation of electromagnetic waves through the ionosphere. The relation between the electric displacement vector and the electiric field in a dielectric can now be expressed as $\mathbf{D} = \varepsilon_0\mathbf{E} + \mathbf{P} = \varepsilon_0(1 + \chi)\mathbf{E}$. $1 + \chi$ is called the dielectric constant of the medium and denoted by ε_r. $\varepsilon_0\varepsilon_r$ is called the electric permittivity of the medium and denoted by ε. In vacuum, $\chi = 0$ (there is nothing to polarize) and $\varepsilon_r = 1$, $\varepsilon = \varepsilon_0$. For a medium with ε depending upon the position **r**, Gauss' law assumes the form $\nabla . (\varepsilon.\mathbf{E}) = \rho_{ext}$

Review Questions with Answers on Electrostatics

[1] To what law in mechanics is Coulomb's law analogous?

Ans. Coulomb's law is analogous to the inverse square law of gravitation which states that two point masses m_1, m_2 separated from each other by a distance of d exert an attractive force upon each other of amount $F = \dfrac{Gm_1m_2}{d^2}$. This force is directed along the line joining the two masses. Coulomb's law is related to this by replacing mass with charge with the only difference being the constant of proportionality and the fact that mass is only of one sign in gravitation where as charge can be of both positive and negative signs. Thus, the force between two charges is attractive or repulsive according as the charges are of the opposite sign or of the same sign whereas the gravitational force is always attractive.

[2] Units of permittivity of free space.

Ans. The force between two charges in free space is $F = \dfrac{q_1q_2}{4\pi\varepsilon_0.d^2}$.

Thus ε_0 has units of *Coulomb*^{2}/(*Newton.m*^{2})

[3] Discuss two applications based on the electric force on a charged particle.

Ans. One application is to the construction of the cathode ray oscilloscope. Here we have two vertical plates and two horizontal plates, all of which are conducting. A fluorescent screen is placed in front of the plates and illuminated by a cathode ray. We apply a voltage waveform $v(t)$ to the vertical plates and this produces an electric field that causes the spot produced by the cathode ray on the screen to get deflected up and down in accordance with $v(t)$. To the horizontal plates, we apply a time base which is a linearly increasing ramp voltage that returns to its zero value after reaching its maximum. This causes the spot on the fluorescent screen to get deflected horizontally back and forth. The overall effect of both these voltages is to produce a scaled plot of the waveform $v(t)$ on the screen and thus view the waveform as if on a graph. The cathode ray particles are electrons which are negatively charged and their movement on the screen is due to the force of the electric field on them produced by the voltage waveform applied to the plates. This is a very important application of the electric force on a charged particle. Another application is to the separation of impurities from an ore. Suppose we have an ore which consists of positively charged ions and impurities which consist of negatively charged ions. We are giving this application without reference to any specific ore but with the knowledge that this is a very common phenomenon. We want to separate out the ion from the impurity. We then simply pass the impure ore through a chamber with two plates lying opposite to each other and charged oppositely by means of a voltage applied to them through a battery. The positively charged ions get attracted towards the side having the negatively charged plate and the negatively charged impurity get attracted towards the side having the positively charged plate. The resulting stream which comes out of the chamber has been split into two parts, one part consisting of the positively charged ore and the other part consisting of the negatively charged impurity.

[4] Given three sinusoidally time varying linearly polarized vector fields

$$\mathbf{X}(t) = (A_1\hat{x} + B_1\hat{y} + C_1\hat{z}).\cos(\omega_c t)$$
$$\mathbf{Y}(t) = (A_2\hat{x} + B_2\hat{y} + C_2\hat{z}).\cos(\omega_c t)$$
$$\mathbf{Z}(t) = (A_3\hat{x} + B_3\hat{y} + C_3\hat{z}).\sin(\omega_c t)$$

find the the conditions on A_i, B_i, C_i, $i = 1, 2, 3$ that guarantee that (i) $\mathbf{Y}(t)$ is polarized perpendicular to the plane $lX + mY + nZ = 0$, (ii) $\mathbf{X}(t) + \mathbf{Y}(t) + \mathbf{Z}(t)$ is circularly polarized in the X–Y plane, (iii) $\mathbf{X}(t) - \mathbf{Y}(t) + \mathbf{Z}(t)$ is linearly polarized.

(i) For **Y** (t) to be polarized perpendicular to the plane having normal (l, m, n), we require that

$$(l\hat{x} + m\hat{y} + n\hat{z}) \times \mathbf{Y}(t) = 0$$

This gives us three equations

$$mC_2 - nB_2 = 0,\ nA_2 - lC_2 = 0,\ lB_2 - mA_2 = 0$$

out of which, only two are linearly dependent. That is $B_2 = \frac{mA_2}{l}, C_2 = \frac{nA_2}{l}$.

Next note that

$$\mathbf{U}(t) = \mathbf{X}(t) + \mathbf{Y}(t) + \mathbf{Z}(t) =$$

$$\hat{x}\ ((A_1 + A_2)\cos\omega_c t + A_3.\sin\omega_c t) + \hat{y}\ ((B_1 + B_2)\cos\omega_c t + B_3.\sin\omega_c t) + \hat{z}\ ((C_1 + C_2)\cos\omega_c t + C_3.\sin\omega_c t)$$

The first condition for this vector to fall in the X—Y plane is that the z component should be zero, i.e. $C_2 = -C_1$, $C_3 = 0$. Under this condition, we can write

$$\mathbf{U}(t) = R_1.\cos(\omega_c t + \phi_1)\ \hat{x} + R_2.\cos(\omega_c t + \phi_2)\ \hat{y}$$

where

$$R_1 = \sqrt{(A_1 + A_2)^2 + A_3^2},\quad R_2 = \sqrt{(B_1 + B_2)^2 + B_3^2}$$

$$\phi_1 = -\tan^{-1}(A_3/(A_1 + A_2)), \phi_2 = -\tan^{-1}(B_3/(B_1 + B_2))$$

$\mathbf{U}(t)$ will be circularly polarized provided $R_2 = R_1$ and $|\phi_2 - \phi_1| = \pi/2$. The second condition is equivalent to $\cos(\phi_2) = \pm \sin(\phi_1)$, that is

$$\frac{B_1 + B_2}{R_2} = \pm\frac{A_3}{R_1}$$

or equivalently, in view of the first condition, $B_1 + B_2 = \pm A_3$. Substituting this into the relation $R_1^2 = R_2^2$ gives $(A_1 + A_2) = \pm B_3$. Thus, the conditions for $\mathbf{U}(t)$ to be circularly polarized in the X—Y plane are

$$C_1 + C_2 = 0,\ C_3 = 0,\ B_1 + B_2 = \pm A_3,\ A_1 + A_2 = \pm B_3$$

Finally, we note that

$$\mathbf{V}(t) = \mathbf{X}(t) - \mathbf{Y}(t) + \mathbf{Z}(t) = \hat{x}\ ((A_1 - A_2)\cos(\omega_c t) + A_3.\sin\omega_c t) + \hat{y}((B_1 - B_2)\cos(\omega_c t) + B_3.\sin(\omega_c t)) + \hat{z}\ ((C_1 - C_2)\cos(\omega_c t) + C_3.\sin(\omega_c t))$$

It is clear that one way in which this field can be linearly polarized is that the coefficients of $\sin(\omega_c t)$ be equal to zero, i.e. $A_3 = B_3 = C_3 = 0$. Another way in which this field can be linearly polarized is that the coefficient of $\cos(\omega_c t)$ be zero, i.e. $A_1 = A_2$, $B_1 = B_2$, $C_1 = C_2$.

[5] Consider a circular ring of charge Q of radius R lying in the X—Y plane with centre coinciding with the origin. Write down the approximate equation of motion for a charge $-q$ executing motion in the vicinity of the origin along the z axis.

Ans. If q is at (0, 0, z) it experiences a force

$$F_z = -\frac{qz}{4\pi\varepsilon_0(R^2+z^2)^{3/2}}.$$

For z small in magnitude compared to R, the equation of motion is

$$m\frac{d^2z}{dt^2} = -\frac{qz}{4\pi\varepsilon_0 R^3}$$

The motion is thus approximately simple harmonic with an angular frequency of

$$\omega = \left(\frac{q}{4m\pi\varepsilon_0 R^3}\right)^{1/2}$$

[6] A sphere of radius a and dielectric constant ε_r has uniform charge density of ρ_0. Find the potential at the centre of the sphere and on the surface of the sphere.

Ans. By using Gauss' law in integral form applied to a sphere of radius r, we find for the radial component of the electric field,

$$E_r(r)4\pi r^2 = \frac{4}{3\varepsilon}\pi r^3\rho_0, r < a$$

and

$$E_r(r)4\pi r^2 = \frac{4}{3\varepsilon_0}\pi a^3, r > a$$

These give

$$E_r(r) = \rho_0 r/3\varepsilon, r < a, E_r(r) = \rho_0 a^3/3r^2\varepsilon_0, r > a$$

The potential at $r = \infty$ is zero. Here, $\varepsilon = \varepsilon_0\varepsilon_r$. Thus,

$$V(0) = \int_0^\infty E_r(r)dr = \int_0^a E_r(r)dr + \int_a^\infty E_r(r)dr$$

$$= \rho_0 a^2/6\varepsilon + \rho_0 a^2/3\varepsilon_0 = \rho_0 a^2(2\varepsilon_r + 1)/6\varepsilon_0\varepsilon_r$$

$$V(a) = \int_a^\infty E_r(r)dr = \rho_0 a^2/3\varepsilon_0$$

[7] An electric field exists in the room in which you are working such that the lines of force are horizontal and at right angles to one wall. As you walk toward the wall from which the lines of force emerge into the room, are you walking toward points of higher or lower or equal potential?

Ans. If V_0 denotes the potential of the wall and V_1 the potential at a distance d from the wall then the electric field normal to the wall directed away from it is given by $E = (V_0 - V_1)/d$. Thus, if the electric field points away from the wall, $E > 0$ and $V_0 > V_1$ which means that the potential increases as we move toward the wall. On the other hand, if the electric field points toward the wall, then $E < 0$ and $V_0 < V_1$ which means that the potential decreases as we move toward the wall. This argument works in the special case when the field is uniform. When the field is nonuniform but directed away from the wall, then $V_0 - V_1 = \int_0^d E(x)dx$ with $E(x) > 0$ and this implies that $V_0 > V_1$. Conversely, when the field is nonuniform but directed toward the wall, then $V_0 - V_1 = \int_0^d E(x)dx$ with $E(x) < 0$ and this implies $V_0 < V_1$. So the final result remains unaltered.

[8] A charge is uniformly distributed throughout a sphere of radius a. Taking the potential at ∞ as zero, find the potential at $r = b < a$.

Ans. From Gauss' law, we find for the electric field $E(r)$, which is directed radially outward,

$$E(r)4\pi r^2 = 4\pi\rho r^3/3\varepsilon_0, \; r < a$$
$$E(r)4\pi r^2 = 4\pi\rho a^3/3\varepsilon_0, \; r > a$$

Thus,

$$E(r) = \rho r/3\varepsilon_0, \; r < a, \; E(r) = \rho a^3/3r^2\varepsilon_0, \; r > a$$

Then,

$$V(b) = V(b) - V(\infty) = \int_b^\infty E(r)dr = \int_b^a \rho r dr/3\varepsilon_0 + \int_a^\infty \rho a^3 dr/3r^2\varepsilon_0$$
$$= \rho(a^2 - b^2)/6\varepsilon_0 + \rho a^2/3\varepsilon_0 = \rho a^2/2\varepsilon_0 - \rho b^2/6\varepsilon_0$$

[9] Find the electric field at point (0, 0, h) due to the rectangle described by $-a < x < a$, $-b < y < b$, $x = 0$ and carrying uniform charge $\rho_s C/m^2$.

Ans. We shall determine the potential V(0, 0, z) and take the negative derivative at $z = h$ to obtain the electric field. Coulomb's law combined with superposition gives

$$V(0,0,z) = \frac{\rho_S}{4\pi\varepsilon_0}\int_{-a}^{a}\int_{-b}^{b}\frac{dxdy}{\sqrt{x^2+y^2+z^2}}$$

We first evaluate the integral

$$I(x) == \int_{-b}^{b} \frac{dy}{\sqrt{x^2 + z^2 + y^2}}$$

Let $x^2 + z^2 = p^2$ and $y = p.\tan\theta$. Then,

$$I(x) = \int_{-\tan^{-1}(b/p)}^{\tan^{-1}(b/p)} \sec(\theta) d\theta = \log(\sec\theta + \tan\theta)|_{-\tan^{-1}(-b/p)}^{\tan^{-1}(b/p)}$$

$$= \frac{1}{2} \log \frac{\sqrt{b^2 + p^2} + b}{\sqrt{b^2 + p^2} - b}$$

Then,

$$\int_{-a}^{a} \int_{-b}^{b} \frac{dxdy}{\sqrt{x^2 + y^2 + z^2}} = \int_{-a}^{a} I(x) dx = \frac{1}{2} \int_{-a}^{a} \log(\sqrt{b^2 + z^2 + x^2} + b) dx$$

$$-\frac{1}{2} \int_{-a}^{a} log(\sqrt{b^2 + z^2 + x^2} - b) \, dx$$

Thus, the potential can be expressed as

$$V(0, 0, z) = \frac{\rho s}{8\pi\varepsilon_0} \int_{-a}^{a} \log(\sqrt{b^2 + z^2 + x^2} + b) dx - \frac{\rho s}{8\pi\varepsilon_0} \int_{-a}^{a} \log$$

$$\log, (\sqrt{b^2 + z^2 + x^2} - b) dx$$

The negative derivative of this is

$$E_z(0, 0, z) = -\frac{\partial V(0,0,z)}{\partial z} = -\frac{\rho s}{4\pi\varepsilon_0} \int_{-a}^{a} \frac{dx}{\sqrt{b^2 + z^2 + x^2} (\sqrt{b^2 + x^2 + z^2} + b)}$$

$$+ \frac{\rho s}{4\pi\varepsilon_0} \int_{-a}^{a} \frac{dx}{\sqrt{b^2 + z^2 + x^2} (\sqrt{b^2 + z^2 + x^2} - b)}$$

Let

$$J_1 = \int_{-a}^{a} \frac{dx}{\sqrt{q^2 + x^2} (\sqrt{q^2 + x^2} + b)}$$

$$J_2 = \int_{-a}^{a} \frac{dx}{\sqrt{q^2 + x^2} (\sqrt{q^2 + x^2} - b)}$$

where $q^2 = b^2 + z^2$. Let $x = q.\tan\theta$. Then $q^2 + x^2 = q^2\sec^2\theta$ and $dx = q.\sec^2\theta.d\theta$. This gives

$$J_1 = \int_{-\theta_1}^{\theta_1} \frac{d\theta}{q + b.\cos\theta}, \theta_1 = \tan^{-1}(a/q)$$

Make another change of variables: $\tan(\theta/2) = u$. Then,

$$J_1 = \int_{-\tan(\theta_{1/2})}^{\tan(\theta_{1/2})} \frac{2du}{(1+u^2)\left(q + b\left(\frac{1-u^2}{1+u^2}\right)\right)}$$

$$= \int_{-\tan(\theta_{1/2})}^{\tan(\theta_{1/2})} \frac{2du}{q + b + (q-b)u^2}$$

$$= \frac{2}{\sqrt{q^2 - b^2}} \tan^{-1}\left(u\sqrt{\frac{q^2 - b^2}{q+b}}\right)\Bigg|_{-\tan(\theta_{1/2})}^{\tan(\theta_{1/2})}$$

$$= \frac{2}{\sqrt{q^2 - b^2}}\left(\tan^{-1}\left(\tan(\theta_1/2)\sqrt{\frac{q-b}{q+b}}\right) + \tan^{-1}\left(\tan(\theta_1/2)\sqrt{\frac{q-b}{q+b}}\right)\right)$$

J_2 is obtained by replacing b with $-b$ in J_1:

$$J_2 = \frac{2}{\sqrt{q^2 - b^2}}\left(\tan^{-1}\left(\tan(\theta_1/2)\sqrt{\frac{q+b}{q-b}}\right) + \tan^{-1}\left(\tan(\theta_1/2)\sqrt{\frac{q+b}{q-b}}\right)\right)$$

We have

$$\mathbf{E}_z(0,0,z) = \frac{\rho_s}{4\pi\varepsilon_0}(J_2 - J_1)$$

Where we substitute $q = \sqrt{b^2 + z^2}$ and in place of $\tan(\theta_1/2)$, we substitute

$$\sqrt{\frac{\sqrt{q^2 + a^2} - q}{\sqrt{q^2 + a^2} + q}}.$$

[10] A charge distribution with spherical symmetry has density $\rho =$

ρ_0, $r < R$ and $\rho = 0$, $r > R$. Determine using Poisson's equation the potential everywhere and the energy stored in the region $r < R$.

Ans. Poisson's equation for the potential having radial symmetry is

$$\frac{1}{r}\frac{d^2}{dr^2} rV = -\rho_0 / \varepsilon_0, \ r < R$$

$$\frac{1}{r}\frac{d^2}{dr^2} rV = 0, \ r > R$$

Integrating the first equation, we get

$$rV(r) = -\rho_0 r^3/6\varepsilon_0 + Ar + \mathbf{B}, \ r < R$$

and since V must be nonsingular at $r = 0$, it must follow that

$$V(r) = -\rho_0 r^2/6\varepsilon_0 + A, \ r < R$$

Likewise, integrating the second equation gives

$$V(r) = C/r + D, \ r > R$$

and since our reference zero potential is at $r = \infty$, we must have $D = 0$. Matching the electric fields and potentials at $r = R + 0$ and $r = R - 0$ gives

$$V'(R+) = V'(R-), \ V(R+) = V(R-)$$

so that

$$\rho_0 R/3\varepsilon_0 = C/R^2, \ -\rho_0 R^2/6\varepsilon_0 + A = C/R$$

so that

$$C = \rho_0 R^3/3\varepsilon_0, \ A = \rho_0 R^2/3\varepsilon_0 + \rho_0 R^2/6\varepsilon_0 = \rho_0 R^2/2\varepsilon_0$$

The electric field for $r < R$ is

$$E(r) = -V'(r) = \rho_0 r/3\varepsilon_0$$

It is directed radially outward. The energy stored in the region $r < R$ is given by (See Solved Problem 8)

$$W = \frac{\varepsilon_0}{2}\int_0^R E(r)^2 4\pi r^2 dr = \frac{2\pi\rho_0^2}{9\varepsilon_0}\int_0^R r^4 dr$$

$$= \frac{2\pi\rho_0^2 R^5}{45\varepsilon_0} \text{ Joules}$$

[11] A spherical capacitor has an inner conductor of radius a carrying charge Q and maintained at zero potential. If the outer conductor contracts from a radius b to c under internal forces, find the work performed by the electric field as a result of the contraction.

Ans. From Gauss' law in integral form the electric field (which is radial) for $a < r < b$ satisfies

$$E(r)4\pi r^2 = Q/\varepsilon_0$$

Thus,

$$E(r) = \frac{Q}{4\pi\varepsilon_0 r^2}, a < r < b$$

This gives for the total field energy stored in between the plates,

$$W = (\varepsilon_0/2)\int_a^b E^2(r)4\pi r^2 dr = \frac{Q^2}{8\pi\varepsilon_0}\int_a^b \frac{dr}{r^2} = \frac{Q^2}{4\pi\varepsilon_0}\left(\frac{1}{a}-\frac{1}{b}\right)$$

The field energy after contraction is given by

$$W' = \frac{Q^2}{4\pi\varepsilon_0}\left(\frac{1}{a}-\frac{1}{c}\right)$$

The work performed by the electric field equals the decrease in the field energy. It is given by

$$W - W' = \frac{Q^2}{8\pi\varepsilon_0}\left(\frac{1}{c}-\frac{1}{b}\right) = \frac{Q^2(b-c)}{8\pi\varepsilon_0 bc}$$

[12] Consider two hollow cylindrical conducting surfaces of radii a and b respectively placed a distance d apart. Assume the two surfaces are parallel to each other and that their centres are located d metres apart. Find the capacitance between the two surfaces.

Ans. The two surfaces are equipotentials. So, we shall find an arrangement of charges for which the two surfaces become equipotentials. Consider two infinitely long line charges having linear charge densities σ and $-\sigma$ Coulombs per unit length respectively. Assume that the first line charge is along the z axis and the second at $x = c$, $y = 0$ parallel to the z axis. The potential by superposition at $(x, y, 0)$ due to this arrangement is

$$V(x, y) = \frac{\sigma}{2\pi}\log\left(\sqrt{x^2+y^2}\right) - \frac{\sigma}{2\pi}\log\left(\sqrt{(x-c)^2+y^2}\right)$$

Consider the equipotential surface $V = -V_0 < 0$. It is given by

$$\frac{x^2+y^2}{(x-c)^2+y^2} = \exp(-4\pi V_0/\sigma) = \lambda$$

say. This can be simplified to get

$$(1-\lambda)(x^2+y^2) + 2c\lambda x - \lambda c^2 = 0$$

or

$$x^2+y^2+\frac{2c\lambda x}{1-\lambda}-\frac{\lambda c^2}{1-\lambda}=0$$

This is a circle of radius R where

$$R^2 = \frac{\lambda^2 c^2}{(1-\lambda)^2}+\frac{\lambda c^2}{1-\lambda} = \frac{c^2\lambda}{(1-\lambda)^2}$$

or

$$R = \frac{c.\exp(-2\pi V_0/\sigma}{1-\exp(-4\pi V_0/\sigma)} = \frac{c}{2.\sinh(2\pi V_0/\sigma)}$$

The centre of this circle is $\left(-\frac{c\lambda}{1-\lambda}, 0\right)$ in the $x-y$ plane. Likewise, the equipotential surface $V_1 > 0$ is given by

$$\frac{x^2+y^2}{(x-c)^2+y^2} = \exp(4\pi V_1/\sigma) = \lambda_1$$

This is the same as

$$(\lambda_1 - 1)(x^2+y^2) - 2c\lambda_1 x + \lambda_1 c^2 = 0$$

or

$$x^2+y^2-\frac{2c\lambda_1 x}{\lambda_1-1}+\frac{\lambda_1 c^2}{\lambda_1-1}=0$$

This is a circle of radius R', where

$$R'^2 = \frac{\lambda_1^2 c^2}{(\lambda_1-1)^2}-\frac{\lambda_1 c^2}{\lambda_1-1} = \frac{c^2\lambda_1}{(\lambda_1-1)^2}$$

or

$$R' = \frac{c}{2.\sinh(2\pi V_1/\sigma)}$$

with centre at $\left(\frac{c\lambda_1}{\lambda_1 - 1}, 0\right)$. We choose the potentials V_0, V_1 so that the former equipotential surface encloses the line charge at the origin and the latter equipotential surface encloses the line charge at $x = c$, $y = 0$. The first condition means that

$$\frac{c\lambda}{1-\lambda} < R = \frac{c.\exp(-2\pi V_0 / \sigma)}{1 - \exp(-4\pi V_0 / \sigma)}$$

which is the same as requiring that $V_0 > 0$, which is indeed the case. The second condition means that

$$\left|\frac{c\lambda_1}{\lambda_1 - 1} - c\right| < R' = \frac{c.\exp(2\pi V_1 / \sigma)}{\exp(4\pi V_1 /.\sigma) - 1}$$

or

$$\frac{c}{\lambda_1 - 1} < \frac{c\sqrt{\lambda_1}}{\lambda_1 - 1}$$

same as requiring that $V_1 > 0$ which is the case. We now choose V_0, V_1, c so that $R = a$, $R' = b$ and the distance between the centres of the two equipotential cylinders is d. This means that

$$2.\sinh(2\pi V_0 / \sigma) = \frac{c}{a}, 2.\sinh(2\pi V_1 / \sigma) = \frac{c}{b},$$

$$c\left(\frac{\lambda_1}{\lambda_1 - 1} + \frac{\lambda}{1 - \lambda}\right) = d$$

This last condition is the same as

$$d / c = \left(\frac{\exp(2\pi V_1 / \sigma)}{2\sinh(2\pi V_1 / \sigma)} + \frac{\exp(-2\pi V_0 / \sigma)}{2\sinh(2\pi V_0 / \sigma)}\right)$$

which in view of the two previous conditions, reduces to

$$d / c = \frac{b}{c}\exp(2\pi V_1 / \sigma) + \frac{a}{c}\exp(-2\pi V_0 / \sigma)$$

or

$$d = b.\exp\ (2\pi V_1/\sigma) + a.\exp\ (-2\pi V_0/\sigma)$$

Define

$$\alpha = \exp(2\pi V_1/\sigma),\ \beta = \exp(2\pi V_0/\sigma)$$

Then, the conditions can be expressed as

$$c/a = \beta - 1/\beta,\ c/b = \alpha - 1/\alpha,\ d = b\alpha + a/\beta$$

The potential difference between the two cylindrical surfaces is $V_1 + V_0$ and by Gauss' law, the charge enclosed per unit length on the first cylinder is σ while on the second, it is $-\sigma$. The capacitance between the two cylinders is therefore

$$C = \frac{\sigma}{V_1 + V_0} = \frac{2\pi}{\log(\alpha\beta)}$$

Taking the ratio of the first two equations gives

$$\frac{b}{a} = \frac{\alpha}{\beta}\frac{\beta^2 - 1}{\alpha^2 - 1}$$

Substituting for α from the third expression $\alpha = \frac{d}{b} - \frac{a}{b\beta}$ into this expression and simplifying gives the quadratic

$$ad\beta^2 + (b^2 - a^2 - d^2)\beta + ad = 0$$

The solution to this equation is

$$\beta = \frac{1}{2ad}(d^2 + a^2 - b^2 \pm \sqrt{(b^2 - a^2 - d^2)^2 - 4a^2d^2})$$

Since $\beta > 1$, we choose the plus sign. It can be rewritten as

$$\beta = \frac{1}{2ad}(d^2 + a^2 - b^2 \pm \sqrt{(a^2 + b^2 - d^2)^2 - 4a^2b^2})$$

and hence

$$\alpha\beta = \frac{d\beta}{b} - \frac{a}{b} = \frac{d\beta - a}{b}$$

$$= \frac{1}{2ab}(d^2 - a^2 - b^2 + \sqrt{(d^2 - a^2 - b^2)^2 - 4a^2b^2})$$

Now if $\cosh^{-1}(p) = q$, then

$$p = \cosh q = \frac{1}{2}(\exp(q) + \exp(-q))$$

and hence

$$\exp(2q) - 2p\exp(q) + 1 = 0$$

or

$$\exp(q) = p + \sqrt{p^2 - 1}$$

It follows that

$$\log(\alpha\beta) = \cosh^{-1}((d^2 - a^2 - b^2)/2ab)$$

and hence

$$C = \frac{2\pi}{\cosh^{-1}((d^2 - a^2 - b^2)/2ab)}$$

[13] An electron is projected with an initial velocity of u_0 m/s into the uniform field between the parallel plates shown in Fig. 2.1. The spacing between the plates is d metres. If the electron just misses the upper plate as it emerges from the field, find the electric field intensity and its velocity as it emerges from the field. Assume that the top plate is positively charged and the bottom plate negatively charged and that the length of the plates is L m.

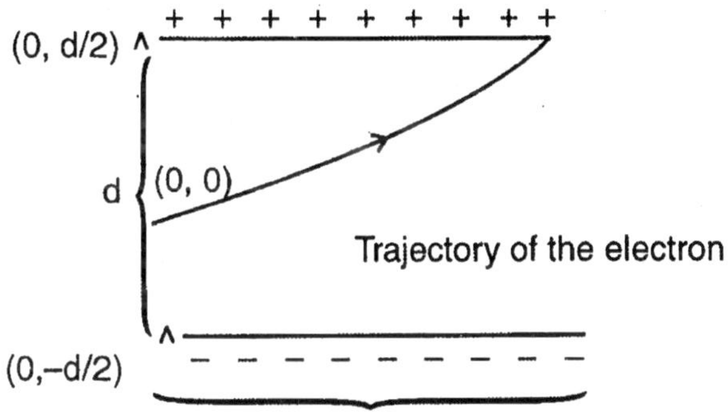

Fig. 2.1

Ans. Let the x axis be parallel to the length of the plates. and the point of start of the electron be (0, 0). Then the upward acceleration of the electron is $a_y = \frac{eE}{m}$ and after time t, the position of the electron is given by

$$x = u_0 t,\ y = \frac{a_y t^2}{2}$$

When $x = L$, we have $t = L/u_0$ and at this time, $y = \frac{a_y L^2}{2u_0^2}$. This must correspond to $y = d/2$, that is $\frac{a_y L^2}{2u_0^2} = d/2$ or $\frac{eEL^2}{2mu_0^2} = d/2$, or $E = \frac{dmu_0^2}{eL^2}$.

At this time, the electron velocity is $\mathbf{v} = u_x\hat{x} + v_y\hat{y}$ where

$$\upsilon_x = u_0, v_y = a_y t = \frac{eEL}{mu_0} = \frac{du_0}{L}$$

[14] A surface has its equation given by **r** = **r**(u, v) with u, v being two free parameters. The surface has its potential specified by V (u, v). Explain how you would numerically determine the surface charge density $\sigma(u,v)$.

Ans. The first step is to partition the surface into little pixels by drawing the lines of constant u and constant v. The pixel regions are given by δS_k and inside each pixel, we choose a point $(u_k,v_k) \in \delta S_k$, $k = 1, 2, \ldots, N$. We are assuming that there are N pixels in number. The pixels are assumed to be so small that we may assume the surface charge density over each pixel to be a constant, say the surface charge density on δS_k is σ_k. We then calculate the potential due to the charge on the kth pixel at a given point **r** by applying Coulomb's law with superposition. Call this potential as $\sigma_k V_k(\mathbf{r})$. Here, $V_k(\mathbf{r})$ is a known function. The potential at the point (u_k,v_k) on the surface is by superposition given by

$$V(\mathbf{r}(u_k, v_k)) = \sum_{j=1}^{N} \sigma_j V_j(\mathbf{r}(u_k, v_k)), k = 1, 2, \ldots N$$

The variables on the left hand side are known and hence by solving these linear equations, we can determine the surface charge densities $\{\sigma_K\}$. This gives us the approximate charge density on the surface. It can also be adapted to volumes where the volume charge density $\rho(\mathbf{r})$ is to be determined when the potential in a region is specified. Let $\mathbf{V}$ be the region where the charge density $\rho(\mathbf{r})$ is to be obtained and let the potential $V(\mathbf{r})$, $\mathbf{r} \in \mathbf{V}'$ be specified. We divide the region $\mathbf{V}$ into little pixels, say cuboids $\mathbf{V}_j$, $j = 1, 2, \ldots, N$ and assume a uniform charge distribution of density ρ_j within $\mathbf{V}_j$. We calculate the potential due to the jth pixel. Let it be denoted by $\rho_j, V_j(\mathbf{r})$. $V_j(\mathbf{r})$ is a known function defined as the potential produced by a unit volume density of charge inside the jth pixel. We also divide the region V' into N little pixels and choose points $\mathbf{r}'_j$, $j = 1, 2, \ldots, N$ one inside each pixel. Then we have the equations

$$V(\mathbf{r}'_k) = \sum_{j=1}^{N} \rho_j V_j(\mathbf{r}_k), k = 1, 2, \ldots, N$$

Solving this simultaneous set of N linear equations, we obtain the charge densities $\{\rho_j\}$ within the different pixels.

[15] Explain how you would apply the method of moments to calculate

the surface charge density on a solid cylinder of radius R and length L when the potential of the cylinder is maintained at a constant value V_0 relative to ∞.

Ans. We assume that the base of the cylinder is the region $z = 0$, $x^2 + y^2 \le R^2$ and the top surface is $z = L$, $x^2 + y^2 \le R^2$. We divide the length of the cylinder into N equal little cylinders, the kth on ρ extending from $z = kd$ to $z = (k + 1)\, d$, $k = 0, 1, \ldots, N - 1$, $d = L/N$. Assume the surface charge density on the sides of the kth little cylinder to be σ_k. We shall then calculate the potential at the centre of the rth cylinder which has co-ordinates $(0, 0, (r + 1/2)d))$ due to the charge on each cylinder. The potential due to the charge on the kth cylinder at $(0, 0, z)$ equals

$$V(0, 0, z) = \frac{\sigma_k}{4\pi\varepsilon_0} \int_0^{2\pi} \int_{kd}^{(k+1)d} \frac{R\, d\phi\, d\xi}{\sqrt{(z-\xi)^2 + R^2}}$$

$$= \frac{\sigma_k R}{2\varepsilon_0} \log\left(\frac{(z-kd) + \sqrt{R^2 + (z-kd)^2}}{(z-(k+1)d) + \sqrt{R^2 + (z-(k+1)d)^2}} \right)$$

Thus, the potential at the centre of the rth cylinder due to the charge on the kth cylinder is given by

$$V(0, 0, (r + 1/2)d) = \sigma_k V_{k,r}$$

where

$$V_{k,r} = \frac{R}{2\varepsilon_0} \log\left(\frac{(r-k+1/2)d + \sqrt{R^2 + (r-k+1/2)^2 d^2}}{((r-k-1/2)d + \sqrt{R^2 + (r-k-1/2)^2 d^2}} \right)$$

We must set this equal to V_0. The equations to be solved are thus

$$V_0 = \sum_{k=0}^{N-1} \sigma_k V_{k,r}, \quad r = 0, 1, \ldots, N-1$$

which yield $\{\sigma_k\}$, the approximate surface charge density on the cylinder.

13. Some Solved Problems in Electrostatics

1. Space is filled with a dielectric of permittivity $\varepsilon(\rho)$, where ρ is the distance of the point from the z axis. Assuming that the charge density in space varies as $\sigma(\rho)$, calculate the electric field and potential using Gauss' law assuming that the potential and field decay down to zero as $\rho \to \infty$ and $\varepsilon(\rho)$ remains bounded.

Solution: From the symmetry of the problem, the electric field has only a ρ component and its magnitude depends only on ρ. Likewise the potential is a function of ρ only. Thus, $V = V(\rho)$, $\mathbf{E} = E_\rho(\rho)$, ρ with $E_\rho = -\frac{dV}{d\rho}$. Gauss' law for the dielectric $\nabla.(\varepsilon.\mathbf{E}) = \sigma$, when expressed in cylindrical co-ordinates with the appropriate symmetry reads

$$\frac{1}{\rho}\frac{d}{d\rho}(\varepsilon.\rho.E_\rho) = \sigma(\rho)$$

Integration from ρ to ∞ assuming that $\varepsilon.\rho.E_\rho \to 0$ as $\rho \to \infty$ since ε remains bounded and $E_\rho \to 0$ gives

$$E_\rho = -\frac{1}{\rho\varepsilon(\rho)}\int_\rho^\infty \rho.\sigma(\rho)d\rho$$

In fact, here, we are assuming that $\rho\, E_\rho \to 0$ as $\rho \to \infty$.
Integration of the field from ρ to ∞ gives on noting that $V(\rho) \to 0$ when $\rho \to \infty$ gives

$$V(\rho) = -\int_\rho^\infty d\rho' \frac{1}{\rho'\varepsilon(\rho')}.\int_{\rho'}^\infty \rho''.\sigma(\rho'')d\rho''$$

2. A cylinder of radius R has its axis coinciding with the z axis. The cylinder has a uniform charge of density σ. Calculate the electric field and potential both inside and outside the cylinder.

Solution: Gauss' law reads, $\frac{1}{\rho}\frac{d}{d\rho}\ \rho.E_\rho = \sigma.\varepsilon_0, 0 < \rho < R$ and $= 0$ for $\rho > R$.
Thus,

$$E_\rho = \frac{\sigma.\rho}{2\varepsilon_0} + A/\rho, \rho < R$$

$$E_\rho = B/\rho, \rho > R$$

Since the electric field must remain finite at $\rho = 0$, i.e. on the z axis, it follows that $A = 0$. The normal component of the electric field must be continuous at the boundary since there is no surface charge. Thus,

$$B/R = \frac{\sigma.R}{2\varepsilon_0}$$

This gives

$$E_\rho(\rho) = \frac{\sigma.\rho}{2\varepsilon_0}, \rho < R, \quad E_\rho(\rho) = \frac{\sigma.R^2}{2\varepsilon_0.\rho}, \rho > R$$

Integration of the equation $E_\rho = -\dfrac{dV}{d\rho}$ from 0 to ρ gives

$$V(\rho) - V(0) = -\int_0^\rho E_\rho(\rho)d\rho$$

Thus,

$$V(\rho) = V(0) - \frac{\sigma\rho^2}{4\varepsilon_0}, \rho < R,$$

$$V(\rho) = V(0) - \frac{\sigma R^2}{4\varepsilon_0} - \frac{\sigma.R}{2\varepsilon_0}\log(\rho / R), \rho > R$$

3. Determine the capacitance of a parallel plate capacitor consisting of two plates of area A separated by a distance d and the region between the plates consisting of a dielectric material of permittivity $\varepsilon(z)$, where z is the distance from the top plate. The bottom plate may be assumed to have a charge $-Q$ and the top plate a charge $+Q$. Neglect edge effects, thereby assuming the electric field between the plates has only a z component and its magnitude depends only upon z. Capacitance is defined as the ratio of the charge on the top plate to the potential difference between the plates.

Solution: Let $E_z(z)$ denote the electric field. It is directed along the z direction, i.e. from the top to the bottom plate. Applying Gauss' law in integral form to a pill box with one surface inside the conducting body of the top plate and the other surface falling in the region between the plates, it follows on noting that the electric field inside the conducting body of the top plate is zero, that $E_z(z) = \sigma/\varepsilon(z)$, where $\sigma = Q/A$ is the surface charge density on the top plate. The potential difference between the top and the bottom plate is therefore,

$$V = \int_0^d E_z(z)dz = \sigma \int_0^d \frac{dz}{\varepsilon(z)}$$

The capacitance is then, $Q/V = \sigma.A/V = \dfrac{A}{\int_0^d \frac{dz}{\varepsilon(z)}}$. In particular, if the permittivity does not vary with z, the capacitance equals $\frac{\varepsilon A}{d}$.

4. Determine the capacitance of a cylindrical capacitor consisting of two concentric cylinders of radii $a < b$ and having length L. Neglect edge effects and assume that the medium between the cylinders is filled

with a dielectric of permittivity $\varepsilon(\rho)$, where ρ is the distance from the common axis of the two cylinders.

Solution: Assume that the axis is along the z axis. The electric field has only a $\hat{\rho}$ component and this depends upon ρ only. Call this $E_\rho(\rho)$. Gauss' law $\nabla \cdot (\varepsilon \cdot \mathbf{E}) = 0$ in cylindrical co-ordinates gives

$$\frac{1}{\rho}\frac{d}{d\rho}(\rho.\varepsilon(\rho)E_\rho) = 0, a < \rho < b$$

The solution to this equation is

$$E_\rho(\rho) = \frac{\alpha}{\rho.\varepsilon(\rho)}, a < \rho < b$$

where α is a constant. The normal component to the displacement vector at any conducting surface gives the surface charge density on the surface. Thus, the surface charge density on the inner cylinder is $\varepsilon(a)\ E_\rho(a) = \frac{\alpha}{a}$.

The total charge on the inner cylinder is given by $Q = \frac{2\pi a L\alpha}{a} = 2\pi L\alpha$.

The potential difference between the cylindrical surfaces is given by

$$V = \int_a^b E_\rho(\rho)d\rho = \alpha.\int_a^b \frac{d\rho}{\rho.\varepsilon(\rho)}$$

The capacitance

$$C = \frac{Q}{V} = \frac{2\pi L}{\int_a^b \frac{d\rho}{\rho.\varepsilon(\rho)}}$$

In particular when ε is a constant, we get $C = \frac{2\pi\varepsilon L}{\log(b/a)}$

5. Determine the general solution to the Laplace's equation in Cartesian, cylindrical and spherical polar co-ordinates. Use the method of separation of variables. Show by means of examples how this solution can be used to solve boundary value problems in electrostatics.

Solution: First we note that the potential V in a region free of charge satisfies the equation $\nabla^2 V = 0$, provided that the permittivity does not vary from point to point. If the space is free of charge but has a permittivity ε dependent upon position, then the appropriate equation for the potential is $\nabla.(\varepsilon\nabla V) = 0$. If the space in addition has charge of volume density ρ, the appropriate equation gets modified to $\nabla.(\varepsilon\nabla V) = -\rho$. We consider the first case here: Laplace's equation in the Cartesian system is

$$\frac{\partial^2 V}{\partial x^2} + \frac{\partial^2 V}{\partial y^2} + \frac{\partial^2 V}{\partial z^2} = 0$$

We adopt the separation of variables: $V(x, y, z) = X(x)\,Y(y)\,Z(z)$. Plugging this into the equation and dividing by V gives

$$\frac{X''(x)}{X(x)}+\frac{Y''(y)}{Y(y)}+\frac{Z''(z)}{Z(z)}=0$$

This can hold good iff each term in this sum is a constant with the sum of the constants adding to zero, i.e.

$$\frac{X''(x)}{X(x)}=-\alpha^2,\quad \frac{Y''(y)}{Y(y)}=-\beta^2,\quad \frac{Z''(z)}{Z(z)}=\gamma^2$$

where $\gamma^2 = \alpha^2 + \beta^2$. The solution can thus be expressed as a superposition

$$V(x,y,z)=\int(A_1(\alpha,\beta).\cos(\alpha.x)++B_1(\alpha,\beta).\sin(\alpha.x))\times$$
$$(A_2(\alpha,\beta).\cos(\beta y)+B_2(\alpha,\beta).\sin(\beta y))\times$$
$$(A_3(\alpha,\beta).\exp(\sqrt{\alpha^2+\beta^2}\,z)+B_3(\alpha,\beta).\exp(-\sqrt{\alpha^2+\beta^2}\,z))d\alpha.d\beta$$

Consider for example the problem of determining the potential inside a cuboid of sidelengths a, b, c with the faces being $x = 0$, a, $y = 0$, b, $z = 0$, c. The potential on all the faces except $z = c$ face is zero and that on the $z = c$ face is $V_0(x, y)$. Using the above separation of variables, it follows that cos (αx) and cos (βy) terms cannot appear since the potential on the planes $x = 0$ and $y = 0$ is zero. The potential on the plane $x = a$ is zero implies that $\alpha = n\pi/a$. The potential on the plane $y = b$ is zero implies that $\beta = m\pi/b$. The potential on the plane $z = 0$ is zero implies that $B_3 = -A_3$. Thus letting $\gamma_{m,n} = \sqrt{(n\pi/a)^2+(m\pi/b)^2}$, we obtain

$$V(x,y,z)=\sum_{m,n=1}^{\infty} c_{m,n}\sin(n\pi x/a).\sin(m\pi y/b).\sinh(\gamma_{m,n}z)$$

The constants $c_{m,n}$ are determined from the boundary condition on $z = c$ plane:

$$V_0(x,y)=\sum_{m,n=1}^{\infty} c_{m,n}\sinh(\gamma_{m,n}c)\sin(n\pi x/a).\sin(m\pi y/b)$$

Thus from the Fourier series property,

$$c_{m,n}.\sinh(\gamma_{m,n}c)=\frac{4}{ab}\int_0^a\int_0^b V_0(x,y)\sin(n\pi x/a).\sin(m\pi y/b)dxdy$$

and this completes the solution to our problem.

We now look at Laplace's equation in cylindrical co-ordinates. The equation is

$$\frac{1}{\rho}\frac{\partial}{\partial\rho}\left(\rho\frac{\partial V}{\partial\rho}\right)+\frac{1}{\rho^2}\frac{\partial^2 V}{\partial\phi^2}+\frac{\partial^2 V}{\partial z^2}=0$$

Adopt the separation of variables: $V(\rho, \phi, z) = R(\rho).\Phi(\phi).Z(z)$. Plugging this into the differential equation and dividing by V gives us

$$\frac{R'(\rho)}{\rho.R(\rho)}+\frac{R''(\rho)}{R(\rho)}+\frac{\Phi''(\phi)}{\rho^2.\Phi(\phi)}=-\frac{Z''(z)}{Z(z)}$$

The left side is a function of ρ and ϕ only while the right side is a function of z only. Thus, both sides must equal a constant, say k^2. We get $Z(z) = A_1\cos(kz) + B_1\sin(kz)$. The left side gives

$$\frac{R'(\rho)}{\rho.R(\rho)}+\frac{R''(\rho)}{R(\rho)}+\frac{\Phi''(\phi)}{\rho^2.\Phi(\phi)}=k^2$$

which can be rewritten as

$$\frac{\rho.R'(\rho)}{R(\rho)}+\frac{\rho^2.R''(\rho)}{R(\rho)}-k^2\rho^2=-\frac{\Phi''(\phi)}{\Phi(\phi)}$$

The left side here is a function of ρ only while the right side ϕ only. Hence, both sides must equal a constant, say m^2. The solution for Φ is $\Phi(\phi) = A_2.\cos(m\phi) + B_2.\sin(m\phi)$. Since the potential must be the same at $\phi = 0$ and $\phi = 2\pi$, it follows that m must be an integer. Without loss of generality, we can assume it to be positive. Finally the left side of the above equation gives

$$\rho^2.R''(\rho) + \rho.R'(\rho) - (k^2\rho^2 + m^2)\,R(\rho) = 0$$

The solutions to these equations are $J_m(ik\rho)$, $Y_m(ik\rho)$, where J_m, and Y_m are the modified Bessel's functions of order m. Y_m has a singularity at the origin while J_m is non-singular. We can form linear combinations of these functions to get a real basis $\{H_{1,m}(k\rho), H_{2,m}(ik\rho)\}$ for the space of solutions for the above equation. These are called the Hankel functions. The general solution in cylindrical co-ordinates can now be expressed as a superposition

$$V(\rho,\phi,z)=\sum_m \int (A_{1,m}(k).\cos(kz)+B_{1,m}(k).\sin(kz)).(A_{2,m}(k).\cos(m\phi)$$

$$+B_{2,m}(k).\sin(m\phi))\times(A_{3,m}(k).H_{1.m}(k\rho)+B_{3,m}.H_{2,m}(k\rho))dk$$

Consider for example the problem of determining the potential in the region between two concentric cylinders, the inner one having radius a and the outer one radius b. The lengths of the cylinders are L and we may assume the common axis to be the z axis. The potential on the wall $\rho = a$ is specified as $V_1(\phi, z)$. The potential on the wall $\rho = b$ is specified as $V_2(\phi, z)$ and finally, the potential on the $z = 0$ and $z = L$ surfaces are specified as zero. The last two boundary conditions imply that no $\cos(kz)$ term can appear and that k must be of the form $n\pi/L$ with n as an integer which may be taken as positive. Thus the solution to the potential is

$$V(\rho,\phi,z)=\sum_{m,n=1}^{\infty} \sin(n\pi z/L).(A_{m,n}.\cos(m\phi)+B_{m,n}.\sin(m\phi))\times$$

$$(C_{m,n}H_{1,m}(n\pi\rho/L)+D_{m,n}H_{2,m}(n\pi\rho/L))$$

The coefficients $A_{m,n}$, $B_{m,n}$, $C_{m,n}$, $D_{m,n}$ are determined from the double Fourier series for $V(a, \phi, z)$ and $V(b, \phi, z)$ which have been specified. The details are left to the student to be worked out. We can also look at the problem when the potential is not a function of z, i.e. the two-dimensional Laplace equation in polar co-ordinates:

$$\frac{1}{r}\frac{\partial}{\partial r}r\frac{\partial V}{\partial r}+\frac{1}{r^2}\frac{\partial^2 V}{\partial \phi^2}=0$$

Substituting $V(r,\phi) = R(r)\Phi(\phi)$ into this equation and dividing by V gives

$$\frac{r^2 R''(r)}{R(r)}+\frac{rR'(r)}{R(r)}=-\frac{\Phi''(\phi)}{\Phi(\phi)}$$

The left side is a function of r only and the right side of ϕ only. Hence, both sides equal a constant m^2 which from the single valuedness of Φ must be an integer. We can take it as a positive integer. Then, $\Phi(\phi) = A.\cos(m\phi) + B.\sin(m\phi)$. The R equation is

$$r^2 R''(r) + rR'(r) - m^2R(r) = 0$$

The solutions are $R(r) = r^{\pm m}$. The general solution is

$$V(r,\phi)=\sum_{m=0}^{\infty} (A_m r^m + B_m r^{-m}).(C_m.\cos(m\phi)+D_m.\sin(m\phi))$$

If the origin is included, then $B_m = 0$. If $r = \infty$ is included, then $A_m = 0$. If both the origin and $r = \infty$ are excluded, then both terms $r^{\pm m}$ will appear. Consider for example, the problem of determining the potential in the region $r \le R$ with the potential at $r = R$ specified as $V_0(\phi)$. The potential has the expansion

$$V(r,\phi) = \sum_{m=0}^{\infty} r^m (A_m.\cos(m\phi) + B_m.\sin(m\phi))$$

The boundary condition gives

$$V_0(\phi) = V(R,\phi) = \sum_{m=0}^{\infty} R^m (A_m.\cos(m\phi) + B_m.\sin(m\phi))$$

so that by the Fourier series expansion,

$$A_m = \frac{1}{\pi R^m} \int_0^{2\pi} V_0(\phi).\cos(m\phi)d\phi, m \ge, 1, \; A_0 = \frac{1}{2\pi} \int_0^{2\pi} V_0(\phi)d\phi$$

$$B_m = \frac{1}{\pi R^m} \int_0^{2\pi} V_0(\phi).\sin(m\phi)d\phi, m \ge, 1$$

Likewise, if we are interested in the potential in the region $r > R$, then

$$V(r,\phi) = \sum_{m=0}^{\infty} R^{-m} (A_m.\cos(m\phi) + B_m.\sin(m\phi))$$

with

$$A_m = \frac{R^m}{\pi} \int_0^{2\pi} V_0(\phi).\cos(m\phi)d\phi, m \ge, 1, \; A_0 = \frac{1}{2\pi} \int_0^{2\pi} V_0(\phi)d\phi$$

$$B_m = \frac{R^m}{\pi} \int_0^{2\pi} V_0(\phi).\sin(m\phi)d\phi, m \ge, 1,$$

Finally, if we are interested in the potential in the region $a < r < b$ with $V(a, \phi) = V_1(\phi)$, $V(b,\phi) = V_2(\phi)$, then

$$V(r,\phi) = \sum_{m-0}^{\infty} (A_m r^m + B_m r^{-m}).(C_m.\cos(m\phi) + D_m.\sin(m\phi)$$

The boundary conditions give

$$C_m(A_m a^m + B_m a^{-m}) = \frac{1}{\pi} \int_0^{2\pi} V_1(\phi).\cos(m\phi)d\phi, m \ge, 1, C_0(A_0 + B_0)$$

$$= \frac{1}{2\pi} \int_0^{2\pi} V_1(\phi)d\phi$$

$$D_m(A_m a^m + B_m a^{-m}) = \frac{1}{\pi} \int_0^{2\pi} V_1(\phi).sin(m\phi)d\phi, m \ge, 1$$

and two similar equations with a replaced by b and V_1 replaced by V_2. These four equations can be solved for the products A_mC_m, B_mC_m, A_mD_m, B_mD_m to give the solution as

$$V(r,\phi) = \sum_{m=0}^{\infty} A_m C_m r^m \cos(m\phi) + A_m D_m r^m \sin(m\phi) + B_m C_m r^{-m} \cos(m\phi)$$

$$+ B_m D_m r^{-m} \sin(m\phi)$$

Finally, we look at Laplace's equation in spherical polar co-ordinates, perform the separation of variables and write out the general solution.

$$\frac{1}{r}\frac{\partial^2 rV}{\partial r^2} + \frac{1}{r^2 \sin\theta}\frac{\partial}{\partial\theta}\sin\theta\frac{\partial V}{\partial\theta} + \frac{1}{r^2\sin^2\theta}\frac{\partial^2 V}{\partial\phi^2} = 0$$

Let $V(r, \theta, \phi) = R(r)\Theta(\theta).\Phi(\theta)$. Plugging this into the differential equation and dividing by V and multiplying by r^2 results in

$$-\frac{r^2 R''(r)}{R(r)} - \frac{2rR'(r)}{R(r)} = \frac{1}{\sin\theta.\Theta}\frac{d}{d\theta}\sin\theta.\frac{d\Theta}{d\theta} + \frac{1}{\sin^2\theta}\frac{\Phi''(\phi)}{\Phi(\phi)}$$

The left side here is a function of r only while the right side of (θ, ϕ) only. Thus both sides must equal a constant say λ. The right side results in the differential equation

$$\frac{\sin\theta}{\Theta}\frac{d}{d\theta}\sin\theta.\frac{d\Theta}{d\theta} - \lambda.\sin^2\theta = -\frac{\Phi''(\phi)}{\Phi(\phi)}$$

Again the left side here is a function of θ only while the right side of ϕ only. Thus both sides here must once again equal a constant, say m^2. The solution for Φ is $\exp(\pm im\phi)$. m must be an integer for the potential to be single valued. The θ equation reduces to

$$\frac{1}{\sin\theta}\frac{d}{d\theta}(\sin\theta.\frac{d\Theta}{d\theta}) - \frac{m^2}{\sin^2\theta}\Theta = \lambda\Theta$$

It can be shown that the allowed values of λ so that the solutions for Θ are physically meaningful are $\lambda = -l(l+1)$ where l is a non-negative integer $\geq |m|$. The above equation is called the Legendre differential equation and its solutions for each pair (l, m) are the modified Legendre polynomials $P_{l,m}(\cos\phi)$ in the variable $\cos\phi$. The R equation is

$$r^2R''(r) + 2rR'(r) - l(l+1)\,R(r) = 0$$

Its solutions are r^l and r^{-l-1}. The general solution is thus given by

$$\sum (A_{1,l,m}\exp(im\phi)+A_{2,l,m}\exp(-im\phi)).P_{l,m}(\cos\theta).\left(C_{l,m}r^l+\frac{D_{l,m}}{r^{l+1}}\right)$$

If the origin is included in the region then $D_{l,m} = 0$ since the potential must be non-singular at $r = 0$. In case the problem has azimuthal symmetry, $m = 0$ is the only allowed value and the solution is of the form

$$V(r,\theta)=\sum_{l=0}^{\infty} P_l(\cos\theta)\left(C_l r^l+\frac{D_l}{r^{l+1}}\right)$$

where $P_l(\cos\theta)$ is the solution to the Legendre equation

$$\frac{1}{\sin\theta}\frac{1}{d\theta}\left(\sin\theta\frac{d\Theta}{d\theta}\right)+l(l+1)\,\Theta=0$$

It is left as an exercise to the student to transform the Legendre equation in $x = \cos\theta$ and obtain its solution in the form of a power series in x. Consider for example the problem of determining the potential inside a sphere of radius R when the potential on the surface is specified as $V_0(\theta)$. The problem has azimuthal symmetry and since the origin is included, it can be expanded as

$$V(r,\theta)=\sum_{l=0}^{\infty} A_l r^l P_l(\cos\theta)$$

From the Sturm Liouville theory, the Legendre polynomials are orthogonal: $\int_{-1}^{1} P_l(x)P_k(x)dx = C_l\delta_{l,k}$, where $\{C_l\}$ are some normalization constants which we shall not need to specify since we are interested only in the method not in the final solution. The boundary condition gives

$$V(R,\theta)=\sum_{l=0}^{\infty} A_l R^l P_l(\cos\theta)$$

so that

$$A_l=\frac{1}{C_l R^l}\int_0^{\pi} V(R,\theta).P_l(\cos\theta).\sin(\theta)d\theta$$

Suppose on the other hand, we are interested in determining the solution outside the sphere i.e. for $r > R$. Since the potential must vanish at ∞, we must not include the r^l terms. Thus

$$V(r,\theta) = \sum_{l=0}^{\infty} \frac{A_l}{r^{l+1}} P_l(\cos\theta)$$

and

$$A_l = \frac{R^{l+1}}{C_l} \int_0^{\pi} V(R,\theta) P_l(\cos\theta) \sin(\theta).d\theta$$

6. A parallel plate capacitor consists of two parallel plates of area A and separated by a distance d. The lower plate is located at the $z = 0$ plane and the top plate at $z = d$ plane. The region between the plates from $z = 0$ to $z = a$ is filled with a dielectric of permittivity ε_1 and the region from $z = a$ to $z = d$ is filled with dielectric of permittivity ε_2. Determine the capacitance between the plates and the fields E, P, D in the two regions. Also determine the surface density of polarization charge on the $z = a$ surface.

Solution: Let Q be the charge on the bottom plate and $-Q$ the charge on the top plate. The equation $\nabla.\mathbf{D} = 0$ in the region between the two plates implies that $\frac{dD(z)}{dz} = 0$, where $D(z)$ is the z component of the electric displacement vector. Note that neglecting edge effects, E, D, P all have only a z component. Thus, $D(z)$ is a constant. Gauss' law in integral form applied to a pillbox with one face inside the bottom conductor and the opposite face in the region between the plates gives $D = \sigma = Q/A$. The electric field E_1 in the region $0 < z < a$ is $E_1 = D/\varepsilon_1 = \sigma/\varepsilon_1$. In the region $a < z < d$, the electric field is given by $E_2 = D/\varepsilon_2 = \sigma/\varepsilon_2$. The polarization vector P_1 in the region $0 < z < a$ is given by $P_1 = D - \varepsilon_0 E_1 = \sigma(1 - \varepsilon_0/\varepsilon_1)$ and the polarization vector P_2 in the region $a < z < d$ is given by $P_2 = D - \varepsilon_0 E_2 = \sigma(1 - \varepsilon_0/\varepsilon_2)$. The surface density of polarization charge on the $z = a$ plane is $\sigma_{pol} = \varepsilon_0(E_2 - E_1) = P_1 - P_2 = \sigma\varepsilon_0(1/\varepsilon_2 - 1/\varepsilon_1)$. The potential difference between the plates is $V = E_1 a + E_2(d - a) = \sigma a/\varepsilon_1 + \sigma(d - a)/\varepsilon_2$. The capacitance between the plates is thus

$$C = Q/V = (\sigma A)/V = \frac{A}{a/\varepsilon_1 + (d-a)/\varepsilon_2}$$

7. Consider two concentric spherical surfaces of radii $a < b$ with the origin as the centre. The space between the two surfaces is filled with dielectric of permittivity $\varepsilon(r)$ where r is the distance from the centre. Find the capacitance between the two surfaces.

Ans. Gauss' law for the electric displacement $D_r(r)$ is $\frac{1}{r^2}\frac{d}{dr}(r^2D_r(r)) = 0$. This gives $D_r(r) = \frac{A}{r^2}$, where A is a constant. The condition $D_r(a) = \sigma$ where σ is the charge density on the inner surface gives $A = \sigma a^2$. Now, $Q = \sigma.4\pi a^2$ where Q is the charge on the inner surface. Thus, $A = \frac{Q}{4\pi}$. The electric field between the plates is $E_r(r) = \frac{D_r(r)}{\varepsilon(r)} = \frac{A}{\varepsilon(r)r^2}$.
This gives for the potential difference between the plates

$$V = \int_a^b E_r(r)dr = A\int_a^b \frac{dr}{\varepsilon(r)r^2}$$

and hence the capacitance is

$$C = \frac{Q}{V} = 4\pi\left(\int_a^b \frac{dr}{\varepsilon(r)r^2}\right)^{-1}$$

8. Derive an expression for the energy density of the electrostatic field in terms of the electric field.

Ans. The field energy can be expressed as $U = \frac{1}{2}\int \rho(\mathbf{r})V(\mathbf{r})d^3r$. Using Poisson's equation, $\nabla^2 V = -\rho/\varepsilon$, we get

$$U = -\frac{\varepsilon}{2}\int V\nabla^2 V d^3\mathbf{r} = -\frac{\varepsilon}{2}\int \nabla.(V\nabla V)d^3r + \frac{\varepsilon}{2}\int |\nabla V|^2\, d^3\mathbf{r}$$

The first integral on the right, by Gauss' theorem, transforms into a surface integral $\int_S V\,\frac{\partial V}{\partial n}\,dS$, where S is a spherical surface of infinite radius. Assuming that the potential at $r = \infty$ decays at least as fast as $1/r$ and the electric field at $r = \infty$ at least as fast as $1/r^2$, it follows that the surface integral is zero since $dS = r^2 d\Omega$, where $d\Omega$ is the infinitesimal solid angle. Thus,

$$U = \frac{\varepsilon}{2}\int |\nabla V|^2\, d^3\mathbf{r} = \int \frac{\varepsilon}{2}|\mathbf{E}|^2\, d^3\mathbf{r}$$

where $\mathbf{E} = -\nabla V$ is the electric field. This means that the electrostatic field energy density is $W = \frac{\varepsilon}{2}|\mathbf{E}|^2$.

9. This problem will introduce you to the method of images. Consider an infinite conducting plane coinciding with the xy plane. A charge Q is placed at a distance d above it, say at $\mathbf{d} = (0, 0, d)$. We want to determine the potential field $V(x, y, z)$, the charge density induced on the plane and the force exerted by the plane on the charge and vice versa.

Ans. We've got to solve Poisson's equation in the region $z > 0$ ($\nabla^2 V\mathbf{r}$) $= -\frac{Q}{\varepsilon}\delta(\mathbf{r} - \mathbf{d})$, with the boundary condition that V is constant, say zero on the xy plane. This boundary condition corresponds to the fact that inside a perfect conductor no electric field can exist at equilibrium and hence all points within the conductor must be at the same potential. This common potential can be taken as the zero reference. Even a slight electric field inside a perfect conductor will cause charges to flow immediately from one side to another thereby neutralizing the electric field. The boundary condition can be met by replacing the conducting plane by an equal and opposite charge $-Q$ at $-\mathbf{d} = (0, 0, -d)$. The potential due to the two charges on the xy plane is zero since $V = \frac{Q}{r_1} - \frac{Q}{r_2}$ is zero because $r_1 = r_2$ on the xy plane. That is every point on the xy plane is equidistant from the two charges. Further, the presence of $-Q$ at $-\mathbf{d}$ is not going to affect the singularity of $\nabla^2 V$ at $\mathbf{d}$. Specifically, if

$$V(\mathbf{r}) = \frac{Q}{4\pi\varepsilon\,|\mathbf{r}-\mathbf{d}|} - \frac{Q}{4\pi\varepsilon|\mathbf{r}+\mathbf{d}|}$$

then $\nabla^2 V(\mathbf{r}|) = -\frac{Q}{\varepsilon}\delta(\mathbf{r} - \mathbf{d})$ in the region $z > 0$ since $-Q$ is present in the region $z < 0$. Expressed in full, we can express the potential in the region $z > 0$ as

$$V(x, y, z) = \frac{Q}{4\pi\varepsilon\sqrt{x^2 + y^2 + (z-d)^2}} - \frac{Q}{4\pi\varepsilon\sqrt{x^2 + y^2 + (z+d)^2}}$$

The normal component to the electric field on the surface multiplied by ε gives the surface charge density:

$$\sigma(\rho) = \sigma(x, y) = -\varepsilon\frac{\partial V(x, y, z)}{\partial z}\Big|_{z=0+} = -\frac{Qd}{2\pi(x^2 + y^2 + d^2)^{3/2}}$$

$$= -\frac{Qd}{2\pi(\rho^2 + d^2)^{3/2}}$$

where $\rho = \sqrt{x^2 + y^2}$ is the distance of the point on the conducting plane from the origin. The total charge on the plane is

$$\int_0^\infty \sigma(\rho)2\pi\rho.d\rho = -Q$$

This result could have been deduced directly from Gauss law. Consider a closed hemispherical surface, consisting of the infinite xy plane and the hemisphere of infinite radius with origin as centre and extending in the region $z < 0$. The normal component of the electric field on the hemisphere vanishes while on the xy plane, it is given by

$$E_z(\rho) = -\frac{\partial V}{\partial z} = -\frac{Qd}{2\pi\varepsilon(\rho^2 + d^2)^{3/2}}$$

Gauss' law gives for this closed surface S, $\varepsilon \int_S \mathbf{E}.\mathbf{n}dS = Q_{encl} = -Q$ since the surface encloses the charge $-Q$ at $-\mathbf{d}$. This proves that $\varepsilon. \int E_z(\rho) 2\pi\rho.d\rho = -Q$ even without performing the integration. We now want to calculate the force of the plane on the charge. Note that the charge Q induces the charge $-Q$ on the plane distributed as $\sigma(\rho)$ which in turn exerts an attractive force on Q. This force will be shown to equal the force between Q and the fictitious charge $-Q$ at $-\mathbf{d}$. By symmetry, the attractive force has only a z component which is

$$F = \int \frac{Q\sigma(\rho)d}{4\pi\varepsilon(\rho^2 + z^2)^{3/2}} 2\pi\rho.d\rho$$

$$= -\frac{Q^2 d^2}{4\pi\varepsilon} \int_0^\infty \frac{\rho.d\rho}{(\rho^2 + d^2)^3} = -\frac{Q^2}{16\pi\varepsilon.d^2}$$

which coincides with the claim made. We shall next compute the work done in transporting the charge from $(0, 0, d)$ to ∞ along the z axis against the attractive force produced by the oppositely charged conducting plane. It is given by the integral of the above expression from d to ∞:

$$W = \int_d^\infty \frac{Q^2}{16\pi\varepsilon.r^2} dr = \frac{Q^2}{16\pi\varepsilon.d}$$

Finally, using the method of images developed here, we shall obtain the Green's function for the potential problem in the half plane $z > 0$ with specified boundary condition for the potential on the xy plane. The Green's function equation to be solved is

$$\nabla^2 G(x,y,z|x', y', z') = \delta\,(x - x')\,\delta\,(y - y')\,\delta\,(z - z'),\ z, z' > 0$$

with boundary condition

$$G(x,y,0|x', y', z') = 0$$

By analogy with the method of images just developed, the solution is simply the formula for the potential obtained above with $\mathbf{d}$ replaced by

(x', y', z') and $-Q/\varepsilon$ replaced by unity. The solution is

$$G(x,y,z \mid x',y',z') = -\frac{1}{4\pi}\left(\frac{1}{\sqrt{(x-x')^2+(y-y')^2+(z-z')^2}} - \frac{1}{\sqrt{(x-x')^2+(y-y')^2+(z+z')^2}}\right)$$

This can be used to solve for example the problem of determining the potential due to any charge distribution in the region $z > 0$ when the potential on the xy plane is specified as $V_0(x, y)$. It is given by

$$V(x,y,z) = -\int_{z>0} \rho(x,y,z)G(x,y,z \mid x',y',z')dxdydz/\varepsilon + \int V_0(x,y) G(x,y,0 \mid x',y',z')dxdy$$

We do not continue on this any more.

10. This is another problem dealing with the method of images. Consider a conducting sphere of radius R with centre at the origin of co-ordinates. We place a charge Q at $(0, 0, d)$ where $d > R$ and wish to determine the potential outside the sphere. We also want to determine the force exerted on the charge by the conducting sphere and vice versa, the surface charge density induced on the sphere and the work done to transport the charge Q from $(0, 0, d)$ to ∞ along the z axis. First assume that an image charge $-Q'$ is placed at $(0, 0, d')$ where $d' < R$. We want to determine the image charge Q' as well as its distance d' from the origin so that the combination of the two charges gives zero potential on the surface of the sphere. The potential on the surface of the sphere at polar co-ordinates (R, θ, ϕ) due to the two charges is given by

$$V(R,\theta) = \frac{1}{4\pi\varepsilon_0}\left(\frac{Q}{\sqrt{d^2+R^2-2dR\cos\theta}} - \frac{Q'}{\sqrt{d'^2+R^2-2d'R\cos\theta}}\right)$$

For this to be zero, we must have

$$Q'^2(d^2 + R^2 - 2dR\cos\theta) = Q^2(d'^2 + R^2 - 2d'R\cos\theta)$$

for all $\theta \in [0, \pi)$. This leads us to the equations $dRQ'^2 = d'RQ^2$, $Q'^2(d^2 + R^2) = Q^2(d'^2 + R^2)$. Eliminating Q' gives $d'R(d^2 + R^2) = dR(d'^2 + R^2)$ or $d'(d^2 + R^2) = d(d'^2 + R^2)$ or $dd'(d - d') = R^2(d - d')$ so that $dd' = R^2$ or $d' = R^2/d$ and $Q'^2 = Q^2d'/d = Q^2R^2/d^2$ or $Q' = QR/d$. Thus, the potential at any point (r, θ, ϕ) outside the sphere is given by

$$V(r,\theta) = \frac{Q}{4\pi\varepsilon_0}((d^2 + r^2 - 2dr\cos\theta)^{-1/2} - (R/d)(R^4/d^2 + r^2 - $$

$$-2(R^2 r/d)\cos\theta)^{-1/2})$$

$$= \frac{Q}{4\pi\varepsilon_0}((d^2 + r^2 - 2dr\cos\theta)^{-1/2} - (R^2 + r^2 d^2/R^2 - 2rd\cos\theta)^{-1/2})$$

The normal component of the electric field on the surface of the sphere is

$$E_r(R,\theta) = -\frac{\partial V(r,\theta)}{\partial r}\Big|_{r=R} = \frac{Q}{4\pi\varepsilon_0 R}\frac{(R^2 - d^2)}{(R^2 + d^2 - 2dR\cos\theta)^{3/2}}$$

This multiplied by ε_0 is the surface charge density on the sphere:

$$\sigma(\theta) = \frac{Q(R^2 - d^2)}{4\pi R(R^2 + d^2 - 2Rd\cos\theta)^{3/2}}$$

This charge is negative and we can check the validity of this result by showing that the surface integral of this density is $-Q' = -QR/d$ as it should be in accordance with Gauss' law:

$$2\pi\int_0^\pi \sigma(\theta)R^2\sin(\theta)d\theta = \frac{Q(R^2 - d^2)R}{2}\int_0^\pi \frac{\sin(\theta)d\theta}{(R^2 + d^2 - 2Rd\cos\theta)^{3/2}}$$

$$= \frac{Q(R^2 - d^2)R}{2}\int_{(R-d)^2}^{(R+d)^2} \xi^{-3/2}(d\xi/2Rd)$$

$$= -\frac{Q(R^2 - d^2)}{2d}\xi^{-1/2}\Big|_{(R-d)^2}^{(R+d)^2} = \frac{Q(R^2 - d^2)}{2d}((d-R)^{-1} - (d+R)^{-1}) =$$

$$= -QR/d = -Q'$$

We can also compute the force exerted by the conducting sphere on the charge Q. It will turn out to equal the force between Q and its image charge $-Q'$. To see this, we note that the force is directed along the z axis and is attractive. Coulomb's law with the appropriate resolution gives

$$F = \frac{Q}{4\pi\varepsilon_0}\int_0^\pi \frac{\sigma(\theta)R^2\sin\theta d\theta.2\pi.(d - R\cos(\theta))}{(d^2 + R^2 - 2dR\cos\theta)^{3/2}}$$

$$= \frac{Q^2(R^2 - d^2)R}{8\pi\varepsilon_0} \int_0^{\pi} \frac{\sin(\theta)(d - R\cos\theta)d\theta}{(d^2 + R^2 - 2dR\cos\theta)^3}$$

On making the change of variables, $d^2 + R^2 - 2dR\cos\theta = \xi$, this integral becomes

$$F \frac{Q^2(R^2 - d^2)R}{8\pi\varepsilon_0} \int_{(d-R)^2}^{(d+R)^2}]xi^{-3}(d/2 - R^2/2d + \zeta/2d)(d\zeta/2dR)$$

$$= \frac{Q^2(R^2 - d^2)}{16\pi\varepsilon_0 d} \left\{\frac{1}{4}(d - R^2/d)((d-R)^{-4} - (d+R)^{-4}) + \frac{1}{2d}((d-R)^{-2}\right.$$

$$\left. -(d+R)^{-2})\right\} = \frac{Q^2(R^2 - d^2)}{16\pi\varepsilon_0 d} \left\{\frac{d^2 - R^2}{4d} \frac{(d+R)^4 - (d-R)^4}{(d^2 - R^2)^4} + \frac{2R}{(d^2 - R^2)^2}\right\}$$

$$= \frac{Q^2(R^2 - d^2)}{16\pi\varepsilon_0 d} \left\{\frac{2R(d^2 + R^2)}{(d^2 - R^2)^3} + \frac{2R}{(d^2 - R^2)^2}\right\}$$

$$\frac{Q^2(R^2 - d^2)}{16\pi\varepsilon_0 d} \frac{4Rd^2}{(d^2 - R^2)^3} = -\frac{Q^2 Rd}{4\pi\varepsilon_0 (d^2 - R^2)^2} = -\frac{QQ'}{4\pi\varepsilon_0 (d - d')^2}$$

This can be verified by letting $d' = R^2/d$ and $Q' = QR/d$. We can also calculate the total force needed to transport the charge Q from d to ∞ along the z axis. It is given by the integral of the above quantity with a sign reversed:

$$W = \int_d^{\infty} \frac{Q^2 Rr}{4\pi\varepsilon_0 (r^2 - R^2)^2} dr = \frac{Q^2 R}{8\pi\varepsilon_0 (d^2 - R^2}$$

11. Consider the situation shown in Fig. 2.2. We are considering a grid inside a material and four points of the grid numbered 1, 2, 3, 4 counterclockwise. Above the horizontal, the dielectric constant is ε_r and below the horizontal, it is unity. The central point in this grid is numbered 0. Let the potential at the point numbered i be V_i. We want to write down an approximate expression for the potential at point 0 in terms of the potentials at 1, 2, 3, 4 based on the discretization of the modified Laplace's equation for a dielectric medium. The modified equation is $\nabla.(\varepsilon\nabla V) = 0$ which for our two dimensional problem, boils down to

$$\frac{\partial}{\partial x}\varepsilon\frac{\partial V}{\partial x} + \frac{\partial}{\partial y}\varepsilon\frac{\partial V}{\partial y} = 0$$

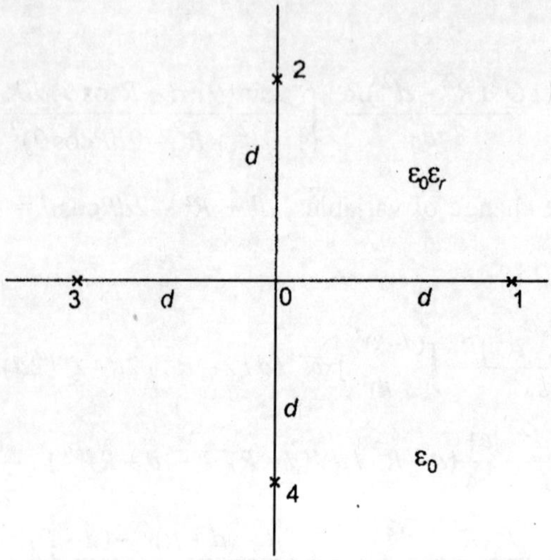

Fig. 2.2

Approximation to $\varepsilon \left.\dfrac{\partial V}{\partial x}\right|_{x+dx,y}$ above the x axis at (0,0 is $\varepsilon_r (V_1 - V_0)/d$.

Approximation to $\varepsilon \left.\dfrac{\partial V}{\partial x}\right|_{x+y}$ just above the x axis at (0,0) is $\varepsilon_r (V_0 - V_3)/d$. Therefore, the approximation to $\dfrac{\partial}{\partial x} \varepsilon \dfrac{\partial V}{\partial x}$ just above the x axis is the difference of these two divided by d: $\varepsilon_r(V_1 - 2V_0 + V_3)/d^2$. Likewise, approximation to $\dfrac{\partial}{\partial x} \varepsilon \dfrac{\partial V}{\partial x}$ just below the x axis is $(V_1 - 2V_0 + V_3)/d^2$. Taking the average of these two gives the approximation to $\dfrac{\partial}{\partial x} \varepsilon \dfrac{\partial V}{\partial x}$ at (0,0) as

$$\frac{(1+\varepsilon_r)}{2d^2}(V_1 - 2V_0 + V_3)$$

Now, approximation to $\varepsilon \left.\dfrac{\partial V}{\partial x}\right|_{x,y+dy}$ at (0, 0) is $\varepsilon_r(V_2 - V_0)/d$.

Approximation to $\varepsilon \left.\dfrac{\partial V}{\partial x}\right|_{x,y}$ at (0,0) is $(V_0 - V_4)/d$. Taking the difference of these two and dividing by d gives the approximation to $\frac{\partial}{\partial y}\varepsilon\frac{\partial V}{\partial y}$ at (0,0) as

$$(\varepsilon_r(V_2 - V_0) - (V_0 - V_4))/d^2 = (\varepsilon_r V_2 + V_4 - (\varepsilon_r + 1)\ V_0)/d^2$$

Modified Laplace's equation with these approximations reads

$$(1 + \varepsilon_r)\ (V_1 - 2V_0 + V_3)/2 + \varepsilon_r V_2 + V_4 - (\varepsilon_r + 1)V_0 = 0$$

or

$$V_0 = \frac{V_1 + V_3}{4} + \frac{\varepsilon_r V_2 + V_4}{2(1+\varepsilon_r)}$$

12. The time averaged potential of a neutral Hydrogen atom is given by $\Phi\ (r) = \frac{q}{4\pi\varepsilon_0}\exp(-ar)\ (1 + ar/2)/r$. Find the distribution of charge that gives this potential.

Ans. We use the Poisson's equation $\nabla^2\Phi = -\rho/\varepsilon_0$. Since the potential depends only upon the radius, we have

$$\nabla^2\Phi = \frac{1}{r}\frac{d^2}{dr^2}r\Phi(r) = \frac{qa^3}{8\pi\varepsilon_0}\exp(-ar)$$

Thus,

$$\rho(r) = -\varepsilon_0\nabla^2\Phi(r) = -\frac{qa^3}{8\pi}\exp(-ar)$$

Charge density decreases exponentially with radial distance.

13. (Thomson's theorem): If a number of surfaces are fixed in position and a given total charge is placed on each surface, then the electrostatic energy in the region bounded by the surfaces is a minimum when the charges are placed such that every surface is an equipotential.

Ans. Let the surfaces bounding the volume V be $S_1, \ldots, S_n$. Let $\sigma_j(\mathbf{r})$ denote the surface charge density on the jth surface S_j. Then the electrostatic energy can be expressed as

$$W = \frac{1}{8\pi\varepsilon_0}\sum_{j,k}\int_{S_j\times S_k}\frac{\sigma_j(\mathbf{r})\sigma_k(\mathbf{r}')}{|\mathbf{r}-\mathbf{r}'|}dS(\mathbf{r})dS(\mathbf{r}')$$

We want to minimize W subject to the constraint $\sum_j \int_s \sigma_j(\mathbf{r})dS(r) = Q$.

The appropriate cost function to be minimized after the introduction of Lagrange multipliers is

$$L = W + \lambda\left(\sum_j \int_{S_j} \sigma_j(\mathbf{r})dS(\mathbf{r}) - Q\right)$$

Taking the variation of W with respect to the functions $\sigma_j(\mathbf{r})$, $j = 1, 2, \ldots, n$ gives

$$\delta L = \frac{1}{4\pi\varepsilon_0}\sum_{j,k}\int_{S_j\times S_k}\frac{\delta\sigma_j(\mathbf{r})\sigma_k(\mathbf{r}')}{|\mathbf{r}-\mathbf{r}'|}dS(\mathbf{r})dS(\mathbf{r}')$$

$$+\lambda\sum_j\int_{S_j}\delta\sigma_j(\mathbf{r})dS(\mathbf{r})$$

Equating to zero the coefficient of $\delta\sigma_j(\mathbf{r})$, we get

$$\frac{1}{4\pi\varepsilon_0}\sum_k\int_{S_k}\frac{\sigma_k(\mathbf{r}')}{|\mathbf{r}-\mathbf{r}'|}dS(\mathbf{r}') + \lambda = 0$$

This equation is the same as $V(\mathbf{r}) = -\lambda$ where $V(\mathbf{r})$ is the (Coulomb) potential at $\mathbf{r}$ due to all the charges distributed on the surfaces.

14. A conducting plane coincides with the xy plane except for the fact that a hole of radius a centred at the origin is cut in the plane and the potential at each point in this hole is a constant V volts. The potential on the surface at other points is obviously zero. Assume that a charge Q is placed at $(0, 0, d)$. Find the potential at all points in the region $z > 0$.

Ans. The Green's function for this problem has already been evaluated in Problem 9. It is given by

$$G(\mathbf{r}\,|\,\mathbf{r}') = -\frac{1}{4\pi}((x-x')^2 + (y-y')^2 + (z-z')^2)^{-1/2} - ((x-x')^2 +$$

$$+(y-y')^2 + (z+z')^2)^{-1/2})$$

For the surface, $n = -\hat{z}$ and we find

$$\frac{\partial G(\mathbf{r}\,|\,\mathbf{r}')}{\partial \mathbf{n}} = -\frac{\partial G(\mathbf{r}\,|\,\mathbf{r}')}{\partial z} =$$

$$-\frac{1}{4\pi}\left(\frac{z-z'}{((x-x')^2+(y-y')^2+(z-z')^2)^{3/2}} + \frac{1}{4\pi}\frac{z+z'}{((x-x')^2+(y-y')^2+(z+z')^2)^{3/2}}\right.$$

In terns of cylindrical co-ordinates, $x = \rho.\cos\phi$, $y = \rho.\sin\phi$, $z = z$, $x' = \rho'.\cos\phi'$, $y' = \rho.\sin\phi'$, $z' = z'$, we find that

$$\frac{\partial G(\mathbf{r}\,|\,\mathbf{r}')}{\partial \mathbf{n}}\Big|_{z=0} = \frac{1}{2\pi}\frac{z'}{((x-x')^2+(y-y')^2+z'^2)^{3/2}}$$

$$= \frac{1}{2\pi}\frac{z'}{(\rho^2+\rho'^2-2\rho.\rho'\cos(\phi-\phi')+z'^2)^{3/2}}$$

This gives

$$\Phi(\rho',\phi',z') = \int_{z=0}\Phi(\mathbf{r})\frac{\partial G(\mathbf{r}\,|\,\mathbf{r}')}{\partial \mathbf{n}}dS$$

$$= \frac{V}{2\pi}\int_{0<\phi<2\pi,0<\rho<a}\frac{z'\rho d\rho d\phi}{(\rho^2+\rho'^2-2\rho.\rho'.\cos(\phi-\phi')+z'^2)^{3/2}}$$

We shall find the leading terms in the expansion for the potential assuming that $\rho'^2 + z'^2 >> a^2$. Let $\rho'^2 + z'^2 = r'^2$. Then,

$$(\rho^2+\rho'^2+z'^2-2\rho\rho'.\cos(\phi-\phi'))^{-3/2} = r'^{-3}(1+(\rho^2-2\rho\rho'\cos(\phi-\phi'))/r'^2)^{-3/2}$$

$$= r'^{-3}\left(1-\frac{3}{2r'^2}(\rho^2-2\rho\rho'\cos(\phi-\phi'))+\frac{3.5}{2^2 2!}(\rho^2-2\rho\rho'\cos(\phi-\phi'))^2/r'^4+\ldots\right)$$

Plugging this into the expression for the potential gives

$$\Phi(\rho',\phi',z') = \frac{Vz'}{2\pi}\left\{\frac{1}{r'^3}\int\rho d\rho d\phi - \frac{3}{2r'^5}\int\rho(\rho^2-2\rho\rho'\cos(\phi-\phi'))d\rho d\phi\right.$$

$$\left.+\frac{15}{8r'^7}\int\rho(\rho^2-2\rho\rho'\cos(\phi-\phi'))^2 d\rho d\phi\right\}$$

The integrals are over the region $\{0<\rho<a,\ 0<\phi<2\pi\}$. The different terms occurring here are evaluated easily:

$$\int\rho.d\rho.d\phi = \frac{a^2}{2}2\pi = \pi a^2$$

$$\int\rho(\rho^2-2\rho.\rho'.\cos(\phi-\phi'))d\rho.d\phi = \frac{a^4}{4}2\pi = \frac{\pi a^4}{2}$$

$$\int\rho(\rho^2-2\rho.\rho'.\cos(\phi-\phi'))^2 d\rho.d\phi = \int\rho(\rho^4+4\rho^2\rho'^2\cos^2(\phi-\phi')$$

$$-4\rho^3\rho'\cos(\phi-\phi'))d\rho d\phi = \frac{a^6}{6}2\pi + 4\frac{a^4}{4}\rho'^2\int_0^{2\pi}\cos^2(\phi)d\phi = \frac{\pi a^6}{3}+\pi a^4\rho'^2$$

This gives the following expansion:

$$V(\rho',\phi',z')=\frac{Vz'a^2}{2r'^3}-\frac{3Vz'a^4}{8r'^5}+\frac{15Vz'}{16r'^7}\left(\frac{a^6}{3}+a^4\rho'^2\right)+\ldots$$

15. Explain the variational approach to the solution of Poisson's equation with (a) the Dirichlet boundary conditions and (b) the Neumann boundary conditions.

Ans. The variational approach means that the solution to the boundary value problem should be derivable by minimizing a certain functional. Consider first the problem of minimizing the functional

$$I(\phi)=\frac{1}{2}\int_V(\nabla\phi,\nabla\phi)dV+\int_V g\phi dV$$

where g is fixed function and V is a volume bounded by a surface S. The minimization is to be carried out subject to the constraint that ϕ is specified on the boundary S of V. We carry out a variation from ϕ to $\phi+\delta\phi$ with the small change $\delta\phi$ zero on S so as to guarantee that the perturbed function will also satisfy the boundary conditions. The change in I is

$$I(\phi+\delta\phi)-I(\phi)=\delta I(\phi)=\int_V(\nabla\delta\phi,\nabla\phi)dV+\int_V g\delta\phi dV$$

$$=\int(\nabla,\delta\phi\nabla\phi)dV-\int\delta\phi\nabla^2\phi dV+\int_V g\delta\phi dV$$

$$=\int_S\delta\phi\frac{\partial\phi}{\partial \mathbf{n}}dS-\int\delta\phi\nabla^2\phi dV+\int_V g\delta\phi dV$$

Since $\delta\phi=0$ on S, the above variation reduces to

$$\delta I(\phi)=\int_V(g-\nabla^2\phi)\delta\phi dV$$

and if this must vanish for all variations $\delta\phi$ inside V, we must have

$$\nabla^2\phi=g$$

inside V. Thus, the function ϕ that minimizes I subject to fixed boundary conditions on S satisfies the Poisson equation $\nabla^2\phi=g$ inside V. This shows how the solution to the Poisson equation with Dirichlet boundary conditions can be obtained from a variational principle. We now look at the problem of minimizing

$$I(\phi)=\frac{1}{2}\int_V(\nabla\phi,\nabla\phi)dV+\int_V g\phi dV-\int_S f\phi dS$$

Taking the variation gives

$$\delta I(\phi) = \int_V (\nabla\delta\phi, \nabla\phi)dV + \int_V g\delta\phi - \int_S f\delta\phi dS$$

$$= \int_V (\nabla, \delta\phi\nabla\phi)dV - \int_V \delta\phi\nabla^2\phi dV + \int_V g\delta\phi dV - \int_S f\delta\phi dS$$

$$= \int \delta\phi\frac{\partial\phi}{\partial\mathbf{n}}dS - \int_V \delta\phi\nabla^2\phi dV + \int_V g\delta\phi dV - \int_S f\delta\phi dS$$

This is zero for all variations $\delta\phi$ provided

$$\frac{\partial\phi}{\partial\mathbf{n}} = f \text{ on } S$$

and

$$\nabla^2\phi = g \text{ in } V$$

That is provided ϕ satisfies the Poisson equation with source g inside V along with the Neumann boundary condition on S.

16. We consider another problem involving the method of images. Two conducting planes are respectively defined by $\{y = 0, x \geq 0\}$ and $\{x = 0, y \geq 0\}$. These form the first quadrant of in the xy plane. All values of z are included. A line charge of linear charge density λ is placed at (x_0, y_0) parallel to the z axis. We want to calculate the potential in the first quadrant, i.e. $x > 0$, $y > 0$.

Ans. The potential due to a line charge is given by $-\frac{\lambda}{2\pi}\log(\rho)$ where ρ is the distance from the line charge. This can be seen via Gauss' law applied to a cylindrical surface of radius ρ with central axis coinciding with the line. Applying Gauss' law to this surface gives $E_\rho(\rho) = \frac{\lambda}{2\pi\rho}$ and the negative integral of this gives the potential. We now place an image line charge of linear density $-\lambda$ at $(-x_0, y_0)$, another image line charge of linear density $-\lambda$ at $(x_0, -y_0)$ and finally an image line charge of linear density λ at $(-x_0, -y_0)$. The overall effect of these four line charges (one original line charge and three image line charges) is to produce a zero potential at the planes $x = 0$ and $y = 0$. The potential at any point is given by superposition

$$V(x,y) = -\frac{\lambda}{4\pi}(\log\frac{((x-x_0)^2+(y-y_0)^2)(((x+x_0)^2+(y+y_0)^2)}{((x+x_0)^2+(y-y_0)^2)((x-x_0)^2+(y+y_0)^2)}$$

17. Find the Green's function for the two-dimensional Dirichlet problem corresponding two concentric cylindrical surfaces one being $\rho = a$, the

other $\rho = b$, with $b > a$. Also repeat for the region inside the single boundary $\rho = a$.

Ans. The relevant equation to be solved is

$$\frac{1}{\rho}\frac{\partial}{\partial\rho}\rho\frac{\partial G(\rho,\phi\,|\,\rho',\phi')}{\partial\rho}+\frac{1}{\rho^2}\frac{\partial^2 G(\rho,\phi\,|\,\rho',\phi')}{\partial\phi^2}=\frac{1}{\rho}\delta(\rho-\rho')\delta(\phi-\phi'),$$

$$a<\rho,\rho'<b, 0<\phi,\phi'<2\pi$$

We make an expansion of the Green's function as a Fourier series in ϕ. We shall assume that ρ', ϕ' are fixed so that G can be regarded as a function of (ρ, ϕ) alone:

$$G(\rho,\phi)=\sum_{n=0}^{\infty} A_n(\rho).\cos(n\phi)+B_n(\rho).\sin(n\phi)$$

with $B_0 = 0$. Then,

$$\nabla^2 G=\sum_{n=o}^{\infty}\left(\frac{1}{\rho}\frac{d}{d\rho}(\rho.A_n'(\rho))-n^2A_n(\rho)/\rho^2\right)cos(n\phi)+$$

$$\left(\frac{1}{\rho}\frac{d}{d\rho}(\rho.B_n'(\rho)\right)-n^2B_n(\rho)/\rho^2)sin(n\phi)$$

We also note that

$$\delta(\phi-\phi')=\frac{1}{2\pi}+\frac{1}{\pi}\sum_{n=1}^{\infty}\cos(n\phi).\cos(n\phi')+\sin(n\phi).\sin(n\phi')$$

since $\frac{1}{\sqrt{2\pi}}, \frac{1}{\sqrt{\pi}}\cos(n\phi), \frac{1}{\sqrt{\pi}}\sin(n\phi), n = 1, 2, ...\}$ forms an orthonormal basis for L^2 ([0, 2π]). Equating coefficients of cos $(n\phi)$ and sin $(n\phi)$ on both sides of the equation gives

$$\frac{1}{\rho}\frac{d}{d\rho}\rho\frac{dA_0(\rho)}{d\rho}=\frac{1}{2\pi\rho}\delta(\rho-\rho')$$

$$\frac{1}{\rho}\frac{d}{d\rho}\rho\frac{dA_n(\rho)}{d\rho}-n^2A_n(\rho)/\rho^2=\delta(\rho-\rho')\cos(n\phi')/\pi\rho$$

$$\frac{1}{\rho}\frac{d}{d\rho}\rho\frac{dB_n(\rho)}{d\rho}-n^2B_n(\rho)/\rho^2=\delta(\rho-\rho')\sin(n\phi)/\pi\rho$$

The right sides of these equations are zero in the regions $a < \rho < \rho'$ and $\rho' < \rho < b$.
Thus, we get

$$A_0(\rho) = \lambda_{1,0}\log(\rho) + \mu_{1,0}$$
$$A_n(\rho) = \lambda_{1,n}\rho^n + \mu_{1,n}\rho^{-n}$$
$$B_n(\rho) = \nu_{1,n}\rho^n + \sigma_{1,n}\rho^{-n}$$

in the range $a < \rho < \rho'$ and

$$A_0(\rho) = \lambda_{2,0}\log(\rho) + \mu_{2,0}$$
$$A_n(\rho) = \lambda_{2,n}\rho^n + \mu_{2,n}\rho^{-n}$$
$$B_n(\rho) = \nu_{2,n}\rho^n + \sigma_{2,n}\rho^{-n}$$

in the range $\rho' < \rho < b$. By integrating the differential equations for $A_n(\rho)$, $B_n(\rho)$ from $\rho'-$ to $\rho'+$ after multiplying these by ρ, we get

$$\rho'(A_0'(\rho'+) - A_0'(\rho'-)) = \frac{1}{2\pi}$$
$$\rho'(A_n'(\rho'+) - A_n'(\rho'-)) = \cos(n\phi')/\pi$$
$$\rho'(B_n'(\rho'+) - B_n'(\rho'-)) = \sin(n\phi')/\pi$$

Again integrating the same equations after multiplication by ρ from a to ρ followed by an integration from $\rho'-$ to $\rho'+$ gives the equations

$$A_n(\rho'+) - A_n(\rho'-) = 0, \; B_n(\rho'+) - B_n(\rho'-) = 0$$

Finally, we note that the boundary condition that $G(\rho, \phi)$ vanish at $\rho = a$ and $\rho = b$ results in the equations

$$A_n(a) = A_n(b) = B_n(a) = B_n(b) = 0$$

These equations translate to

$$\lambda_{1,0} - \lambda_{2,0} = \frac{1}{2\pi}$$
$$n(\lambda_{1,n}\rho'^n - \mu_{1,n}\rho'^{-n}) - n(\lambda_{2,n}\rho'^n - \mu_{2,n}\rho'^{-n}) = \cos(n\phi')/\pi, \; n \geq 1$$
$$n(\nu_{1,n}\rho'^n + \sigma_{1,n}\rho'^{-n}) - n(\nu_{2,n}\rho'^n - \sigma_{2,n}\rho'^{-n}) = \sin(n\phi')/\pi, \; n \geq 1$$
$$(\lambda_{1,n}\rho'^n + \mu_{1,n}\rho'^{-n}) = (\lambda_{2,n}\rho'^n + \mu_{2,n}\rho'^{-n}), \; n \geq 1$$
$$(\nu_{1,n}\rho'^n + \sigma_{1,n}\rho'^{-n}) = (\nu_{2,n}\rho'^n + \sigma_{2,n}\rho'^{-n}), \; n \geq 1$$
$$\lambda_{1,0}\log(a) + \mu_{1,0} = 0$$
$$\lambda_{2,0}\log(b) + \mu_{2,0} = 0$$
$$\lambda_{1,n}a^n + \mu_{1,n}a^{-n} = 0, \; n \geq 1$$
$$\nu_{1,n}a^n + \sigma_{1,n}a^{-n} = 0, \; n \geq 1$$
$$\lambda_{2,n}b^n + \mu_{2,n}b^{-n} = 0, \; n \geq 1$$
$$\nu_{2,n}b^n + \sigma_{2,n}b^{-n} = 0, \; n \geq 1$$

We leave it to the student to obtain the solution for $\{\lambda_{i,n}, \mu_{i,n} : i = 1, 2, n = 0,1,...\}$, $\{\nu_{i,n}, \sigma_{i,n}, i = 1, 2, n = 1,2,....\}$ and hence for the functions $A_n(\rho)$, $B_n(\rho)$ and finally express the Green's function in terms of these. It involves only algebra and we omit the details here.

18. A sphere of radius R has its centre at the origin of co-ordinates. The permittivity of the medium inside the sphere is ε. This sphere is placed in a constant electric field $\mathbf{E} = E_0\hat{z}$. That is at great distances from the sphere the electric field is $E_0\hat{z}$. Determine the electric potential and electric field both inside and outside the sphere.

Ans. Outside the sphere, since we have azimuthal symmetry, we can expand the potential in terms of the Legendre polynomials. We will have a term $Ar\ P_1(\cos\theta) = Ar\cos\theta = Az$ corresponding to a constant field at large distances. A must be chosen as $-E_0$. The other terms must decay with increasing r. Thus, the potential will have the form

$$V(r,\theta) = -E_0 r\cos\theta + \frac{A_1}{r^2}\cos\theta + \sum_{l=2}^{\infty} \frac{A_l}{r^{l+1}} P_l(\cos\theta), r > R$$

Inside the sphere, we must have the expansion

$$V(r,\theta) = \sum_{l=0}^{\infty} B_l r^l P_l(\cos\theta) = B_0 + B_1 r\cos\theta + \sum_{l=2}^{\infty} B_l r^l P_l(\cos\theta)$$

The normal component of $\mathbf{D}$ must be continuous at the boundary. The normal component of $\mathbf{D}$ to the sphere for $r = R+$ is given by

$$D_r(R+) = -\varepsilon \frac{\partial V(r,\theta)}{\partial r}\Big|_{r=R+} = \varepsilon E_0 \cos\theta + \frac{2\varepsilon A_1}{R^3}\cos\theta$$
$$+\varepsilon \sum \frac{(l+1)A_l}{R^{l+2}} P_l(\cos\theta)$$

The normal component of $\mathbf{D}$ to the sphere for $r = R-$ is given by

$$D_r(R-) = -\varepsilon_0 \frac{\partial V(r,\theta)}{\partial r}\Big|_{r=R-} = -\varepsilon_0 B_1 \cos\theta - \varepsilon_0 \sum_{l=2}^{\infty} l B_l R^{l-1} P_l(\cos\theta)$$

Equating these two expressions gives

$$-\varepsilon_0 B_1 = \varepsilon E_0 + \frac{2\varepsilon A_1}{R^3}$$

$$\varepsilon(l+1)A_l / R^{l+2} = -\varepsilon_0 B_l R^{l-1},\ l = 2, 3, ...$$

The other boundary condition is the continuity of the tangential component of **E** at the boundary. The tangential component of the electric field at $r = R+$ has only a $\hat{\theta}$ component. It is given by

$$E_\theta(R+) = -\frac{1}{R}\frac{\partial V(R+,\theta)}{\partial\theta}$$

Likewise the tangential component of the electric field at $r = R-$ also has only a $\hat{\theta}$ component which is given by

$$E_\theta(R-) = -\frac{1}{R}\frac{\partial V(R-,\theta)}{\partial\theta}$$

If these two expressions are to be equal, we must have

$$\frac{\partial V(R+,\theta)}{\partial\theta} = \frac{\partial V(R-,\theta)}{\partial\theta}$$

and this relation must hold for all θ. Integration shows that this is compatible with the condition that the potential be continuous at the boundary, i.e.

$$V(R+,\ \theta) = V(R-,\ \theta)$$

and so we need only to match the potentials. The continuity of the tangential component to the electric field will then be automatically guaranteed. This condition gives

$$B_0 = 0,\ B_1 R = -E_0 R + \frac{A_1}{R^2},\ B_l R^l = \frac{A_l}{R^{l+1}},\ l = 2,3,...$$

The two equations connecting $\{A_l : l \geq 2\}$ and $\{B_l : l \geq 2\}$ clearly imply $A_l = B_l = 0,\ l \geq 2$. The two equations connecting A_1, B_1 give on elimination of B_1,

$$\varepsilon E_0 + \frac{2\varepsilon A_1}{R^3} = -\varepsilon_0\left(-E_0 + \frac{A_1}{R^3}\right)$$

which gives

$$A_1 = -\frac{R^3 E_0(\varepsilon - \varepsilon_0)}{2\varepsilon + \varepsilon_0}$$

Plugging this into the second equation connecting B_1 and A_1, we get

$$B_1 = -E_0 + \frac{A_1}{R^3} = -\frac{3\varepsilon E_0}{2\varepsilon + \varepsilon_0}$$

This completes the solution to our problem:

$$V(r,\theta) = \left(\frac{A_1}{r^2} - E_0 r\right)\cos\theta, r > R$$

$$V(r,\theta) = B_1 r\cos\theta,\ r < R$$

19. Find the potential in the region $0 < \rho < R$, $0 < z < L$, $0 < \phi < 2\pi$ when the boundary condition is specified as $V(R, \phi, z) = 0$ for $0 < z < L$, $0 < \phi < 2\pi$ and $V(\rho, \phi, 0) = 0$, $V(\rho, \phi, L) = V_0(\rho, \phi)$ for $0 < \rho < R$, $0 < \phi < 2\pi$.

Ans. The potential in the cylindrical volume satisfies Laplace's equation

$$\frac{1}{\rho}\frac{\partial}{\partial\rho}\rho\frac{\partial V}{\partial\rho}+\frac{1}{\rho^2}\frac{\partial^2 V}{\partial\phi^2}+\frac{\partial^2 V}{\partial z^2}=0$$

Separation of variables $V(\rho, \phi, z) = R(\rho)\,\Phi(\phi)\,Z(z)$ gives

$$(\rho R'(\rho))'/\rho R(\rho) + \Phi''(\phi)/\rho^2\Phi(\phi) + Z''(z)/Z(z) = 0$$

It follows that

$$Z''(z)/Z(z) = -(\rho R'(\rho))'/\rho R(\rho) + \Phi''(\phi)/(\rho^2\Phi(\phi))$$

and hence both sides must equal a constant, say k^2. Thus,

$$Z(z) = A.\sinh(kz) + B.\cosh(kz)$$

since V must vanish when $z = 0$, we require $B = 0$ that is $Z(z)$ is proportional to $\sinh(kz)$. The equation for $R(\rho)$ and $\Phi(\phi)$ is

$$\rho(\rho R'(\rho))'/R(\rho) + k^2\rho^2 = 0 = -\Phi''(\phi)/\Phi(\phi)$$

and hence both sides must equal a constant, say m^2. This gives

$$\Phi(\phi) = C.\cos(m\phi) + D.\sin(m\phi)$$

and since V must be single valued, m is an integer. The equation for $R(\rho)$ is then

$$\rho^2 R''(\rho) + \rho R'(\rho) + (k^2\rho^2 - m^2)\,R(\rho) = 0$$

the solution to which having no singularity at $\rho = 0$ is given by

$$R(\rho) = J_m(k\rho)$$

where $J_m(x)$ is the Bessel function of order m. Since V must vanish when $\rho = R$, we have $k = \alpha_{m,n}/R$, $n = 1, 2, \ldots$, where $\{\alpha_{m,n}\}$ are the roots of $J_m(x) = 0$. The solution for the potential is thus of the form

$$V(\rho,\phi,z)=\sum_{m=0}^{\infty}\sum_{n=1}^{\infty} J_m(\alpha_{m,n}\rho/R).\sinh(\alpha_{m,n}z)(A_{m,n}\cos(m\phi)+B_{m,n}\sin(m\phi))$$

The coefficients $A_{m,n}$, $B_{m,n}$ are determined from the condition $V(\rho, \phi, L) = V_0(\rho, \phi)$. We shall first of all deduce orthogonality relations for the Bessel functions. Fix n and consider the functions

$$f_n(x) = J_m(\alpha_{m,n} x/R),\ n = 1, 2,...$$

We have

$$f_n'(x) = (\alpha_{m,n}/R)J_m'(\alpha_{m,n}x/R),\ f_n''(x) = (\alpha_{m,n}/R)^2 J_m''(\alpha_{m,n}x/R)$$

Thus,

$$(\alpha_{m,n}x/R)^2(R/\alpha_{m,n})^2 f_n''(x) + (\alpha_{m,n}x/R)(R/\alpha_{m,n})f_n'(x) + ((\alpha_{m,n}x/R)^2 - m^2)f_n(x) = 0$$

which simplifies to

$$x^2 f_n''(x) + xf_n'(x) + ((\alpha_{m,n}x/R)^2 - m^2)f_n(x) = 0$$

Likewise

$$x^2 f_q''(x) + xf_q'(x) + ((\alpha_{m,q}x/R)^2 - m^2)f_q(x) = 0$$

We multiply the first equation here by $f_q(x)/x$, the second equation by $f_n(x)/x$, subtract the second from the first and integrate from $x = 0$ to $x = R$. Making use of the identity

$$\int_0^R (x(f_n''f_q - f_q''f_n) + (f_n'f_q - f_q'f_n))dx = \int_0^R (x(f_n'f_q - f_q'f_n)' + (f_n'f_q - f_q'f_n))dx = x(f_n'f_q - f_q'f_n)\Big|_0^R = 0$$

since $f_n(R) = f_q(R) = 0$, we finally obtain the orthogonality equation

$$(\alpha_{m,n}^2 - \alpha_{m,q}^2)\int_0^R x\, f_n(x) f_q(x)dx = 0$$

and since $\alpha_{m,n}$, $\alpha_{m,q}$ are distinct positive numbers for $n \neq q$, we get

$$\int_0^R x\, f_n(x) f_q(x) = 0,\ n \neq q$$

We define

$$C_{m,n} = \int_0^R xf_n^2(x)dx = \int_0^R xJ_m(\alpha_{m,n}x/R)^2 dx$$

Then, we can write the above orthogonality equation as

$$\int_0^R xJ_m(\alpha_{m,n}x/R)J_m(\alpha_{m,q}x/R)dx = C_{m,n}\delta_{n,q}$$

From the equation

$$V_0(\rho,\phi)=\sum_{m=0}^{\infty}\sum_{n=1}^{\infty}J_m(\alpha_{m,n}\rho/R)\sinh(\alpha_{m,n}L)(A_{m,n}\cos(m\phi)+B_{m,n}\sin(m\phi))$$

we deduce using these orthogonality equations that

$$\int_0^R\int_0^{2\pi}V_0(\rho,\phi)\rho.J_m(\alpha_{m,n}\rho/R)\cos(m\phi)d\phi=\pi C_{m,n}A_{m,n}, m\geq 1, n\geq 1$$

$$\int_0^R\int_0^{2\pi}V_0(\rho,\phi)\rho.J_m(\alpha_{m,n}\rho/R)\sin(m\phi)d\phi=\pi C_{m,n}B_{m,n}, m\geq 1, n\geq 1$$

$$\int_0^R\int_0^{2\pi}V_0(\rho,\phi)\rho.J_0(\alpha_{0,n}\rho/R)d\phi=2\pi C_{0,n}A_{0,n}, n\geq 1$$

20. Solve for $V(x, y)$ in the region $0 < x < a$, $0 < y < \infty$ when V satisfies Laplace's equation

$$\frac{\partial^2 V}{\partial x^2}+\frac{\partial^2 V}{\partial y^2}=0$$

with boundary conditions $V = 0$ when $x = 0, a$ and $V(x, 0) = V_0(x)$, $0 < x < a$.

Ans. Using separation of variables, we write $V(x,y) = X(x)Y(y)$. Plugging this into the equation gives

$$\frac{X''(x)}{X(x)}+\frac{Y''(y)}{Y(y)}=0$$

and hence

$$\frac{X''(x)}{X(x)}=-m^2=-\frac{Y''(y)}{Y(y)}$$

The solution is

$$X(x) = A\cos(mx) + B\sin(mx),\ Y(y) = C\exp(-my) + D\exp(my)$$

The condition $V(0,y) = V(a,y) = 0$ gives $A = 0$, $m = n\pi/a$ where $n = 1, 2,\ldots$ Since $V(y) \to 0$ when $y \to \infty$, we must have $D = 0$. Thus, the solution for the potential can be expressed as

$$V(x,y)=\sum_{n=1}^{\infty}A_n\exp(-n\pi y/a)\sin(n\pi x/a), 0<x<a, y>0$$

We determine $\{A_n\}$ from the condition

$$V_0(x) = V(x,0) = \sum_{n=1}^{\infty} A_n \sin(n\pi x/a), 0 < x < a$$

This gives

$$A_n = \frac{2}{a}\int_0^a V_0(x)\sin(n\pi x/a)dx, n = 1,2,...$$

Hence, the complete solution can be expressed as

$$V(x,y) = \int_0^a K(x,y,\xi)V_0(\xi)d\xi$$

where

$$K(x,y,\xi) = \frac{2}{a}\sum_{n=1}^{\infty} \sin(n\pi x/a)\sin(n\pi\zeta/a)\exp(-n\pi y/a)$$

21. Solve Laplace's equation in two dimensions for the potential in the region $0 < \rho < \infty$, $0 < \phi < \phi_0$. Also determine the Green's function for this region with the Dirichlet boundary conditions on the planes $\phi = 0$, $\rho > 0$ and $\phi = \phi_0$, $\rho > 0$.

Ans. Laplace's equation in plane polar co-ordinates is

$$\frac{1}{\rho}\frac{\partial}{\partial\rho}\rho\frac{\partial V}{\partial\rho} + \frac{1}{\rho^2}\frac{\partial^2 V}{\partial\phi^2} = 0$$

Adopting separation of variables $V(\rho, \phi) = R(\rho)\Phi(\phi)$, we get

$$(\rho R'(\rho))'/\rho R(\rho) + \Phi''(\phi)/\rho^2\Phi(\phi) = 0$$

It follows that

$$\rho(\rho R'(\rho))'/R(\rho) = -\Phi''(\phi)/\Phi(\phi)$$

and hence both sides must equal a constant, say m^2. The solution for Φ is

$$\Phi(\phi) = A\cos(m\phi) + B\sin(m\phi), 0 < \phi < \phi_0$$

m may be taken as positive without loss of generality. The equation satisfied by $R(\rho)$ is then

$$\rho^2 R''(\rho) + \rho R'(\rho) - m^2 R(\rho) = 0$$

the solutions to which are $R(\rho) = \rho^{\pm m}$ when $m > 0$. Since V must be finite when $\rho = 0$, it must follow that only ρ^m terms can occur in R. Note that for $m = 0$, the two solutions are 1 and $log(\rho)$. $log(\rho)$ must be excluded in view of finiteness of V when $\rho = 0$. The requirement that V vanish when $\phi = 0$ and $\phi = \phi_0$ gives A = 0 and $m = n\pi/\phi_0$ with $n = 1, 2, ...$ Thus the solution for the potential can be expressed as

$$V(\rho,\phi) = \sum_{n=1}^{\infty} A_n Sin(n\pi\phi/\phi_0)\rho^{n\pi\phi/\phi_0}$$

Suppose we are interested in the solution for the potential in the region $0 < \rho < R$, $0 < \phi < \phi_0$ with the boundary conditions that V vanish when $\phi = 0, \phi_0$ and $V(R,\phi) = V_0(\phi)$. Then we get the additional condition

$$V_0(\phi) = \sum_{n=1}^{\infty} A_n R^{n\pi\phi/\phi_0} Sin(n\pi\phi/\phi_0), 0 < \phi < \phi_0$$

from which $\{A_n\}$ can be determined as

$$A_n R^{n\pi\phi/\phi_0} = \frac{2}{\phi_0}\int_0^{\phi_0} V_0(\phi)\sin(n\pi\phi/\phi_0)d\phi$$

Exercises

[1] An infinite flat sheet carries a constant surface charge of density σ Coul/m^2. Show by integration that the electric field above it is a constant equal to $\frac{\sigma}{\varepsilon}\hat{z}$ assuming the sheet of charge coincides with the xy plane.

[2] Consider the situation shown in Fig. 2.3. The electric field lines make an angle θ_1 with the normal to the interface in medium 1 and an angle θ_2 with the normal in medium 2. Medium 1 has permittivity ε_1 and medium 2 ε_2. Show that $\frac{\tan\theta_1}{\tan\theta_2} = \frac{\varepsilon_1}{\varepsilon_2}$.

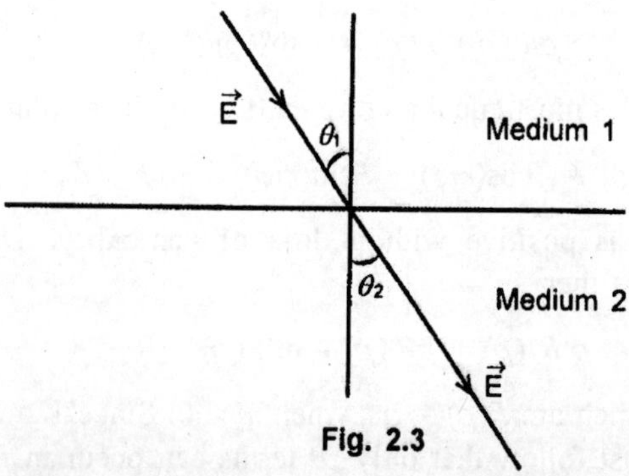

Fig. 2.3

[3] Assume that a medium has electric field **E** in it, conductivity σ and permittivity ε which vary with position and time. Using Gauss' law in the form $\nabla.(\varepsilon\mathbf{E}) = \rho$, the equation of continuity in the form $\nabla.\mathbf{J} + \frac{\partial \rho}{\partial t} = 0$ and Ohm's law in the form $\mathbf{J} = \sigma\mathbf{E}$, deduce that the charge density ρ can be determined in terms of the electric field using the differential equation

$$\frac{\varepsilon}{\sigma}\frac{\partial \rho}{\partial t} + \rho = \sigma\left(\nabla\left(\frac{\varepsilon}{\sigma}\right), \mathbf{E}\right)$$

[4] Using integration, calculate the potential due to a sphere of radius R with centre at the origin and having charge density $\rho(r)$ which is spherically symmetric.

Hint:

$$V(0,0,d) = \frac{1}{4\pi\varepsilon}\int_0^R \int_0^\pi \int_0^{2\pi} \frac{\rho(r) r^2 \sin\theta dr d\theta d\phi}{\sqrt{r^2 + d^2 - 2rd\cos\theta}}$$

Evaluate this integral for $r \le R$ and for $r > R$.

[5] Calculate the potential in the region $\rho \le R$ where ρ is the distance from the z axis when this region is free of charge and $V(R, \phi, z) = -V_0$ for $0 < \phi \le \pi$ and $V(R, \phi, z) = V_0$ for $\pi < \phi < 2\pi$.

Hint: From the solution to Laplace's equation,

$$V(\rho,\phi,z) = \sum_{n=0}^{\infty} \rho^n (A_n \cos(n\phi) + B_n \sin(n\phi))$$

Determine $\{A_n\}$ and $\{B_n\}$ using the Fourier series for the potential $V(R, \phi, z)$ in the variable ϕ.

[6] Consider an infinite grid in the xy plane with the grid wires being located at $x = na$, $y = mb$, $n, m \in \mathbb{Z}$. The wires are maintained at a constant potential. Hence, the potential at a point (x, y, z) above the grid is a periodic function of x and y with periods of a and b respectively.. This means that the potential can be expanded as a double Fourier series

$$V(x,y,z) = \sum_{n,m=-\infty}^{\infty} c_{n,m}(z) \exp(2\pi i(nx/a + my/b))$$

Determine the form of the functions $c_{n,m}(z)$ by using the fact that V satisfies the Laplace equation

$$\frac{\partial^2 V}{\partial x^2}+\frac{\partial^2 V}{\partial y^2}+\frac{\partial^2 V}{\partial z^2}=0$$

and $V \to 0$ as $z \to \infty$.

[7] Assume that the region $R \le r < \infty$ is filled with charge of density $\rho(r)$. Show using Poisson's equation in spherically symmetric situations that for $r > R$,

$$V(r)=\frac{1}{r}\int_R^r (r-\xi)\xi\rho(\xi)d\xi+\frac{B}{r}+A$$

where A, B are constants determined by the conditions $V(\infty) = 0$, $V(R) = V_0$. Assume that

$$lim_{r\to\infty}\frac{1}{r}\int_R^\infty (r-\xi)\xi\rho(\xi)d\xi$$

is finite. Note that when $\int_R^\infty \xi^2\rho(\xi)d\xi < \infty$, this limit equals $\int_R^\infty \xi\rho(\xi)d\xi$ assuming that this integral is finite.

[8] Dipoles of strength $p_1\hat{z}$ and $p_2\hat{z}$ are located respectively at $(0, 0, a_1)$ and $(0, 0, a_2)$. Find the potential and electric field at the origin.

[9] Dipoles of strength $p_j\ (\hat{x}.\cos\sigma_j + \hat{y}.\sin\sigma_j)$, $j = 1, 2,..., n$ are located on the $x - y$ plane at the points $(x_j, y_j, 0)$, $j = 1, 2, ..., n$ respectively. Find the potential and electric field at the origin.

[10] Explain why at equilibrium, charge in a perfect conductor must reside only on the surface.

[11] Explain why at equilibrium, the electric field inside a perfect conductor must be zero.

[12] Calculate the Green's function for the Dirichlet problem corresponding to two concentric spherical surfaces of radii $a < b$. Assume that the common centre coincides with the origin.

Hint: The Green's function satisfies the equation $\nabla^2 G(\mathbf{r}|\mathbf{r}') = \delta(\mathbf{r} - \mathbf{r}')$ for $a < r, r' < b$ with the boundary condition $G(\mathbf{r}|\mathbf{r}') = 0$ when $r = a$ or when $r = b$. Here, $r = |\mathbf{r}|$, $r' = |\mathbf{r}'|$. In terms of spherical co-ordinates, this equation is

$$\frac{1}{r}\frac{\partial^2 rG}{\partial r^2}+\frac{1}{r^2\sin\theta}\frac{\partial}{\partial\theta}\sin\theta\frac{\partial G}{\partial\theta}+\frac{1}{r^2\sin^2\theta}\frac{\partial^2 G}{\partial\phi^2}=\frac{1}{r^2}\delta(r-r')$$
$$\delta(\cos\theta-\cos\theta')\delta(\phi-\phi')$$

If $Y_{lm}(\theta,\phi)$ denote the spherical harmonics (these are eigenfunctions to the angular part of the Laplace's equation), these form an orthonormal basis for functions defined on the unit sphere with respect to the area measure $\sin\theta d\theta d\phi$. This means that

$$\frac{1}{\sin\theta}\frac{\partial}{\partial\theta}\sin\theta\frac{\partial Y_{lm}(\theta,\phi)}{\partial\theta}+\frac{1}{\sin^2\theta}\frac{\partial^2 Y_{lm}(\theta,\phi)}{\partial\phi^2}=-l(l+1)Y_{lm}(\theta,\phi)$$

for $l = 0, 1, 2, \ldots, m = -l, -l+1, \ldots, l$. Thus,

$$\delta(\cos\theta-\cos\theta')\delta(\phi-\phi')=\sum_{l=0}^{\infty}\sum_{m=-l}^{l}Y_{lm}(\theta,\phi)\bar{Y}_{lm}(\theta',\phi'),$$

and hence we try expanding the Green's function as

$$G(\mathbf{r}\,|\,\mathbf{r}')=\sum_{l=0}^{\infty}\sum_{m=-l}^{l}g_{lm}(r)Y_{lm}(\theta,\phi)$$

Substituting this into the above equation gives

$$\frac{1}{r}\frac{d^2}{dr^2}(rg_{lm}(r))-l(l+1)g_{lm}(r)=\frac{1}{r^2}\delta(r-r')$$

These can be solved in the region $a < r < r'$ and $r' < r < b$ using tables of standard functions. The conditions determining the arbitrary constants are obtained by integrating from $r' - 0$ to $r' + 0$ and integrating from a to r followed by an integration from $r' - 0$ to $r' + 0$. We leave the details to the student.

[13] Consider a regular tetrahedron with vertices at (0, 0, 0), (a, 0, 0), $(a/2, a\sqrt{3}/2, 0)$, $\left(\frac{a}{2}, \frac{a}{2\sqrt{3}}, d\right)$ where d is chosen so that the distance of the last point from (0, 0, 0) equals a. Assume that equal charges Q are located at these vertices. Determine (a) The force on any one charge due to the others, (b) The electrostatic potential energy of the configuration.

[14] A line charge of length L is maintained at a constant potential of V_0. The potential at any point (0, 0, z) on this wire is therefore given by the expression $V_0 = V(z) = \frac{1}{4\pi\varepsilon}\int_0^L \frac{\rho(z')dz'}{|z-z'|}$, $0 < z < L$. By discretizing this equation obtain numerically the value of the linear charge density $\rho(z)$, $0 < z < L$.

[15] The electric field in space is given by $\mathbf{E} = A(x)\hat{x} + B(y)\hat{y} + C(z)\hat{z}$. Assuming permittivity ε, determine an expression for the total charge

enclosed in a cuboid having faces $x = 0, x = a, y = 0, y = b, z = 0, z = c$.

[16] An electric dipole $\mathbf{p} = p\hat{z}$ is placed at (a_1, b_1, c_1). The potential due to this dipole at (a_2, b_2, c_2) is V_0. Find the potential at (a_3, b_3, c_3) due to this dipole. Your answer should be independent of p.

[17] Point charges $+Q$ and $-Q$ are located respectively at $(0, 0, d/2)$ and $(0, 0, -d/2)$. Find the potential at (r, θ, ϕ) with neglect of $O(d^4/r^5)$ terms.

Hint:

$$(x^2+y^2+(z+d/2)^2)^{-1/2}-(x^2+y^2+(z-d/2)^2)^{-1/2}=(r^2+d^2/4 + dr.\cos\theta)^{-1/2}-(r^2+d^2/4-dr.\cos\theta)^{-1/2}$$

$$=r^{-1}\left(1-\frac{1}{2}\left(\frac{d^2}{4r^2}+\frac{d\cos\theta}{r}\right)+\frac{1.3}{2^2.2!}\left(\frac{d^2}{4r^2}+\frac{d.\cos\theta}{r}\right)^2-\frac{1.3.5}{2^3 3!}\right.$$

$$=r^{-1}\left(1-\frac{1}{2}\left(\frac{d^2}{4r^2}+\frac{d\cos\theta}{r}\right)+\frac{1.3}{2^2.2!}\left(\frac{d^2}{4r^2}+\frac{d.\cos\theta}{r}\right)^2-\frac{1.3.5}{2^3 3!}\right.$$

$$\left.\left(\frac{d^2}{4r^2}+\frac{d.\cos\theta}{r}\right)^3\right)-r^{-1}(1-\frac{1}{2}\left(\frac{d^2}{4r^2}-\frac{d\cos\theta}{r}\right)+\frac{1.3}{2^2.2!}\left(\frac{d^2}{4r^2}-\frac{d\cos\theta}{r}\right)^2-$$

$$-\frac{1.3.5}{2^3 3!}\left(\frac{d^2}{4r^2}-\frac{d.\cos\theta}{r}\right)^3)+O(d^4/r^5)$$

[18] The electric field in space is given by $\mathbf{E} = -\nabla V$. Write down the equations of motion of a point charge having mass m and charge Q moving in this field and show that the energy

$$\varepsilon=\frac{m}{2}\left(\left(\frac{dx}{dt}\right)^2+\left(\frac{dy}{dt}\right)^2+\left(\frac{dz}{dt}\right)^2\right)+QV(x,y,z)$$

is a constant.

[19] Given an electric field $\mathbf{E}(x, y, z)$ in space, the field lines or lines of force are trajectories that are tangential to the field at each point. Show that in parametric form, $(x(\lambda), y(\lambda), z(\lambda))$, the equation of the field line passing through a point (x_0, y_0, z_0) is given by

$$\frac{dx(\lambda)}{d\lambda}=E_x(x(\lambda),y(\lambda),z(\lambda))$$

$$\frac{dy(\lambda)}{d\lambda} = E_y(x(\lambda), y(\lambda), z(\lambda))$$
$$\frac{dz(\lambda)}{d\lambda} = E_z(x(\lambda), y(\lambda), z(\lambda))$$

for $\lambda \in \mathbb{R}$ with the initial condition $(x(0), y(0), z(0)) = (x_0, y_0, z_0)$. Describe a numerical method for plotting the fields.

[20] Plot the electric lines of force corresponding to (a) a uniformly charged disc of radius R, (b) A line charge of length L having uniform linear charge density, (c) A solid cylinder of charge having radius R and length L, (d) Two equal and opposite point charges Q, $-Q$ placed a distance d apart in the plane.

[21] A solid cylinder of radius R and length L is maintained at a constant potential V_0 throughout its interior. Assuming that the volume charge density within it is given by $\sigma(\rho, z)$, show that

$$V_0 = \frac{1}{4\pi\varepsilon}\int_0^R\int_0^{2\pi}\int_0^L \frac{\sigma(\rho,z)}{\sqrt{\rho'^2+\rho^2-2\rho\rho'\cos(\phi-\phi')+(z'-z)^2}}\rho d\rho d\phi dz$$

Discretize this equation and hence obtain linear equations for the charge density. Your discretization process should involve placing grids at $\rho = n\Delta$, $z = m\Delta'$, $\phi = 2\pi k/K$, $n = 0, 1, \ldots, N$, $m = 0, 1, \ldots, M$, $k = 0, 1, \ldots, K$ and replacing an integral of the form

$$\int_0^R\int_0^{2\pi}\int_0^L \psi(\rho,\phi,z)d\rho d\phi dz$$

by the discrete sum

$$\sum_{n=0}^{N-1}\sum_{m=0}^{M-1}\sum_{k=0}^{K-1} \psi(n\Delta, m\Delta', 2\pi k/K)\Delta.\Delta'.2\pi/K$$

Do this exercise on a computer choosing $V_0 = 1$, $L = R = 1$, $N = M = K = 100$.

[22] Using the formula $U = \int_V \frac{\varepsilon}{2} E^2 dV$ for the energy stored inside a volume V in the electrostatic field, determine the total energy stored in the field produced by a solid spherical charge of radius R having uniform charge density ρ Coul/m^3. Consider both the fields in the region $r \le R$ and $r > R$.

[23] Find the energy stored in the region $0 \le r \le R$ when the electric field is given by

$$\mathbf{E} = Ar\cos\phi\sin\theta\,\hat{r} + Br\cos\phi\cos\theta\,\hat{\theta} + Cr\sin\phi\,\hat{\phi}$$

[24] If $V = A(\rho)B(\phi)C(z)$, derive an expression for the energy stored in the region

$$0 \le \rho \le b,\ z_1 \le z \le z_2,\ 0 \le \phi \le \phi_0$$

Your answer must be expressed entirely in terms of the integrals

$$\int_a^b A'^2(\rho)\rho d\rho,\ \int_a^b \frac{1}{\rho^2}A^2(\rho)d\rho,\ \int_0^{\phi_0} B'^2(\phi)d\phi,$$

$$\int_0^{\phi_0} B^2(\phi)d\phi,\ \int_{z_1}^{z_2} C^2(z)dz,\ \int_{z_1}^{z_2} C'^2(z)dz$$

[25] The region $0 \le \rho \le a$ is filled with charge of uniform density σ_1 and the region $b \le \rho \le c$ is filled with charge of uniform density σ_2. Here, $a < b < c$. Determine the electric field and potential everywhere in space assuming the region $0 \le \rho \le a$ has permittivity ε_1, the region $a < \rho \le b$ has permittivity ε_2, the region $b < \rho \le c$ has permittivity ε_3 and finally the region $c < \rho < \infty$ has permittivity ε_4.

[26] Repeat problem [24] when the cylindrical co-ordinate ρ is replaced by the radial co-ordinate r.

[27] Explain conceptually how you would calculate the capacitance between two conducting surfaces S_1 and S_2.

Hint: Assume that S_1 is maintained at a potential $+V_0$ and S_2 at a potential $-V_0$. Then in the region between S_1 and S_2, we solve Laplace's equation $\nabla^2 V = 0$ with the boundary condition $V = V_0$ on S_1, $= -V_0$ on S_2 and V vanishes at ∞. The charge on S_1 can be calculated as

$$Q = \frac{\varepsilon}{2}\int_{S_1} \frac{\partial V}{\partial \mathbf{n}} dS$$

The charge on S_2 will then be $-Q$ since by Gauss' theorem, the flux of ∇V over a closed surface including a sphere of infinite radius and $S_1 \cup S_2$ must be zero because $\nabla^2 V = 0$ in the volume enclosed by these surfaces. The quantity Q will be some number depending only upon the geometry of the surface times V_0. The ratio $C = Q/2V_0$ is then the capacitance between the two surfaces. Explain how you would do this computation numerically by means of an example.

[28] A uniform sheet of charge having surface density ρ_s is located at $x = a$ metre and another uniform sheet of charge having surface density $-\rho_s$ is located at $x = b$ metre. Assume $a < b$. Find the potential difference between two points (x_1, y_1, z_1) and (x_2, y_2, z_2). Consider the cases $a < x_1, x_2 < b$, x_2, $x_2 < a$, $x_1, x_2 > b$, $a < x_1 < b < x_2$, $x_1 < a < x_2 < b$ separately.

[28] Suppose we have a thin closed conducting surface. The region within the surface contains charge Q distributed in some manner. Show that the electric field outside the sphere is not necessarily zero.

Hint: Use the fact that the potential V outside the surface satisfies Laplace's equation $\nabla^2 V = 0$ with vanishing boundary conditions at ∞ and constant potential, say V_0 on the surface of the conductor. Show that this will not be the zero solution unless $V_0 = 0$. Equivalently, since the conductor is neutral as a whole, the total charge enclosed by a closed surface enclosing the conductor is Q and this must equal ε_0 times the flux of the electric field out of the surface. So the electric field to the exterior of the conductor cannot be zero.

[30] Suppose we have a closed conducting surface and we place a charge Q outside it distributed in some fashion. Show that the electric field in the region inside the surface must be zero.

Hint: The potential in the region inside the surface satisfies Laplace's equation $\nabla^2 V = 0$ with the boundary condition that $V = V_0$ on the surface. The solution to this boundary value problem is $V = V_0$ inside the surface. Hence, the field inside the surface must be zero.

Note: The above two problems show that a conducting surface shields the region interior to it from the electric field produced by any charge distribution outside it while it is unable to shield the region exterior to it from the electric field produced by any charge distribution within it. This fact is seen to be just a consequence of the boundary conditions involved in both the cases. In the former case the boundary is the conducting surface while in the latter case, it is the union of the conducting surface and a sphere of infinite radius.

[31] The electric displacement vector **D** is given by $f(r)\,\hat{r}$ in the region $r \le a$, by $g(r)\,\hat{r}$ in the region $a < r \le b$, and by $h(r)\,\hat{r}$ in the region $r > b$. Determine the charge density in space, total charge enclosed by a surface of radius c. Consider the cases $c \le a$, $a < c \le b$ and $b < c$ separately.

[32] A thin circular loop of radius R lies in the xy plane with the origin as its centre. Assume that the wire carries a linear charge density of σ Coulombs/m. Determine the potential and electric field in integral form. Determine the multipole expansion for the potential and electric field up to the first five terms.

[33] The momentum of a charge Q moving with velocity v along the x axis and having mass m, according to special relativity is given by

$$p = \frac{mv}{\sqrt{1 - v^2/c^2}}$$

Assume that an electrostatic potential $V(x)$ exists in space. Show that Newton's equation $\frac{dp}{dt} = V$ for the motion along the $\hat{x}$ axis gives on integration the following energy conservation equation

$$\frac{mc^2}{\sqrt{1-v^2/c^2}} + QV(x) = E$$

Where E is a constant. Through what potential difference starting from rest must the particle be accelerated for it to attain a speed equal to α fraction of the speed of light?

[34] A disc of radius R lies in the xy plane and carries a surface charge of density $\sigma(\rho)$ Coul/m². Show that the potential at (ρ', ϕ', z') is given by

$$V(\rho',\phi',z') = \frac{1}{4\pi\varepsilon}\int_0^R \int_0^{2\pi} \frac{\sigma(\rho)2\pi\rho d\rho d\phi}{\sqrt{\rho^2+\rho'^2-2\rho\rho'\cos\phi+z'^2}}$$

Determine the components of the electric field in cylindrical co-ordinates (ρ', ϕ', z').

[35] A sphere of radius R carries a surface charge $\sigma(\theta, \phi) = \sigma_0\cos\theta$. Show that the potential at (r, θ, ϕ) is given by

$$V(r,\theta,\phi) = \frac{\sigma_0 R^2}{4\pi\varepsilon_0}\int_0^{\pi}\int_0^{2\pi} \frac{(\cos\alpha\sin\beta\sin\theta+\cos\beta\cos\theta)\sin\beta d\beta d\alpha}{\sqrt{R^2+r^2-2Rr\cos\beta}}$$

Evaluate the integral for both $r < R$ and $r > R$.

Hint: Choose the point (r, θ, ϕ) where the potential is to be evaluated as being along the pole of the co-ordinate system so that the orginal $\theta = 0$ makes and angle θ with the new pole and corresponds to $\phi = 0$. This means that the point (r, θ, ϕ) has new Cartesian co-ordinates $(0, 0, r)$ and the point having Cartesian coordinates $(0, 0, R)$ has new Cartesian co-ordinates $(\sin\theta, 0, \cos\theta)$. Then a point on the sphere in the new system having polar co-ordinates (R, β, α) will make an angle η with the original pole given by $\cos\eta = \cos\alpha\sin\beta\sin\theta + \cos\beta\cos\theta$ and hence the charge density at this point will be $\sigma_0 \cos\eta$. Moreover, this point on the sphere will have a distance from the orginal point (r, θ, ϕ) (which now occupies the position of the new pole) given by $\sqrt{R^2+r^2-2Rr.\cos\beta}$.

[36] A plane $z = a$ carries a charge of uniform surface density σ Coul/m² while the line $\mathbf{r} = \mathbf{b} + \lambda\mathbf{c}$ with λ a variable parameter, carries a

uniform linear charge density σ_0. Determine an expression for the electric field everywhere in space.

[37] Equal point charges Q are placed at the vertices of a regular polygon having n sides with each side having length a. Determine an expression for the electric field everywhere and a simplified expression for the field on the $\hat{z}$ axis.

[37] Solve Laplace's equation in two dimensions

$$\frac{\partial^2 V}{\partial x^2} + \frac{\partial^2 V}{\partial y^2} = 0$$

in the region $0 \le x \le b$, $0 \le y \le a$ with the boundary conditions $V = 0$ when $x = 0$, $x = b$, $0 \le y \le a$ and $V = 0$ when $y = 0$ with $0 \le x \le b$ and $V = V_0$ when $y = a$ with $0 \le x \le b$.

Hint: The solution obtained by separation of variables that meets the given boundary conditions is

$$V(x, y) = \sum_{n=1}^{\infty} c_n \sin(n\pi x / b) \sinh(n\pi y / b)$$

Determine $\{c_n\}$ so that the given boundary condition at $y = a$ is satisfied.

[39] Find the potential in the region $0 \le \rho \le R$, $0 \le z \le L$ when the region is charge free and the potential V at $\rho = R$ is V_0 for $0 \le z \le L$ and V is zero when $z = 0$ and $z = L$.

Hint: $V(\rho, \phi, z)$ satisfies Laplace's equation

$$\frac{1}{\rho}\frac{\partial}{\partial \rho}\rho\frac{\partial V}{\partial \rho} + \frac{1}{\rho^2}\frac{\partial^2 V}{\partial \phi^2} + \frac{\partial^2 V}{\partial z^2} = 0$$

Adopt the separation of variables

$$V(\rho, \phi, z) = R(\rho)\, \Phi(\phi)\, Z(z)$$

to get

$$\frac{(\rho R'(\rho))'}{\rho R(\rho)} + \frac{\Phi''(\phi)}{\rho^2 \Phi(\phi)} + \frac{Z''(z)}{Z(z)} = 0$$

Deduce from the boundary condition $Z(0) = Z(L) = 0$ that

$$Z(z) = \sin(n\pi z/L),\ n = 1, 2, \ldots$$

and

$$\Phi(\phi) = A_m.\cos(m\phi) + B_m.\sin(m\phi)$$

Show that $R(\rho)$ satisfies

$$\rho^2 R''(\rho) + \rho R'(\rho) - ((n\pi\rho/L)^2 + m^2)R(\rho) = 0$$

[40] Repeat the above problem when $V(R, \phi, z) = 0$, $V(\rho, \phi, 0) = 0$, $V(\rho, \phi, L) = V_1$.
Hint: The same separation of variables with the boundary condition $Z(0) = 0$ gives

$$Z(z) = \sinh(\lambda z)$$

and

$$\frac{(\rho R'(\rho))'}{\rho R(\rho)} + \frac{\Phi''(\phi)}{\rho^2 \Phi(\phi)} = -\lambda^2$$

so that

$$\Phi(\phi) = A_m \cos(m\phi) + B_m \sin(m\phi)$$

and

$$\rho^2 R''(\rho) + \rho R'(\rho) + (\lambda^2 \rho^2 - m^2) R(\rho) = 0$$

with solutions $J_m(\lambda\rho)$. The boundary condition $R(R) = 0$ gives $\lambda = \alpha_{m,n}/R, n = 1, 2, \ldots$, where $\{\alpha_{m,n}\}$ are the roots of $J_m(x) = 0$. The solution is therefore of the form

$$V(\rho,\phi,z) = \sum_{m=0}^{\infty} \sum_{n=1}^{\infty} J_m(\alpha_{m,n}\rho / R) \sinh(\alpha_{m,n} z / R)(A_m \cos(m\phi) + B_m \sin(m\phi))$$

Determine the coefficients $\{A_m, B_m\}$ from the boundary condition at $z = L$.

[41] Write down the equations of motion for a charge q moving in the central field produced by a charge Q in a plane using plane polar co-ordinates. Show that the angular momentum is conserved and derive the equation for the trajectory in polar co-ordinates.

[42] If a conductor has inifinitesimal length dz and area A perpendicular to the length and the material of which the conductor is made has resistivity ρ, then it is known that the resistance of the material is $\rho dz/A$. Further resistances in series add up. Thus, if the material has a cross section $A(z)$ at a distance z from one end and L is the length of the conductor, then its resistance equals $R = \int_0^L \rho dz/A(z)$. Determine using this formula, the resistance of a sector of a spherical shell of conductivity σ with radius ranging from $r = r_1$ to $r = r_2$ and $0 \le \phi \le 2\pi$. Assume that the angle subtended by the sector at the origin equals α. See Fig. 2.4.

[43] Consider a plate of length L, width W lying in the xy plane with one side of the length extending from $x = 0$ to $x = L$ at $y = 0$ and one

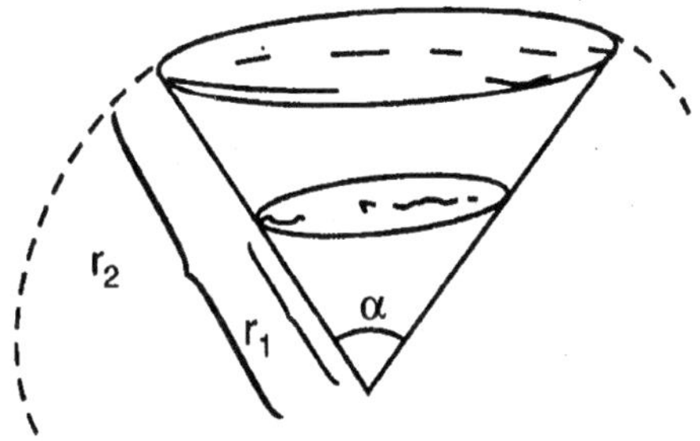

Fig. 2.4

side of the width extending from $y = 0$ to $y = w$ at $x = 0$. Another rectangular plate of breadth w and length $L.\sec\theta$ makes an angle θ with the xy plane. One side of its breadth extends from $y = 0$ to $y = w$ at $x = 0$, $z = d$ and the other side of its breadth extends from $y = 0$ to $y = w$ at $x = L$, $z = d + L.\tan\theta$. Draw a diagram for this geometry and calculate the capacitance between the two plates.

[44] Assume that there are k planes whose equations are given respectively by the equations $l_i x + m_i y + n_i z = p_i$, $i = 1, 2, ..., k$. The ith plane carries a uniform surface charge of density σ_i Coul/m^2. Determine an expression for the electric field at all points in space not falling on any plane.

Hint: Show that the electric field due to the charge on the ith plane can be expressed as

$$\mathbf{E}_i = \frac{\sigma_i}{2\varepsilon} sgn(l_i x + m_i y + n_i z - p_i)(l_i, m_i, n_i)$$

where $sgn(\xi) = +1$ if $\xi > 0$ and $= -1$ if $\xi < 0$.

[45] Explain how you can regard Coulomb's law as an alternative statement of Gauss' law and Gauss' law as an alternative statement of Coulomb's law.

Hint: Derive Coulomb's law from Gauss' law by considering a point charge at the centre of a hypothetical sphere. Then derive Gauss' law from Coulomb's law by considering the flux of the electric field out of a surface enclosing a point charge and not enclosing a point charge respectively.

[46] Derive an expression for the total charge enclosed by a cube having faces $x = 0, a$, $y = 0, b$, $z = 0, c$ when the electric field in space is given by

$$\mathbf{E} = (A_1 x + A_2 y + A_3 z + A_{11} x^2 + A_{22} y^2 + A_{33} y^2 + A_{12} xy + A_{23} yz + A_{31} zx)\hat{x}$$
$$+ (B_1 x + B_2 y + B_3 z + B_{11} x^2 + B_{22} y^2 + B_{33} z^2 + B_{12} xy + B_{23} yz + B_{31} zx)\hat{y}$$

$$+ (C_1xz + C_2y + C_3z + C_{11}x^2 + C_{22}y^2 + C_{33}z^2 + C_{12}xy + C_{23}yz + C_{31}zx)\hat{z}$$

Do this using Gauss' law in integral form as well as by using Gauss' law in differential form.

[47] Derive an expression for the total charge enclosed by a sphere of radius R with the origin as centre when the electric field in space is given by

$$\mathbf{E} = f_1(r)A_1(\theta)\hat{r} + f_2(r)g_2(\theta)\hat{\theta}$$

Do this using both Gauss' law in integral form and in differential form.

[48] Determine an integral expression for the charge enclosed by the cylinder $0 \le \rho \le R$, $0 \le \phi \le 2\pi$, $0 \le z \le L$ when the electric field in space is given by

$$\mathbf{E} = A(\rho, \phi, z)\hat{\rho} + B(\rho, \phi, z)\hat{\phi} + C(\rho, \phi, z)\hat{z}$$

Do this problem using both Gauss' law in integral as well as in differential form.

[49] The region $0 \le r \le R$ is filled with charge of uniform density ρ. Write down the equation of motion of a charge q moving radially in this region and solve it. Consider both the cases when ρ and q have the same sign and when they have opposite signs.

[50] Consider k straight wires passing through the points $\mathbf{r}_j$, $j = 1, 2, \ldots, k$ and having directions along the unit vectors $\mathbf{n}_j$, $j = 1, 2, \ldots, k$ respectively. The jth wire carries a uniform linear charge of density σ_k Coul/m. Determine an expression for the electric field and potential at all points in space not falling on any of the wires.

[51] Determine an approximate expression for the capacitance per unit length between two parallel straight wires of radii a, b respectively separated by a distance d. Assume that $d >> a, b$.

Hint: Assume that the first wire carries a linear charge density σ and the second wire a linear charge $-\sigma$. The difference between the potential on the surface of the first wire and that on the second wire due to the charge on the first wire is approximately

$$V_{11} - V_{21} = -\frac{\sigma}{2\pi}\log(a/d)$$

The same potential difference due to the charge on the second wire is

$$V_{12} - V_{22} = \frac{\sigma}{2\pi}\log(d/b)$$

The total potential difference by superposition is the sum of these two. Deduce from this the formula for the capacitance.

3

CHAPTER

AMPERE'S LAW AND MAGNETOSTATICS

1. INTRODUCTION

Ampere observed the deflection of the needle of a magnetic compass when placed close to a current. He concluded that the current must produce a magnetic field similar to that produced by a permanent magnet. Later on he performed an experiment with two wires carrying current placed close to each other but with adjustable orientations. He measured the force on each wire and concluded that if I_1 and I_2 are two currents in infinitesimal current elements $d\mathbf{l}_1$ and $d\mathbf{l}_2$, then the magnetic field produced by the first element located at $\mathbf{r}_1$ at the location $\mathbf{r}_2$ of the second element must be

$$\mathbf{B}_{21} = \mu_0 I_1 \frac{d\mathbf{l}_1 \times \mathbf{r}_{12}}{r_{12}^3}$$

where μ_0 is a constant of proportionality called the magnetic permeability of free space. The force experienced by the second element was measured by Ampere as

$$d\mathbf{F}_{21} = I_2 d\mathbf{l}_2 \times \mathbf{B}_{21} = \frac{\mu_0 I_1 I_2}{r_{12}^3} d\mathbf{l}_2 \times (d\mathbf{l}_1 \times \mathbf{r}_{12})$$

This is in accordance with the Lorentz force equation according to which, a charge Q travelling with a velocity **v** when placed in a magnetic field **B** experiences a force $Q\mathbf{v} \times \mathbf{B}$. A curious fact about the above result is that the Newtonian action–reaction law is violated, i.e. the force of the magnetic field produced by current element 1 upon element 2 is in general, unequal to the force of the magnetic field produced by element 2 upon element 1: $d\mathbf{F}_{21} \neq d\mathbf{F}_{12}$. Ampere also showed by means of experiments that if a steady current I flows through a surface S bounded by a contour Γ, then the magnetic field **H** produced by this current satisfies the equation

$$\int_{\Gamma} \mathbf{H}.d\mathbf{l} = I$$

Here, $\mathbf{H} = \mathbf{B}/\mu_0$. In words, he stated this result by saying that the line integral of the magnetic field around a closed loop equals the net current cutting through the loop (see Fig. 3.1).

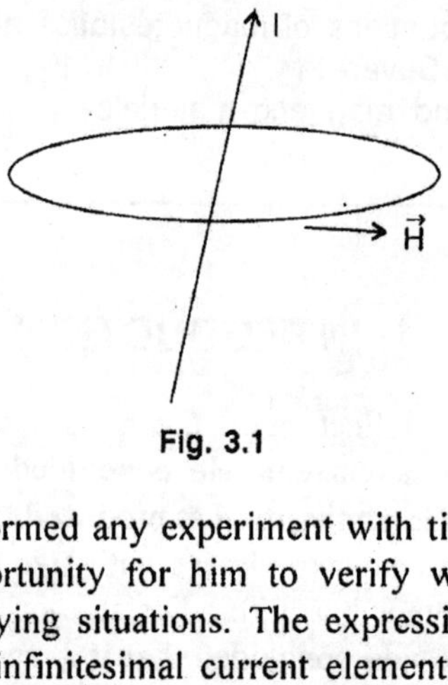

Fig. 3.1

Ampere never performed any experiment with time varying currents, so there was no opportunity for him to verify whether his law was applicable to time varying situations. The expression for the magnetic field produced by an infinitesimal current element can be extended to finite current elements as

$$\mathbf{B}(\mathbf{r}_2) = \frac{\mu_0 I}{4\pi} \int \frac{d\mathbf{r}_1 \times \mathbf{r}_{12}}{r_{12}^3}$$

the integral being over the concerned current carrying wire. This is known as the Biot–Savart law after its discoverers and as we shall soon see here, can be derived starting from Ampere's law and the non-existence of magnetic monopoles.

2. Ampere's Law in Differential Form

Let S denote a surface bounded by the closed curve Γ and let $\mathbf{J}(\mathbf{r})$

be the current density in space. At the point **r**, if a small surface element δS is placed whose normal points along the direction of the unit vector **n**, then the current flowing across this surface is given by $\mathbf{J}(\mathbf{r}).\mathbf{n}\delta S$ (see Fig. 3.2). The current cutting through the surface S is thus $I = \int_S \mathbf{J}.\mathbf{n}dS$ and Ampere's law can be stated as

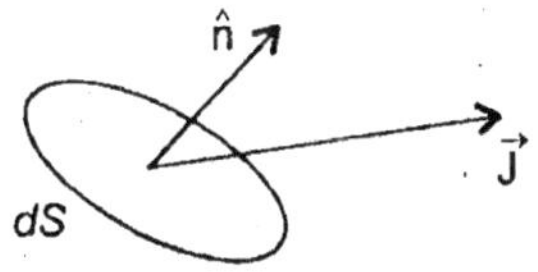

Fig. 3.2.

$$\int_\Gamma \mathbf{H}.d\mathbf{r} = \int_S \mathbf{J}.\mathbf{n}dS$$

By Stokes theorem, the left side of this equation is $\int_S \nabla \times \mathbf{H}.\mathbf{n}dS$ and hence another formulation of Ampere's law is

$$\int (\nabla \times \mathbf{H} - \mathbf{J}).\mathbf{n}dS = 0$$

Since S can be chosen arbitrarily, it follows that

$$\nabla \times \mathbf{H} = \mathbf{J}$$

This is called the differential form of Ampere's law. To determine the magnetic field from the specified current density in space, Ampere's law alone is insufficient. This is because the curl of a vector field does not uniquely specify the vector field. To specify a vector field uniquely (except for an additive constant vector), we require both its curl and divergence. It is known that there do not exist isolated magnetic charges. This means that we cannot have a situation in which the magnetic field lines emanate from a given point without returning back. An isolated north pole would for example, produce a magnetic field that is directed everywhere radially outwards. This situation is analogous to the electric field produced by an isolated positive or negative point charge. It is known that the north and south poles always occur in pairs. A permanent bar magnet for example, has one end as a north pole and the other end as a south pole. If the magnet is split into two halves transversely, then we get two permanent magnets each having a north–south pole pair.

3. The complete equations of magnetostatics

If $\mathbf{B}(\mathbf{r})$ is a magnetic field in space, then the fact that there do not exist any magnetic monopoles implies that the flux of **B** through any closed surface S is zero. That is every magnetic field line that leaves the surface at some point must return into the surface from some other point. Thus,

$$\int_S \mathbf{B}.\mathbf{n}dS = 0$$

If V denotes the volume enclosed by the surface S, then this boils down to $\int_V \nabla.\mathbf{B} = 0$ by Gauss' theorem. Since V can be any volume, it follows that

$$\nabla.\mathbf{B} = 0$$

This is the differential form of the condition that there do not exist any magnetic monopoles in nature. Thus the complete equations of magnetostatics are

$$\nabla \times \mathbf{B} = \mu_0 \mathbf{J},\ \nabla.\mathbf{B} = 0$$

This should be compared to the complete equations of electrostatics:

$$\nabla \times \mathbf{E} = 0,\ \nabla.\mathbf{E} = \rho/\varepsilon_0$$

The vanishing of the divergence of a vector field implies that the field must be the curl of another vector field. Thus,

$$\mathbf{B} = \nabla \times \mathbf{A}$$

where $\mathbf{A}$ is a vector field called the magnetic vector potential. It is uniquely determined up to an additive gradient, i.e. if $\mathbf{A}$ is a vector potential, then so is $\mathbf{A}' = \mathbf{A} + \nabla f$, f being any scalar field. Indeed,

$$\nabla \times \mathbf{A}' = \nabla \times \mathbf{A}$$

since $\nabla \times \nabla f = 0$. We can choose f so that $\nabla.\mathbf{A}' = 0$ Indeed, this means that f is a solution to the equation $\nabla^2 f = -\nabla.\mathbf{A}$. Substituting for the magnetic field $\mathbf{B} = \nabla \times \mathbf{A}$ into the Ampere's law equation gives

$$\nabla \times \nabla \times \mathbf{A}' = \mu_0\mathbf{J}$$

or

$$\nabla(\nabla.\mathbf{A}') - \nabla^2\mathbf{A}' = \mu_0\mathbf{J}$$

or

$$\nabla^2\mathbf{A}' = -\mu_0\mathbf{J}$$

Henceforth, we shall call $\mathbf{A}'$ by $\mathbf{A}$. Then, $\nabla^2\mathbf{A} = -\mu_0\mathbf{J}$. This is a set of three independent Poisson's equations for the three components of the vector potential:

$$\nabla^2 A_x = -\mu_0 J_x,\ \nabla^2 A_y = -\mu_0 A_y,\ \nabla^2 A_z = -\mu_0 J_z$$

The solution to these equations is obtained from analogy with electrostatics where we had the Poisson's equation for the potential with the charge density as the source. The solution obtained was the Coulomb potential. Likewise, here, we have

$$A_x(\mathbf{r}) = \frac{\mu_0}{4\pi}\int \frac{J_x(\mathbf{R})}{|\mathbf{r}-\mathbf{R}|} d^3\mathbf{R}$$

with similar equation for A_y and A_z. Combining all these three into a single vector equation gives

$$\mathbf{A}(\mathbf{r}) = \frac{\mu_0}{4\pi}\int \frac{J(\mathbf{R})}{|\mathbf{r}-\mathbf{R}|} d^3\mathbf{R}$$

To complete the solution to the magnetostatics problem, we take the curl of the above equation and obtain the magnetic field:

$$\mathbf{B}(\mathbf{r}) = \nabla \times \mathbf{A}(\mathbf{r}) = \frac{\mu_0}{4\pi}\int \nabla\left(\frac{1}{|\mathbf{r}-\mathbf{R}|}\right) \times \mathbf{J}(\mathbf{R}) d^3\mathbf{R}$$

$$= \frac{\mu_0}{4\pi}\int \frac{J(\mathbf{R}) \times (\mathbf{r}-\mathbf{R})}{|\mathbf{r}-\mathbf{R}|^3} d^3\mathbf{R}$$

This specializes to the case when the only current source is a thin wire carrying a current I. Then we replace $\mathbf{J}(\mathbf{R})d^3\mathbf{R}$ by $Id\mathbf{R}$ or $Id\mathbf{l}$ to get the Biot–Savart law:

$$\mathbf{A}(\mathbf{r}) = \frac{\mu_0 I}{4\pi}\int \frac{d\mathbf{R}}{|\mathbf{r}-\mathbf{R}|}$$

The line integral being along the wire. The corresponding expression for the magnetic field is

$$\mathbf{B}(\mathbf{r}) = \frac{\mu_0 I}{4\pi}\int \frac{d\mathbf{R} \times (\mathbf{r}-\mathbf{R})}{|\mathbf{r}-\mathbf{R}|^3}$$

which is the Biot–Savart law.

It should be mentioned here that the equations of electrostatics and magnetostatics are decoupled from one another. That is, the magnetic field does not appear anywhere in the equations of electrostatics and likewise, the electric field appears nowhere in the equations of magnetostatics. This is true only because electrostatics and magnetostatics deal with time independent charge and current distributions. A time independent charge distribution in space produces only an electric field and likewise a time independent current density in space produces only a magnetic field. In the time varying situation, the equations of electrostatics and magnetostatics need to be modified. The modification turns out to be such that the electric and magnetic fields get coupled to each other and the modified equations are four in number called the Maxwell equations. The modification causes a changing magnetic field to produce an electric field and likewise a changing electric field to produce a magnetic field. It is surprising that despite these modifications, Gauss' law and the no magnetic monopole conditions remain unaffected.

4. Magnetization and magnetic materials

A magnetic material consists of a volume distribution of magnetic moments that are normally randomly aligned, but on the application of an external magnetic field tend to orient themselves in one direction either along the field direction or in the opposite direction thus enhancing or reducing the effective magnetic field. This is analogous to the electrostatic situation where the dipoles align themselves in a direction so as to reduce the effective electric field. Each atom can be regarded as a little current loop due to the electron revolving around it and this constitutes a little magnetic dipole. Just as the dipole moment per unit volume, namely the polarization vector is used to describe a dielectric material, likewise the magnetic dipole moment per unit volume or the magnetization is used to describe a magnetic material. We have, in one of the exercises computed the magnetic vector potential due to a single magnetic dipole as $\frac{\mu_0}{4\pi}\frac{\mathbf{m}\times\mathbf{r}}{\mathbf{r}^3}$. Assume that we have a volume distribution of dipoles of density $\mathbf{M}(\mathbf{r})$ dipoles per unit volume at $\mathbf{r}$. The vector potential due to this distribution by the superposition principle is

$$\mathbf{A}(\mathbf{r}) = \frac{\mu_0}{4\pi}\int_V \frac{\mathbf{M}(\mathbf{R})\times(\mathbf{r}-\mathbf{R})}{|\mathbf{r}-\mathbf{R}|^3}d^3\mathbf{R}$$

$$= \frac{\mu_0}{4\pi}\int_V \mathbf{M}(\mathbf{R})\times\nabla_R\frac{1}{|\mathbf{r}-\mathbf{R}|}d^3\mathbf{R}$$

$$= -\frac{\mu_0}{4\pi}\int_V \nabla_R\times\frac{\mathbf{M}(\mathbf{R})}{|\mathbf{r}-\mathbf{R}|}d^3\mathbf{R} + \frac{\mu_0}{4\pi}\int_V \frac{\nabla_R\times\mathbf{M}(\mathbf{R})}{|\mathbf{r}-\mathbf{R}|}d^3\mathbf{R}$$

We now make use of the vector identity that if S is the bounding surface for a volume V, then for any vector field $\mathbf{F}$,

$$\int_V \nabla\times\mathbf{F}dV = \int_S \mathbf{n}\times\mathbf{F}dS$$

to express the above equation as

$$\mathbf{A}(\mathbf{r}) = \frac{\mu_0}{4\pi}\int_S \frac{\mathbf{M}(\mathbf{R})\times\mathbf{n}}{|\mathbf{r}-\mathbf{R}|}dS(\mathbf{R}) + \frac{\mu_0}{4\pi}\int_V \frac{\nabla_R\times\mathbf{M}(\mathbf{R})}{|\mathbf{r}-\mathbf{R}|}d^3\mathbf{R}$$

It follows from this expression that $\mathbf{J}_m = \nabla \times \mathbf{M}$ equals the current density corresponding to the volume distribution of magnetic dipoles and $\mathbf{J}_{sm} = \mathbf{M} \times \mathbf{n}$ the surface current density due to magnetic dipoles. Now the total current density in the magnetic medium can be split into an external component $\mathbf{J}_c$ corresponding to conduction current and a magnetization component $\mathbf{J}_m$: $\mathbf{J} = \mathbf{J}_c + \mathbf{J}_m$. Thus Ampere's law can be formulated as

$$\nabla \times \mathbf{B}/\mu_0 = \mathbf{J}_c + \mathbf{J}_m = \mathbf{J}_c + \nabla \times \mathbf{M}$$

or equivalently

$$\nabla \times (\mathbf{B}/\mu_0 - \mathbf{M}) = \mathbf{J}_c$$

This is expressed in a more convenient form by introducing

$$\mathbf{H} = \mathbf{B}/\mu_0 - \mathbf{M}$$

so that

$$\nabla \times \mathbf{H} = \mathbf{J}_c$$

In most materials, to a good degree of approximation, the magnetization **M** is proportional to the magnetic field **H** existing in the medium. Thus, $\mathbf{M} = \chi_m \mathbf{H}$. This gives

$$\mathbf{H} = \mathbf{B}/\mu_0(1 + \chi_m) = \mathbf{B}/\mu$$

χ_m is called the magnetic susceptibility and $\mu = \mu_0 (1 + \chi_m)$ the magnetic permeability of the medium. Sometimes these two constants depend upon the field strength itself and this dependence is indicated by means of a curve called the **B–H** curve. This curve is also called a hysterisis curve for the following reason. As **H** is increased, the increase in **B** follows a certain curve; on the other hand, when **H** is decreased, the decrease in **B** follows another curve. In other words, this means that the variation in **B** as a function of **H** is not memoryless. The value of **B** for a given value of **H** depends upon the path by which the current value of **H** has been arrived at.

Magnetic Scalar Potential: Consider now a situation in which there is no conduction current, that is the current is only due to magnetization caused by circulating atomic currents. The relevant equations of magnetostatics to be solved are then

$$\nabla \times \mathbf{H} = \mathbf{0},\ \nabla.\mathbf{B} = 0,\ \mathbf{B} = \mu_0 (\mathbf{H} + \mathbf{M})$$

Letting $\rho_m = -\nabla.\mathbf{M}$, (the fictitious magnetic charge), our equations can then be expressed as

$$\nabla^2\Phi_m = -\rho_m$$

where $\mathbf{H} = -\nabla\Phi_m$. The situation is analogous to electrostatics. Φ_m is called the magnetic scalar potential. It is defined whenever there is no conduction current, for then, we are guaranteed that **H** is a gradient of a scalar field. The solution to the equation for the magnetic scalar potential is analogous to the Coulomb solution:

$$\Phi_m(\mathbf{r}) = \frac{1}{4\pi}\int \frac{\rho_m(\mathbf{r}')}{|\mathbf{r}-\mathbf{r}'|}dV' = \frac{-1}{4\pi}\int \frac{\nabla'.\mathbf{M}(\mathbf{r}')}{|\mathbf{r}-\mathbf{r}'|}dV'$$

$$= \frac{-1}{4\pi}\int \nabla'.\frac{\mathbf{M}(\mathbf{r}')}{|\mathbf{r}-\mathbf{r}'|}dV' + \frac{1}{4\pi}\int \mathbf{M}(\mathbf{r}').\nabla'\frac{1}{|\mathbf{r}-\mathbf{r}'|}dV'$$

The integrals are taken over entire space. The first integral transforms by Gauss' theorem into a surface integral over a sphere of imprint radius and hence vanishes.
Thus

$$\overline{\Phi}_{-m}(\vec{r})\frac{1}{4\pi}\int_{IR^3} \vec{M}(\vec{r}').\nabla'\frac{1}{|\vec{r}-\vec{r}'|}dV'$$

$$= \frac{1}{4\pi}\int_{IR^3} \frac{\mathbf{M}(\vec{r}').(\vec{r}-\vec{r}')}{|\vec{r}-\vec{r}'|^3}dV'$$

This means that $\dfrac{\mathbf{M}(\vec{r}')dV'.(\vec{r}-\vec{r}')}{4\pi|\vec{r}-\vec{r}'|^3}$ can be interpreted as the magnetic scalar potential at $\vec{r}$ due to a magnetic moment $\vec{m} = \vec{M}(\vec{r}')dV'$ at $\vec{r}'$. Hence the magnetic scalar potential due to a magnetic moment $\vec{m}$ at the origin equals

$$\frac{\vec{m}.\vec{r}}{4\pi r^3}$$

If S is a surface bounding a volume V then the magnetic scalar potential due to magnetic moments within V is given by

$$\overline{\Phi}_m(\vec{r}) = \frac{1}{4\pi}\int_V \frac{\vec{M}(\vec{r}').(\vec{r}-\vec{r}')}{|\vec{r}-\vec{r}'|^3}dV'$$

$$= \frac{1}{4\pi}\int_V \vec{M}(\vec{r}').\vec{\nabla}'\frac{1}{|\vec{r}-\vec{r}'|}dV'$$

$$= \frac{1}{4\pi}\int_V \vec{\nabla}'.\left(\frac{\vec{M}(\vec{r}')}{|\vec{r}-\vec{r}'|}\right)dV' - \frac{1}{4\pi}\int_V \frac{\vec{\nabla}'\vec{M}(\vec{r}')}{|\vec{r}-\vec{r}'|^3}dV'$$

$$= \frac{1}{4\pi}\int_S \frac{\vec{M}(\vec{r}').\hat{n}'}{|\vec{r}-\vec{r}'|}dS' - \frac{1}{4\pi}\int_V \frac{\vec{\nabla}'.\vec{M}(\vec{r}')}{|\vec{r}-\vec{r}'|^3}dV'$$

Thus $\vec{m} = \vec{M}\cdot\hat{n}$ has the interpretation of surface density of magnetic charge.

Review Questions on Magnetic Fields

[1] What are the units of magnetic flux density? How is magnetic flux density defined in terms of force on a current element?

Ans. Magnetic flux density is denoted by the symbol **B**. To determine its units, we note that the force on a charge Q moving with velocity **v** is $\mathbf{F} = Q\mathbf{v} \times \mathbf{B}$. It follows that **B** has units of Newton.sec/Coulomb.m. But 1 Newton equals 1kg.m/sec^2. Thus, **B** has units of kg/Coul.sec. These units could also be derived using Faraday's law that $\nabla \times \mathbf{E} = -\frac{\partial B}{\partial}$. Since **E** has units of Newton/Coul and ∇ has units of m^{-1}, it follows that **B** has units of Newton.sec/Coul.m.

[2] Explain the Hall effect and its use.

Ans. Consider a p type semiconducting bar in which the majority carriers are the holes. The holes are moving in the $\hat{x}$ direction under the action of a voltage. Now assume that the cross section of the bar has an area A and that we apply a magnetic field B_y along the y direction. This field exerts a force $F_z = q.v_xB_y$ on the p type conductor along the $\hat{z}$ direction causing the holes to get deflected along the $\hat{z}$ direction and hence deposit on the top surface of the bar. This deposit produces an electric field $-E_z$ whose magnitude is directed along the negative z direction. After a sufficient number of holes have got deposited on the top surface, the electric force along the negative z direction balances the magnetic force along the z direction causing the holes to move along a straight path. The condition for this to happen is that $E_z = v_xB_y$ according to the Lorentz force equation. When this happens, we can measure the potential difference between the top and bottom surfaces of the bar and hence determine E_z. Since B_y is known, we can calculate $v_x = E_z/B_y$, the velocity of the holes and by measurement of the current $I = q.p.v_x.A$, where p is the hole density, we get $p = I/q.v_xA$. The Hall effect thus enables us to measure the hole density in a semiconductor.

5. Solved Problems

[1] Determine the magnetic field due to a circular loop of radius R carrying a current I on the axis passing through the centre at a distance d from the plane of loop.

Solution: Assume that the loop lies in the X–Y plane with centre coinciding with the origin. Take a point $\mathbf{R} = (R.\cos\phi, R.\sin\phi, 0)$ on the loop. Then $d\mathbf{R} = R.(-\sin\phi.\cos\phi, 0)d\phi$ and

$$d\mathbf{R} \times (\mathbf{r} - \mathbf{R}) = d\phi(-R.\sin\phi, R.\cos\phi, 0) \times (-R.\cos\phi, -R.\sin\phi, d)$$
$$= d\phi.(d.R.\cos\phi, d.R.\sin\phi, R^2)$$

Also $|\mathbf{r} - \mathbf{R}| = (R^2 + d^2)^{1/2}$. This gives

$$\mathbf{B}(r) = \frac{\mu_0 I}{4\pi}\int_0^{2\pi} \frac{(d.R.\cos\phi, d.R.\sin\phi, R^2)d\phi}{(R^2 + d^2)^{3/2}}$$

The x and y components are clearly zero since $\int_0^{2\pi} \cos\phi d\phi = \int_0^{2\pi} \sin d\phi d\phi = 0$. Thus the magnetic field on the axis has only a z component and it is given by

$$B_z = \frac{\mu_0 I R^2}{2(R^2 + d^2)^{3/2}}$$

In particular, putting $d = 0$, we obtain the magnetic field at the centre of the loop as

$$B_z(0,0,0) = \frac{\mu_0 I}{2R}$$

[2] Consider a coil wound around a cylindrical core consisting of N turns per unit length. Let L be the length of the core. Determine the magnetic field on the axis at a distance z from one end. Assume that a current I flows through the coil.

Solution: Assume that the base of the cylinder is on the x-y plane with centre coinciding with the origin so that the top end of the cylinder is in the plane parallel to the x-y plane at a distance L from the x-y plane. The magnetic field at (0, 0, z) due to the circular loop in the interval $[\xi, \xi + d\xi]$, where ξ is measured from the bottom end is given by the expression derived above

$$dB_z = \frac{\mu_0.N.d\xi.I.R^2}{2(R^2+(z-\xi)^2)^{3/2}}$$

The integral of this from $\xi = 0$ to $\xi = L$ gives the total magnetic field B_z at $(0, 0, z)$. Set $z - \xi = R.\tan\theta$. Then, $d\xi = -R.\sec^2\theta.d\theta$ and we get

$$B_z = \frac{\mu_0.N.I}{2}\int_{-\theta_1}^{\theta_2} \cos\theta.d\theta = \frac{\mu_0 NI}{2}(\sin\theta_2 + \sin\theta_1)$$

where

$$\theta_1 = \tan^{-1}((L-z)/R),\ \theta_2 = \tan^{-1}(z/R)$$

The student is advised to show using a picture what the angles θ_1 and θ_2 actually correspond to on a figure.

[3] Determine the magnetic field produced by a straight wire of length L carrying a current I. Assume that the wire extends from $z = 0$ to $z = L$ along the z axis. By the Biot–Savart law, the magnetic field at (x, y, z) is given by

$$\mathbf{B} = \frac{\mu_0 I}{4\pi}\int_0^L \frac{(0,0,d\xi)\times(x,y,z-\xi)}{(x^2+y^2+(z-\xi)^2)^{3/2}}$$

$$= \frac{\mu_0 I}{4\pi}\int_0^L \frac{(-yd\xi, xd\xi, 0)}{(x^2+y^2+(z-\xi)^2)^{3/2}}$$

This expression and also our intuition shows that the magnetic field has only a $\hat{\phi}$ component and this component is given by

$$B_\phi = -B_x\sin\phi + B_y\cos\phi = \frac{\mu_0 I}{4\pi}(y.\sin\phi + x.\cos\phi)\int \frac{d\xi}{\rho^2+(z-\xi)^2)^{3/2}}$$

where $\rho^2 = x^2 + y^2$. Since $x = \rho.\cos\phi$ and $y = \rho.\sin\phi$, it follows that

$$B_\phi = \frac{\mu_0 I\rho}{4\pi}\int_0^L \frac{d\xi}{(\rho^2+(z-\xi)^2)^{3/2}}$$

$$= \frac{\mu_0 I}{4\pi\rho}\int_{-\theta_1}^{\theta_2} \cos\theta.d\theta$$

where we have adopted the change of variables $z - \xi = \rho.\tan\theta$ so that $\theta_1 = \tan^{-1}((L-z)/\rho)$, $\theta_2 = \tan^{-1}(z/\rho)$. The integral evaluates to

$$B_\phi = \frac{\mu_0 I}{4\pi\rho}(\sin\theta_1 + \sin\theta_1) = \frac{\mu_0 I}{4\pi\rho}\left(\frac{L-z}{\sqrt{\rho^2+(L-z)^2}} + \frac{z}{\sqrt{\rho^2+z^2}}\right)$$

[4] Determine the magnetic field produced by a circular loop of radius R carrying current I. Assume that the loop lies in the X–Y plane with its centre coinciding with the origin.

Solution: We shall be content with just determining the magnetic vector potential due to the circular loop. From the solution to the Poisson's equation derived in the previous section, we have in terms of cylindrical co-ordinates,

$$\mathbf{A}(\rho,\phi,z) = \frac{\mu_0 I}{4\pi}\int_0^{2\pi} \frac{(-R.\sin\alpha, R.\cos\alpha, 0)}{((\rho.\cos\phi - R.\cos\alpha)^2 + (\rho.\sin\phi - R.\sin\alpha)^2 + z^2)^{1/2}}\, d\alpha$$

The three components appearing here are the Cartesian components. Thus,

$$A_\rho = A_x.\cos\phi + A_y.\sin\phi = \frac{\mu_0 I}{4\pi}\int_0^{2\pi} \frac{R.\sin(\phi-\alpha)}{(\rho^2 + R^2 + z^2 - 2\rho.R.\cos(\phi-\alpha))^{1/2}}\, d\alpha = 0$$

as we expect from our intuition. Further,

$$A_\phi = -A_x.\sin\phi + A_y.\cos\phi = \frac{\mu_0 I}{4\pi}\int_0^{2\pi} \frac{R.\cos(\alpha-\phi)}{(\rho^2 + R^2 + z^2 - 2\rho.R.\cos(\alpha-\phi))^{1/2}}\, d\alpha$$

We let $\rho^2 + z^2 = r^2$ and make an expansion in powers of $\dfrac{\rho R}{\rho^2 + R^2}$ assuming $r^2 >> \rho.R$, i.e. either $\rho >> R$ or $z >> R$. This amounts to saying that the magnetic field is being determined at distances far compared to the loop dimensions.

$$(\rho^2 + R^2 + z^2 - 2\rho.R.\cos(\alpha - \phi))^{-1/2} = (r^2 + R^2 - 2\rho.R.\cos(\alpha - \phi))^{-1/2} =$$

$$(r^2 + R^2)^{-1/2}\left(1 - \frac{2\rho.R}{r^2 + R^2}.\cos(\alpha - \phi)\right)^{-1/2} =$$

$$(r^2 + R^2)^{-1/2}\left(1 + \frac{\rho.R}{r^2 + R^2}.\cos(\alpha - \phi) + (3/8)\left(\frac{2\rho.R}{r^2 + R^2}\right)^2.\cos^2(\alpha - \phi) + (5/16)\left(\frac{2\rho.R}{r^2 + R^2}\right)^3 \cos^3(\alpha - \phi) + \ldots\right)$$

Thus,

$$A_\phi = \frac{\mu_0 IR}{4\pi\sqrt{r^2 + R^2}}\left(\frac{\rho R}{r^2 + R^2}\int_0^{2\pi}\cos^2(\alpha - \phi)d\alpha + (5/2)\left(\frac{\rho R}{r^2 + R^2}\right)^3 \int_0^{2\pi}\cos^4(\alpha - \phi)d\alpha + \ldots\right)$$

Evaluating the integrals

$$\int_0^{2\pi} \cos^2(\alpha - \phi) d\alpha = \pi$$

$$\int_0^{2\pi} \cos^4(\alpha - \phi) d\alpha = 3\pi / 4$$

we get

$$A_\phi = \frac{\mu_0 I}{4}\left(\frac{\rho R^2}{(r^2 + R^2)^{3/2}} + \frac{15}{8}\frac{\rho^3 R^4}{(r^2 + R^2)^{7/2}} + \ldots\right)$$

[5] Determine the magnetic field produced by a small current loop at distances far from the loop.

Solution: The loop equation is $\mathbf{R}(\lambda)$ and the field is to be determined at $\mathbf{r}$, where $r >> R$. We shall determine the vector potential at large distances and differentiate to get the magnetic field.

$$\mathbf{A}(\mathbf{r}) = \frac{\mu_0 I}{4\pi} \int_{\text{loop}} \frac{d\mathbf{R}}{|\mathbf{r} - \mathbf{R}|}$$

Now

$$|\mathbf{r} - \mathbf{R}|^{-1} = (r^2 + R^2 - 2\mathbf{r}.\mathbf{R})^{-1/2} = r^{-1}\left(1 + \frac{\mathbf{r}.\mathbf{R}}{r^2} - \frac{R^2}{2r^2} + (3/2)(\mathbf{r}.\mathbf{R})^2 / r^4\right) + \ldots$$

The vector potential contributions corresponding to the different terms are obtained by using the equations

$$\int_{loop} d\mathbf{R} = \mathbf{0}$$

$$\int \mathbf{r}.\mathbf{R} d\mathbf{R} = \sum_{i,j} x_i \int X_i dX_j \, \hat{e}_j$$

where (x_1, x_2, x_3) and (X_1, X_2, X_3) are respectively the Cartesian components of $\mathbf{r}$ and $\mathbf{R}$ and $\hat{e}_1$, $\hat{e}_2$, $\hat{e}_3$ are respectively the unit vectors along the three Cartesian axes. We use the expression

$$\int_{\text{loop}} X_i dX_j = \frac{1}{2} \int_{\text{loop}} (X_i dX_j - X_j dX_i) = a_{ij}$$

This is an antisymmetric tensor of rank 3. We can write the contribution

$$\frac{\mu_0 I}{4\pi r^3} \int \mathbf{r}.\mathbf{R} d\mathbf{R} = \frac{\mu_0 I}{4\pi r^3} \sum_{i,j=1}^{3} a_{ij} x_i \hat{e}_j$$

This is the magnetic dipole contribution to the magnetic vector potential. In tensor notation, this contribution can be expressed with the obvious meanings as

$$\frac{\mu_0 I}{4\pi r^3}\mathbf{a.r}$$

The magnetic dipole tensor may be defined as $\mathbf{m} = I\mathbf{a}$. Then, this contribution is expressible in the alternative form

$$\frac{\mu_0 \mathbf{m.r}}{4\pi r^3}$$

Suppose for example the loop is planar, say falling in the x–y plane. Then, there are just two nonzero contributions to the magnetic dipole tensor. They are

$$m_{12} = I.\int_{\text{loop}} X_1 dX_2 - X_2 dX_1,\ m_{21} = I.\int_{\text{loop}} X_2 dX_1 - X_1 dX_2 = -m_{12}$$

Then, the dipole contribution to the magnetic vector potential is

$$\frac{\mu_0 (m_{21}x_2, m_{12}x_1, 0)}{4\pi r^3} = \frac{\mu_0}{4\pi r^3}\mathbf{m}\times\mathbf{r}$$

when $\mathbf{m} = IA\hat{z}$ is now to be interpreted as the magnetic moment vector. This is a well-known result. We now look at the next term, i.e. the magnetic quadrupole contribution. The term is

$$\frac{\mu_0 I}{8\pi r^5}\int_{\text{loop}} (3(\mathbf{r.R})^2 - R^2 r^2)d\mathbf{R}$$

$$= \frac{\mu_0 I}{8\pi r^5}\sum_{i,j} x_i x_j \sum \hat{e}_k \left(\int_{\text{loop}} (3X_i X_j - R^2\delta_{ij})dX_k\right)$$

The term

$$I\int_{\text{loop}} (3X_i X_j - R^2\delta_{ij})dX_k = Q_{ijk}$$

can be termed as the quadrupole moment tensor of the current loop. We can write the quadrupole contribution as

$$\frac{\mu_0}{8\pi r^5}\sum_{i,j,k} Q_{ijk} x_i x_j \hat{e}_k$$

or in tensorial form as

$$\frac{\mu_0}{8\pi r^5}\mathbf{r.Q.r}$$

[6] Determine the motion of a charged particle in a uniform electromagnetic field.

Solution: The Lorentz force equation describes the motion of the charge as

$$\frac{d\mathbf{v}}{dt}=\frac{e}{m}(\mathbf{E}+\mathbf{v}\times\mathbf{B})$$

where **E** and **B** are constant vectors. Assume without loss of generality that **B** is directed along the $\hat{z}$ axis. The equations of motion in component form then are

$$\frac{dv_x}{dt}=\frac{e}{m}(E_x+v_yB)$$

$$\frac{dv_y}{dt}=\frac{e}{m}(E_y-v_xB)$$

$$\frac{dv_z}{dt}=\frac{e}{m}E_z$$

The last equation gives $v_z(t)=\frac{eE_z}{m}t+v_z$ (0). The first two equations involve two variables v_x, v_y. We differentiate the first equation and substitute for $\frac{dv_y}{dt}$ from the second to get

$$\frac{d^2v_x}{dt^2}=\frac{eB}{m}\frac{dv_y}{dt}=\frac{e^2E_yB}{m^2}-\frac{e^2B^2}{m^2}v_x.$$

We define $\omega = eB/m$, $a=\frac{e^2E_yB}{m^2}$ and get

$$\frac{d^2v_x}{dt^2}+\omega^2v_x=a$$

The solution is

$$v_x(t)=A_1.\sin(\omega t)+A_2.\cos(\omega t)+\frac{a}{\omega^2}=A_1.\sin(\omega t)+A_2.\cos(\omega t)+b$$

where $b = E_y/B$ has is a constant having the dimensions of velocity. We obtain v_y from the first equation

$$\frac{dv_x}{dt} = \frac{eE_x}{m} + \omega.v_y$$

or

$$v_y = \frac{1}{\omega}\frac{dv_x}{dt} - \frac{eE_x}{m\omega} = \frac{1}{\omega}\frac{dv_x}{dt} - b$$
$$= A_1.\cos(\omega t) - A_2.\sin(\omega t) - b$$

A_1, A_2 are determined from the initial conditions $v_x(0)$, $v_y(0)$. We shall not go any further but leave the qualitative description of the trajectory of the particle to the student.

[7] Explain how the magnetic field can be determined by three different measurements on the force experienced by a moving charge taken at three different velocities of the charge.

Solution:

$$\mathbf{F}_1 = Q\mathbf{v}_1 \times \mathbf{B},\ \mathbf{F}_2 = Q\mathbf{v}_2 \times \mathbf{B},\ \mathbf{F}_3 = Q\mathbf{v}_3 \times \mathbf{B}$$

This gives

$$\mathbf{F}_1 \times \mathbf{F}_2 = Q^2(\mathbf{v}_1 \times \mathbf{v}_2, \mathbf{B})\mathbf{B}$$
$$\mathbf{F}_2 \times \mathbf{F}_3 = Q^2(\mathbf{v}_2 \times \mathbf{v}_3, \mathbf{B})\mathbf{B}$$
$$\mathbf{F}_3 \times \mathbf{F}_1 = Q^2(\mathbf{v}_3 \times \mathbf{v}_1, \mathbf{B})\mathbf{B}$$

Then,

$\mathbf{A} = \mathbf{F}_1 \times \mathbf{F}_2 + \mathbf{F}_2 \times \mathbf{F}_3 \times \mathbf{F}_3 \times \mathbf{F}_1 = \mathbf{Q}^2 (\mathbf{v}_1 \times \mathbf{v}_2 + \mathbf{v}_2 \times \mathbf{v}_3 + \mathbf{v}_3 \times \mathbf{v}_1, \mathbf{B})\mathbf{B} = \lambda\mathbf{B}$ say. Also

$$\mathbf{F}_2 - \mathbf{F}_1 = \mathbf{Q}\ (\mathbf{v}_2 - \mathbf{v}_1) \times \mathbf{B}$$

and

$$(\mathbf{v}_2 - \mathbf{v}_1) \times \mathbf{A} = \lambda\ (\mathbf{v}_2 - \mathbf{v}_1) \times \mathbf{B}$$

From the two previous equations, it follows that

$$\frac{Q}{\lambda} = \left|\frac{|\mathbf{F}_2 - \mathbf{F}_1|}{|(\mathbf{v}_2 - \mathbf{v}_1) \times \mathbf{A}|}\right|$$

Thus knowing λ, $\mathbf{B}$ can be determined from the equation $\mathbf{B} = \frac{1}{\lambda}\mathbf{A}$.

[8] N infinitesimal current elements of length dl are arranged on a circle of radius R in the X-Y plane, all pointing along the $\hat{z}$ direction and the kth element being located at $(R.\cos(2\pi k/N), R.\sin(2\pi k/N), 0)$, $k = 0, 1, \ldots, N - 1$. Calculate the force on any one element due to the magnetic field produced by the other $N - 1$ elements.

Solution: The magnetic field produced by the kth element at (x, y, z) is given by

$$\frac{\mu_0 Idl}{4\pi}\frac{(0,0,1)\times(x-R.\cos\theta_k, y-R.\sin\theta_k, z)}{((x-R.\cos\theta_k)^2+(y-R.\sin\theta_k)^2+z^2)^{3/2}}$$

where $\theta_k = 2\pi k/N$. The magnetic field at the location $(R, 0, 0)$ of the zeroth element is, therefore

$$\frac{\mu_0 Idl}{4\pi}\frac{(0,0,1)\times(R(1-\cos\theta_k),-R.\sin\theta_k,0)}{R^3((1-\cos\theta_k)^2+\sin^2\theta_k)^{3/2}}=\mathbf{B}_k$$

say. The force on the zeroth element due to this field is then given by

$$d\mathbf{F}_k = I.dl(0, 0, 1)\times\mathbf{B}_k =$$

$$\frac{\mu_0 I^2 dl^2}{4\pi R^2}\frac{(0,0,1)\times(\sin\theta_k, 1-\cos\theta_k, 0)}{(2(1-\cos\theta_k))^{3/2}}$$

$$=\frac{\mu_0 I^2 dl^2}{32\pi R^2}(-\text{cosec}(\theta_k/2), \cot(\theta_k/2), 0)$$

The total force on the zeroth element is the sum of all these vectors from $k = 1$ to $k = N$:

$$\sum_{k=1}^{N} d\mathbf{F}_k = \frac{\mu_0^2 I^2 dl^2}{32\pi R^2}\left(-\sum_{k=1}^{N}\text{cosec}(k\pi/N), \sum_{k=1}^{N}\cot(k\pi/N), 0\right)$$

[9] Find the magnetic field due to a strip lying in the X-Z plane extending from $x = 0$ to $x = b$ and inifinitely along the positive and negative z directions. The strip carries a surface current of $J_s\hat{z}$ Amperes per unit metre.

Solution: First we calculate using Ampere's law the magnetic field produced by an infinitely long thin wire carrying a current I when the wire extends to infinity in both the directions. Assuming that the wire coincides with the $\hat{z}$ axis, it follows from symmetry considerations that the magnetic field has only a $\hat{\phi}$ component. Taking a circular contour of radius ρ whose plane is perpendicular to the wire and with the wire passing through the centre of the circle, it follows from Ampere's law that $\mathbf{H}_\phi.2\pi.\rho = I$. Thus, $\mathbf{H}_\phi = \frac{I}{2\pi\rho}$. This means that if the straight wire points along the direction $\hat{\mathbf{n}}$, and $\mathbf{q}$ is any point on the wire, then the magnetic field at $\mathbf{r}$ can be expressed as

$$\mathbf{B}(\mathbf{r}) = \frac{\mu_0 I(\mathbf{n}\times\mathbf{r})}{2\pi|\mathbf{n}\times\mathbf{r}|^2}$$

Now divide the strip into infinitesimal strips of length $d\xi$ all parallel to the $\hat{z}$ axis. Consider the magnetic field on the x–y plane at $(x, y, 0)$ due to the strip between $[\xi, \xi + d\xi]$. This magnetic field is given by

$$d\mathbf{B} = \frac{\mu_0 J_s d\xi}{2\pi}\frac{(0,0,1)\times(x-\xi,y,0)}{(x-\xi)^2+y^2}$$

$$= \frac{\mu_0 J_s d\xi}{2\pi}\frac{(-y,x-\xi,0)}{(x-\xi)^2+y^2}$$

The total magnetic field due to the entire strip at $(x, y, 0)$ is thus the integral of this from $\xi = 0$ to $\xi = b$. Its components are

$$B_x = -\frac{\mu_0 J_s y}{2\pi}\int_0^b \frac{d\xi}{(x-\xi)^2+y^2}$$

$$B_y = -\frac{\mu_0 J_s}{2\pi}\int_0^b \frac{(x-\xi)d\xi}{(x-\xi)^2+y^2}$$

and $B_z = 0$. We leave the evaluation of these integrals to the student. From the z-symmetry of the problem, this expression for the magnetic field holds even when $z \neq 0$, i.e. at the point (x, y, z). In other words, the magnetic field does not depend upon z.

[10] Find the magnetic field due to a surface current $J_s\hat{z}$ flowing in the x–z plane.

Solution: This is the same as the previous exercise except that the limit of the integrals is taken from $\xi = -\infty$ to $\xi = +\infty$. Noting that

$$\int_{-\infty}^{\infty} \frac{d\xi}{(x-\xi)^2+y^2} = \pi/|y|$$

and

$$\int_{-\infty}^{\infty} \frac{(x-\xi)d\xi}{(x-\xi)^2+y^2} = 0$$

we get

$$B_x = -\frac{\mu_0 J_s}{2},\ y>0, = \frac{\mu_0 J_s}{2}, y<0$$

with other components of the magnetic field being zero.

[11] Consider the current distribution with uniform density $J_0\hat{z} A/m^2$ in the volume between the planes $y = -a$ and $y = 0$ with uniform density $-J_0\hat{z}$ in the volume between the planes $y = 0$ and $y = a$. Find the magnetic field in the region $|y| \le a$.

Solution: The magnetic field due to the thin current sheet in the region $y, y + dy$ which carries a surface current $J_0\hat{z}dy$ A/m at the point $(0,\xi,0)$ is given according to the previous problem by $-\hat{x}\ \dfrac{\mu_0 J_0 dy}{2}$ if $-a < y < \xi < 0$ and by $\dfrac{\mu_0 J_0 dy}{2}\hat{x}$ if $\xi < y < 0$. Further the magnetic field due to the thin current sheet in the region y, y + dy at $(0, \xi, 0)$ is given by $\dfrac{\mu_0 J_0 dy}{2}$ $\hat{x}$ if $\xi < y < 0$ and the field due to the current sheet in y, y + dy at $(0, \xi, 0)$ is given by $\dfrac{-\mu_0 J_0 dy}{2}\ \hat{x}$ if $\xi < 0 < y < a$. Summing up these contributions we get for the total field at $(0, \xi, 0)$ where $-a < \xi < 0$ the following expression:

$$\vec{B} = \left(\frac{-\mu_0 J_0}{2}(\xi + a) + \frac{\mu_0 J_0}{2}(-\xi) - \frac{\mu_0 J_0}{2}a\right)^{\hat{x}}$$

Similarly if $\xi < -a$, the contribution to the magnetic field due to the current in the region $-a < y < 0$ is $\dfrac{\mu_0 J_0 a}{2}\ \hat{x}$ and that due to the current in the region $0 < y < a$ is $\dfrac{-\mu_0 J_0 a}{2}\hat{x}$. The total contribution is 0 likewise for $\xi > a$, the magnetic field is zero. Finally if $0 < \xi < a$, the contribution due to current in $-a < y < 0$ is $\dfrac{-\mu_0 J_0 a}{2}\ \hat{x}$ the contribution due to current in $0 < y < \xi$ is $\dfrac{\mu_0 J_0 \xi}{2}\hat{x}$ and due to current in $\xi < y < a$ is $-\dfrac{\mu_0 J_0 (a-\xi)}{2}\ \hat{x}$, this totals to $-\mu_0 J_0\ (2a + \xi)\ \hat{x}$

Thus

$$\vec{B} = \begin{cases} -\mu_0 J_0 (a-|\xi|)\hat{x}, & |\xi|<a \\ 0, & |\xi|>a \end{cases}$$

[12] Determine the magnetic field and vector potential produced by an infinitely long cylinder of radius R with axis falling along the $\hat{z}$ axis and carrying a current I of uniform density along its axis.

Solution: The current density in the wire is $J_z = \frac{1}{\pi R^2}$. Thus, the current flowing through a circular loop of radius ρ with centre on the z axis is given by $J_z \pi\rho^2 = \frac{I\rho^2}{R^2}$ provided $0 \le \rho \le R$. For $\rho > R$, the current flowing through the loop is I. From the symmetry of the problem, we have only a $\hat{\phi}$ component of the magnetic field. Thus, Ampere's law applied to the circular loop of radius ρ gives

$$H_\phi(\rho).2\pi\rho = \frac{I\rho^2}{R^2},\ 0 \le \rho \le R$$

$$H_\phi(\rho).2\pi\rho = I,\ \rho > R$$

This gives $H_\phi(\rho) = \frac{I\rho}{2\pi R^2}$, $0 \le \rho \le R$ and $H_\phi\ (\rho) = \frac{I}{2\pi\rho}$, $\rho > R$,. or equivalently,

$$B_\phi(\rho) = \frac{\mu_0 I\rho}{2\pi R^2},\ 0 \le \rho \le R$$

$$B_\phi(\rho) = \frac{\mu_0 I}{2\pi\rho},\ \rho > R$$

We assume the magnetic vector potential to be $A_z\ (\rho)\hat{z}$. Its curl is

$$\nabla A_z \times \hat{z} = A_z'(\rho)\hat{\rho} \times \hat{z} = -A_z'(\rho)\hat{\phi}$$

Thus, the condition $\nabla \times \mathbf{A} = \mathbf{B}$ gives

$$-A_z'(\rho) = \frac{\mu_0 I\rho}{2\pi R^2},\ 0 < \rho < R$$

$$-A_z'(\rho) = \frac{\mu_0 I}{2\pi\rho},\ \rho > R$$

or

$$A_z(\rho) = -\frac{\mu_0 I \rho^2}{4\pi R^2} + C, 0 < \rho < R$$

$$A_z(\rho) = -\frac{\mu_0 I}{2\pi}\log(\rho), \rho > R$$

The constant C is chosen by matching the boundary condition at $\rho = R$: $A_z(R-) = A_z(R+)$. This gives

$$C = \frac{\mu_0 I}{4\pi} - \frac{\mu_0 I}{2\pi}\log(R)$$

[13] Determine the magnetic dipole expansion for the vector potential and magnetic field due to a volume current distribution $\mathbf{J}(\mathbf{r})$ which is assumed to be steady.

Solution: The assumption is that the dimensions over which the current is spread out are small compared to the point where the magnetic field is to be computed. We make the approximation

$$|\mathbf{r} - \mathbf{r}'|^{-1} = (r^2 + r'^2 - 2\mathbf{r}.\mathbf{r}')^{-1/2} \approx r^{-1}(1 + \frac{\mathbf{r}.\mathbf{r}'}{r^2} - \frac{r'^2}{2r^2} + \frac{3(\mathbf{r}.\mathbf{r}')^2}{2r^4}$$

In this expression, we have neglected terms smaller than $O(r'^2/r^2)$. The magnetic vector potential is given by

$$\mathbf{A}(\mathbf{r}) = \frac{\mu_0}{4\pi}\int \frac{\mathbf{J}(\mathbf{r}')}{|\mathbf{r} - \mathbf{r}'|} dV'$$

$$\approx \frac{\mu_0}{4\pi r}\int \mathbf{J}(\mathbf{r}')dV' + \frac{\mu_0}{4\pi r^3}\mathbf{J}(\mathbf{r}')\mathbf{r}.\mathbf{r}'.dV'$$

with neglect of terms higher than $O(r'/r)$ in the integrand. Now for steady currents, $\int \mathbf{J}(\mathbf{r}')dV' = \mathbf{0}$. We evaluate the other term:

$$I = \sum_k \int \frac{\partial}{\partial x_k}(J_k(\mathbf{r}')x'_j x'_i)dV'$$

is the volume integral of a divergence and can be transformed into a surface integral over a surface of infinite radius. This surface integral is zero assuming rapid decay of the currents at ∞. Thus, $I = 0$ and we get on noting that $\nabla.\mathbf{J} = 0$ (for steady currents), the equation

$$\sum_k \int J'_k(\delta_{kj}x'_i + \delta_{ki}x'_j)dV' = 0$$

which simplifies to give

$$\int (J'_j x'_i + J'_i x'_j) dV' = 0$$

Thus, the dipole approximation for the magnetic vector potential is

$$A_i(\mathbf{r}) \approx \frac{\mu_0}{4\pi r^3} \sum_j x_j \int J_i(\mathbf{r}') x'_j dV' = \frac{\mu_0}{4\pi r^3} \frac{1}{2} \sum_j x_j \int (J_i(\mathbf{r}') x'_j - J_j(\mathbf{r}') x'_i) dV'$$

For example, letting $i = 1$ gives

$$A_1(\mathbf{r}) \approx \frac{\mu_0}{4\pi r^3} \frac{1}{2} \left(x_2 \int (J'_1 x'_2 - J'_2 x'_1) dV' + x_3 \int (J_i x'_3 - x'_3 J'_1) dV' \right)$$

The magnetic moment of the current distribution is defined to be

$$\mathbf{m} = \frac{1}{2} \int \mathbf{r}' \times \mathbf{J}' dV'$$

Then, the above formula can be expressed as

$$A_1(\mathbf{r}) = \frac{\mu_0}{4\pi r^3} (m_2 x_3 - m_3 x_2)$$

with similar expressions holding for the other components. The final vector form of the expression for the distant vector potential in the dipole approximation is

$$\mathbf{A}(\mathbf{r}) = \frac{\mu_0}{4\pi r^3} \mathbf{m} \times \mathbf{r}$$

We leave the problem of differentiating the vector potential to compute the dipole contribution to the magnetic field to the student.

[14] Derive the Lorentz equations of motion for a charge in an electromagnetic field from an action principle.

Ans. The equations of motion of a charged particle are given by

$$m \frac{d\mathbf{v}}{dt} = q(\mathbf{E} + \mathbf{v} \times \mathbf{B})$$

Here, the electric and magnetic fields can be derived from the scalar and vector potential using the equations

$$\mathbf{E} = -\nabla V - \frac{\partial \mathbf{A}}{\partial t}$$

$$\mathbf{B} = \nabla \times \mathbf{A}$$

We want to derive these equations from an action principle. It is known in mechanics that if $L(x, y, z, v_x, v_y, v_z, t)$ is a function of the three position co-ordinates x, y, z, the three velocity coordinates

$$v_x = \frac{dx}{dt}, v_y = \frac{dy}{dt}, v_z = \frac{dz}{dt}$$

and time t, and we minimize the action functional

$$\int_{t_1}^{t_2} L(x(t), y(t), z(t), v_x(t), v_y(t), v_z(t), t)dt$$

subject to the boundary conditions that $x(t_i) = x_i$, $y(t_i) = y_i$, $z(t_i) = t_i$, $i = 1$, 2, then we end up with the equations of motion

$$\frac{d}{dt}\frac{\partial L}{\partial v_x} = \frac{\partial L}{\partial x}$$

$$\frac{d}{dt}\frac{\partial L}{\partial v_y} = \frac{\partial L}{\partial y}$$

$$\frac{d}{dt}\frac{\partial L}{\partial v_z} = \frac{\partial L}{\partial z}$$

These are called the Euler-Lagrange equations of motion. When motion takes place in a potential field $V(x, y, z, t)$, the equations of motion are

$$m\frac{dv_x}{dt} = -\frac{\partial V}{\partial x}$$

$$m\frac{dv_y}{dt} = -\frac{\partial V}{\partial y}$$

$$m\frac{dv_z}{dt} = -\frac{\partial V}{\partial z}$$

These equations can be derived from the Lagrangian function

$$L(x, y, z, v_x, v_y, v_z, t) = \frac{m}{2}(v_x^2 + v_y^2 + v_z^2) - V(x, y, z, t)$$

which is the difference of the kinetic and potential energies of the particle. We shall show that when the Lagrangian function is chosen as

$$L = \frac{m}{2}(v_x^2 + v_y^2 + v_z^2) - q(V - (\mathbf{v}, \mathbf{A}))$$

where (**a**, **b**) is the scalar product between the vectors **a** and **b**, then the Euler-Lagrange equations give us the correct Lorentz equations of motion.

To see this, we compute

$$\frac{\partial L}{\partial v_x} = mv_x + qA_x$$

Then,

$$\frac{d}{dt}\frac{\partial L}{\partial v_x} = m\frac{dv_x}{dt} + q\left(\frac{\partial A_x}{\partial t} + v_x\frac{\partial A_x}{\partial x} + v_y\frac{\partial A_x}{\partial y} + v_z\frac{\partial A_x}{\partial z}\right)$$

Also

$$\frac{\partial L}{\partial x} = -q\frac{\partial V}{\partial x} + q\left(v_x\frac{\partial A_x}{\partial x} + v_y\frac{\partial A_y}{\partial x} + v_z\frac{\partial A_z}{\partial x}\right)$$

Thus, the x component of the Euler-Lagrange equation expands to

$$m\frac{dv_x}{dt} = -q\left(\frac{\partial V}{\partial x} + \frac{\partial A_x}{\partial t}\right) + q\left(v_y\left(\frac{\partial A_y}{\partial x} - \frac{\partial A_x}{\partial y}\right) - v_z\left(\frac{\partial A_x}{\partial z} - \frac{\partial A_z}{\partial x}\right)\right)$$
$$= qE_x + q(v_yB_z - v_zB_y)$$

as required. We leave the proof for the other components to the student.

[14] Compute the average force and torque on a system of charges in a uniform magnetic field. For the latter, assume bounded motion.

Ans. Let e_1, e_1, ..., e_n be the system of n charges and let their velocities be $\mathbf{v}_1$, ..., $\mathbf{v}_n$. The force on the ith charge is $\mathbf{F}_i = e_i(\mathbf{v}_i \times \mathbf{B})$ and the total torque of these forces about the origin is given by

$$\tau = \sum \mathbf{r}_i \times \mathbf{F}_i = \sum e_i\mathbf{r}_i \times (\mathbf{v}_i \times \mathbf{B})$$
$$= \sum e_i((\mathbf{B}.\mathbf{r}_i)\mathbf{v}_i - (\mathbf{r}_i.\mathbf{v}_i)\mathbf{B})$$

The average force on the system is

$$<\mathbf{p}'> \times \mathbf{B}$$

where

$$<\mathbf{p}'> = <\sum e_i\mathbf{v}_i> = <\frac{d}{dt}\sum e_i\mathbf{r}_i>$$

is the average rate of change of the dipole moment of the system. For bounded motion, this is zero resulting in zero average force. Indeed, bounded motion means that

$$\frac{1}{T}\int_0^T \mathbf{p}'(t)dt = \frac{1}{T}\sum e_i(\mathbf{r}_i(T) - \mathbf{r}_i(0)) \to 0$$

Assuming bounded motion, the average of $\mathbf{r}_i.\mathbf{v}_i$ is zero since

$$\mathbf{r}_i.\mathbf{v}_i = \frac{1}{2}\frac{d}{dt}r_i^2$$

and

$$\frac{1}{T}\int_0^t \mathbf{r}_i.\mathbf{v}_i dt = (r_i^2(T) - r_i^2(0))/2T \rightarrow 0$$

as $T \rightarrow \infty$. Also with $<>$ denoting time average, we have

$$<(\mathbf{B}.\mathbf{r}_i)\mathbf{v}_i> = <(\mathbf{B}.\mathbf{r}_i)\frac{d\mathbf{r}_i}{dt}> = <\frac{d}{dt}((\mathbf{B}.\mathbf{r}_i)\mathbf{r}_i)> - <(\mathbf{B}.\mathbf{v}_i)\mathbf{r}_i>$$
$$= -<(\mathbf{B}.\mathbf{v}_i)\mathbf{r}_i>$$

Thus,

$$<\tau> = \frac{1}{2}\sum e_i(<(\mathbf{B}.\mathbf{r}_i)\mathbf{v}_i> - <(\mathbf{B}.\mathbf{v}_i)\mathbf{r}_i> = \frac{1}{2}\sum e_i <\mathbf{r}_i \times \mathbf{v}_i> \times \mathbf{B}$$
$$= <\mathbf{m}> \times \mathbf{B}$$

where

$$\mathbf{m} = <\frac{1}{2}\sum e_i \mathbf{r}_i \times \mathbf{v}_i>$$

is the average magnetic moment of the system of charges.

[15] If a current, I_1 flows through a coil Γ_1 and a current I_2 flows through a coil Γ_2, then show that the flux due to the current in the first coil that links the second coil can be expressed as $\Phi_{21} = M_{21}I_1$ and that the flux due to the current in the second coil that links the first coil can be expressed as $\Phi_{12} = M_{12}I_2$ where $M_{12} = M_{21}$, where M_{12} depends only upon the geometry of the coils and their relative location. Obtain a formula for the mutual inductance M_{12}.

Ans. Let $\mathbf{A}_1(\mathbf{r})$ and $\mathbf{B}_1(\mathbf{r})$ denote respectively the magnetic vector potential and magnetic field produced by the current in the first coil. Then, we have

$$\mathbf{A}_1(\mathbf{r}) = \frac{\mu I_1}{4\pi}\int_{\Gamma_1}\frac{d\mathbf{r}_1}{|\mathbf{r}-\mathbf{r}_1|}$$

The magnetic flux due to the first coil that links the second one is given by

$$\Phi_{21} = \int_{S_2}\mathbf{B}_1.d\mathbf{S}_2 = \int_{S_2}\nabla \times \mathbf{A}_1.d\mathbf{S}_2 = \int_{\Gamma_2}\mathbf{A}_1.d\mathbf{r}_2$$
$$= \frac{\mu I_1}{4\pi}\int_{\Gamma_1 \times \Gamma_2}\frac{d\mathbf{r}_1.d\mathbf{r}_2}{|\mathbf{r}_1 - \mathbf{r}_2|}$$

This gives

$$M_{21} = \frac{\mu}{4\pi}\int_{\Gamma_1 \times \Gamma_2} \frac{d\mathbf{r}_1 . d\mathbf{r}_2}{|\mathbf{r}_1 - \mathbf{r}_2|}$$

for the mutual inductance. Note that this expression is symmetric with respect to the two coils. Thus, $M_{21} = M_{12}$.

[16] Derive the expression $W_B = \dfrac{B^2}{2\mu}$ for the density of energy stored in the magnetic field.

Ans. There are two ways of approaching this problem. First consider an inductor having N turns of wire and carrying a current I. Assume the length of the inductor is L and that its cross-sectional area is A. It can be shown from Faraday's law that the voltage across the inductor is $v = \alpha\dfrac{dI}{dT}$ with $\alpha = \dfrac{\mu N^2 A}{L}$. This expression follows from the expression for the flux $\phi = \dfrac{\mu N^2 AI}{L}$ obtained by applying Amperes law to a rectangular circuit within one leg passing through the coil as in and the other outside it and assuring a uniform field. Thus, the energy stored is given by $\int vIdt$ = L $I^2/2$. Show that this energy can be expressed as $\dfrac{B^2}{2\mu}$ LA where $B = \dfrac{\mu NI}{L}$ is the magnetic field inside the inductor and LA the volume of the inductor. Hence, deduce the desired expression for the magnetic energy density.

The more rigorous derivation proceeds by noting that if a loop carrying a current I is placed in a magnetic field **B**, then the energy of interaction is given by $I \int \mathbf{B}.d\mathbf{S}$ where the surface integral is taken over the loop area. This follows from the fact that if an element $d\mathbf{l}$ of the loop gets displaced by $d\mathbf{r}$, then the work involved in this displacement is given by

$$\mathbf{F}.d\mathbf{r} = Id\mathbf{l} \times \mathbf{B}.d\mathbf{r} = Id\mathbf{r} \times d\mathbf{l}.\mathbf{B}$$

and $d\mathbf{r} \times d\mathbf{l} = d\mathbf{S}$ is the area vector swept by the loop. Using this, compute the total energy required to bring the loop from ∞ where the magnetic field is zero to its current position. The stored magnetic energy in the loop due to this interaction can thus be expressed as

$$I\int \mathbf{B}.d\mathbf{S} = I\int \nabla\times\mathbf{A}.d\mathbf{S} = I\int_{\Gamma} \mathbf{A}.d\mathbf{r}$$

where Γ is the trajectory of the loop. Now, instead of having a wire, if we have a volume current distribution of density $\mathbf{J}(\mathbf{r})$, and the magnetic field is generated by this volume current distribution, we replace $Id\mathbf{r}$ by $\mathbf{J}dV$ and express the energy as $\int \mathbf{J}.\mathbf{A}dV$. This is the same as

$$\frac{\mu}{4\pi}\int \frac{\mathbf{J}(\mathbf{r}_1).\mathbf{J}(\mathbf{r}_2)}{|\mathbf{r}_1-\mathbf{r}_2|}dV_1dV_2$$

However, in this formula, a factor of 1/2 is to be introduced since we are counting the interaction between two current elements twice. Thus, the final expression for the stored magnetic energy is

$$U = \frac{\mu}{8\pi}\int \frac{\mathbf{J}(\mathbf{r}_1).\mathbf{J}(\mathbf{r}_2)}{|\mathbf{r}_1-\mathbf{r}_2|}dV_1dV_2 = \frac{1}{2}\int \mathbf{J}.\mathbf{A}dV$$

We replace $\mathbf{J}$ here by the expression $\nabla \times \mathbf{B}/\mu$ in accordance with Ampere's law. We use the vector identity

$$\begin{aligned}\nabla.(\mathbf{B}\times\mathbf{A}) &= (\nabla\times\mathbf{B}).\mathbf{A} - \mathbf{B}.\nabla\times\mathbf{A}\\ &= \mathbf{A}.\nabla\times\mathbf{B} - B^2\end{aligned}$$

Plugging this into the expression for the total magnetic energy, we get

$$U = \frac{1}{2\mu}\int \nabla.(\mathbf{B}\times\mathbf{A})dV + \frac{1}{2\mu}\int B^2dV$$

The first integral here by Gauss' theorem is the same as the surface integral $\int \mathbf{B}\times\mathbf{A}.d\mathbf{S}$ taken over the surface of a sphere of infinite radius and is zero assuming that $\mathbf{A}$ and $\mathbf{B}$ decay faster than $1/r$ as $r\to\infty$. This gives $U = \dfrac{1}{2\mu}\int B^2dV$ and hence $W_B = B^2/2\mu$ for the magnetic energy density.

Exercises

[1] Derive an expression for the vector potential and magnetic field produced by a straight wire of length L carrying a current $I(z)$ that varies along the length of the wire. The wire is assumed to extend from $z = 0$ to $z = L$ along $(x, y) = (0, 0)$. Determine the approximate expansion for these two quantities at distances $r >> L$ up to the first three terms.

Hint:

$$A_z(x,y,z) = \frac{\mu}{4\pi}\int_0^L \frac{I(\xi)d\xi}{x^2+y^2+(z-\xi)^2)^{1/2}}$$

Set $\rho^2 = x^2 + y^2$ and $r^2 = \rho^2 + z^2$ and make an expansion of this equation in powers of $1/r$. Also note that Cartesian components of the magnetic field are given by

$$\mathbf{B}(x,y,z) = \frac{\mu}{4\pi}\int_0^L \frac{I(\xi)(-y,x,0)}{(x^2+y^2+(z-\xi)^2)^{3/2}}\,d\xi$$

Evaluate the cylindrical components and show that the only non-vanishing component is the $\hat{\phi}$ component.

[2] Derive an expression for the vector potential and magnetic field produced by a solid cylinder or radius R and length L carrying a current of density $J_z(\rho)$ along the $\hat{z}$ axis. Assume that the cylinder is defined by the region $0 \le \rho \le R$, $0 \le z \le L$.

Hint:

$$A_z(\rho,\phi,z) = \frac{\mu}{4\pi}\int_0^R\int_0^{2\pi}\int_0^L \frac{J_z(\rho')2\pi\rho' d\rho' d\phi' dz'}{(\rho^2+\rho'^2-2\rho\rho'\cos\phi'+(z-z')^2)^{3/2}}$$

Derive from this the expression for the magnetic field.

[3] A rectangular loop has two of its sides along $x = 0$, $z = 0$, $0 \le y \le a$ and $y = 0$, $z = 0$, $0 \le x \le b$. A magnetic field $\mathbf{B}(x, y)$ exists in space. Determine an expression for the force on the current loop and an expression for the torque of this force about the line passing through the point $\mathbf{r}_0 = (x_0, y_0, z_0)$ and having direction defined by the unit vector $\mathbf{n} = (l, m, n)$.

[4] The coil of a dc motor has a rectangular shape with side lengths L and b. The coil has an axis passing through the centres of the sides having length b and can rotate freely about this axis. The coil makes contact with two brushes via a slip ring and a magnetic field exists in the horizontal direction. The brushes supply a current I through the coil. The entire arrangement is shown in Fig. 3.3 Show that when the plane of the

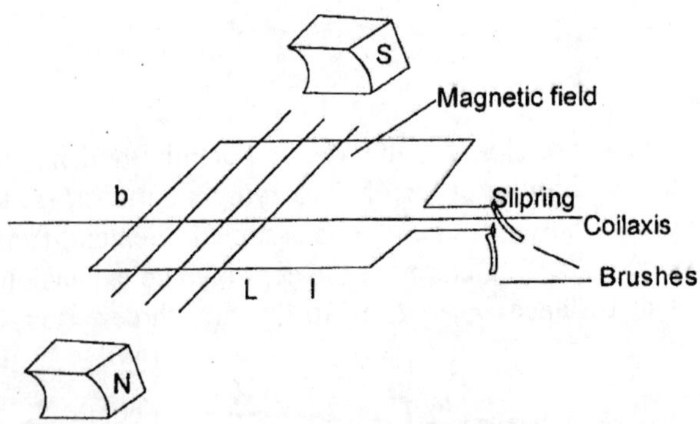

Fig. 3.3

coil makes an angle θ with the horizontal, the torque about the axis is given by $IBbL.\cos(\theta)$. Hence, show that if frictional torque is proportional to $\frac{d\theta}{dt}$ the equation of motion of the coil can be written down as

$$J\frac{d^2\theta}{dt^2}+\gamma\frac{d\theta}{dt}+IBLb\cos\theta=0$$

where J is the moment of inertia of the coil about the axis.

[5] A circular loop of radius R falls in the xy plane with the origin as centre. The loop carries a current $I(\phi)$. Determine the expansion for the magnetic vector potential and field using cylindrical co-ordinates. Evaluate the first five terms of the expansion for the field at $z = 0$, i.e. in the xy plane and at $\rho = 0$, i.e. on the z axis.

Hint: The Cartesian components of **A** are given by

$$\mathbf{A}(\rho,\phi,z)=\frac{\mu R}{4\pi}\int_0^{2\pi}\frac{I(\phi')(-\sin\phi',\cos\phi'0)}{(\rho^2+R^2-2\rho R\cos(\phi'-\phi)+z^2)^{1/2}}d\phi'$$

Evaluate using this, the $\hat{\rho}$, $\hat{\phi}$ and $\hat{z}$ components of **A**.

[5] A loop of wire consists of three straight segments connecting the points $(a, 0, 0)$, $(0, a, 0)$ and $(0, 0, a)$. The loop carries a current I. Determine the expression for the magnetic field in space.

[6] Determine an approximate expression for the flux of the magnetic field through the entire cross section of a cylindrical coil having N turns. The length of the coil is L and its cross-sectional area is A.

Hint: Ampere's law applied to a rectangular loop of length L with one length passing through the centre of the coil (on assuming that the field has only a longitudinal component inside the coil, that this component is a constant and that the field outside the coil is zero) gives $BL = \mu NI$ since NI is the total current passing through the loop. Thus, the flux of the field through the coil (this field cuts N cross-sectional areas each one having area A) is given by

$$\Phi=BA=\frac{\mu N^2IA}{L}=LI$$

where $L=\frac{\mu N^2A}{L}$ is called the coil inductance. Later on, while discussing Faraday's law, we will show that if Φ varies with time. A voltage $\frac{d\phi}{dt}$ is induced through the coil. Thus, the voltage current equation for the inductor is given by $v = L\frac{dI}{dt}$.

[7] Consider a core consisting of a rectangle with a rod connecting the two mid points the longer pair of opposite sides. The arrangement is shown in Fig. 3.4. The legs of the core are labelled as *AB, BC, CD, DE, EF, FA* and the central rod is *CF*. The lengths of the sides *AB, DE, FA* are L_1 each. The lengths of *BC, CD, EF, FA* are L_2 each. The cross-sectional areas of the legs *AB, DE, CF* are A_1 each and the cross-sectional areas of *BC, CD, EF, FA* are A_2 each. The sides *AB, DE* have respectively N_1 and N_2 turns of wire wound around them, and carry respectively currents I_1 and I_2 amperes. Determine the magnetic field in each leg.

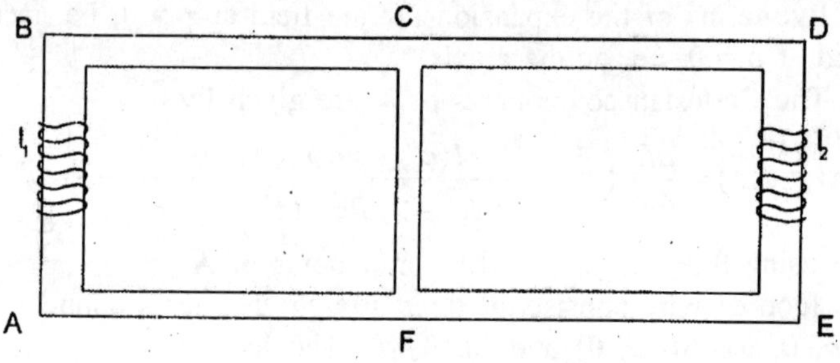

Fig. 3.4

Hint: The magnetic flux Φ through a given leg remains constant. If several legs meet at a junction, and the fluxes through the legs that meet at this junction are Φ_1, ..., Φ_k directed away from the junction, then $\Sigma_{i=1}^{k} \Phi_i = 0$. If Φ is the flux through a leg and the cross-sectional area of the leg is A and its magnetic permeability is μ with the length of the leg L, then the line integral $\int \mathbf{H}.d\mathbf{r}$ along this leg equals $\frac{\Phi L}{\mu A} = R\Phi$ where $R = \frac{L}{\mu A}$ is the reluctance of the leg. The sum of $R\Phi$ around a loop consisting of legs must equal the total magnetomotive force through $\Sigma N_i I_i$ the loop where N_i is the number of turns of wire around the ith leg and I_i is the current with the correct sign through these turns of wire. We thus have an analogue of an electric circuit with the flux playing the role of current, the reluctance playing the role of resistance and the line integral of **H** over a leg playing the role of voltage drop across the leg. The sum of fluxes leaving a junction is zero is the analogue of Kirchhoff's current equation and the line integral of the magnetic field over a closed loop is the total magnetomotive force through the loop is the analogue of Kirchhoff's voltage equation.

[8] A sphere of radius R has uniform magnetization $M_0\hat{z}$. Assume that the centre of the sphere coincides with the origin and that its radius is R. Determine surface density of magnetic charge, the magnetic scalar potential and hence the magnetic field in space.

[9] Consider a closed contour in the xy plane defined by an equation $\rho = \rho(\phi)$, $z = 0$, $0 \le \phi < 2\pi$. Assume that straight wires are passing through points $(x_k, y_k, 0)$, $k = 1, 2, \ldots, p$ within the contour with direction cosines (l_k, m_k, n_k), $k = 1, 2, \ldots, p$ respectively. The kth wire carries a current I_k along the direction (l_k, m_k, n_k). Assume also that current carrying straight wires are passing through the points $(x_k', y_k', 0)$, $k = 1, 2, \ldots, q$ located outside the contour. The kth wire carries a current I_k' and points in the direction (l_k', m_k', n_k'). Determine $\oint_\Gamma \mathbf{H}.d\mathbf{r}$ where Γ is the closed contour. Make a sketch of the above situation.

[10] Consider two circular coils. The first coil has equation $x^2 + y^2 = a^2$, $z = z_1$ and the second coil has equation $x^2 + y^2 = b^2$, $z = z_2$. The first coil carries a current I_1 in the counterclockwise direction looking from the top while the second coil carries a current I_2 in the clockwise direction looking from the top. Determine the magnetic field at an arbitrary point on the $\hat{z}$ axis.

[11] Explain using the differential laws discussed in this chapter why there cannot exist isolated magnetic charges in nature? If a bar magnet of length L is placed along the line $x = y = 0$, $0 < Z \le L$ with the north pole having magnetic charge Q_m Amp meter, find the flux of the magnetic field out of the surface of a sphere of radius $R > L$ with centre at the origin.

[12] A straight wire located along the $\hat{z}$ axis carries current I. Find the flux of the magnetic field produced by this wire through the surface $y_1 \le y \le y_2$, $x = 0$, $z_1 \le z \le (y - y_1)\tan\theta$, where θ is fixed.

[13] Compute the mutual inductance between closely wound coaxial solenoids of length l_1 and l_2 with $l_1 >> l_2$, turns N_1, N_2 and radii $r_1 \approx r_2$.

Hint: The magnetic field produced by the current in coil 1 is approximately constant over the length of coil 2. This field is given by $B_2 = \mu N_1 I_1/L_1$. Thus, the flux of this field linking coil 1 is given approximately by $B_2 N_2 \pi r_2^2$. Do the remaining computations.

[14] Determine the mutual inductance between a rectangular loop of sidelengths a, b and an infinitely long wire that lies in the plane of loop,

is parallel to the side a of the loop and is at a distance d from the side of length a nearer to the loop.

Hint: Assume that the straight wire coincides with the $\hat{z}$ axis and that the rectangular loop is in the yz plane. Assume that the straight wire carries a current I_1. The magnetic field due to this wire at a distance r from it is given by $B(r) = \frac{\mu I}{2\mu r}$. This field is perpendicular to the plane of the loop. Thus, the flux of this field through the loop is given by $\Phi_{21} = \int_d^{d+b} a.B(r)dr$. Do the remaining computations.

[15] A rectangular core of sidelengths a, b and cross-sectional area A has an air gap cut in one of its sides. The length of the air gap is d. The permeability of air is μ_1 and the permeability of the core is μ_2. Assume that a coil of N turns is wound around one of the legs of the core and that this coil carries I amperes. Determine the magnetic field B in the air gap and in the core.

Hint: The magnetic field in the core and in the air gap is the same since the normal component of **B** is continuous. Thus, Ampere's law gives $\frac{Bd}{\mu_1} + \frac{B(2a+2b-d)}{\mu_2} = NI$. Do the remaining calculations.

[16] Explain the following equations starting from first principles: (a) For a volume current distribution of density **J(r)**, the magnetic vector potential and magnetic field are respectively given by

$$\mathbf{A}(\mathbf{r}) = \frac{\mu}{4\pi}\int_V \frac{\mathbf{J}(\mathbf{R})}{|\mathbf{r}-\mathbf{R}|}dV(\mathbf{R})$$

$$\mathbf{B}(\mathbf{r}) = \frac{\mu}{4\pi}\int_V \frac{\mathbf{J}(\mathbf{R})\times(\mathbf{r}-\mathbf{R})}{|\mathbf{r}-\mathbf{R}|^3}dV(\mathbf{R})$$

For a surface current of density **K(r)** on a surface, these quantities are given respectively by

$$\mathbf{A}(\mathbf{r}) = \frac{\mu}{4\pi}\int_S \frac{\mathbf{K}(\mathbf{R})}{|\mathbf{r}-\mathbf{R}|}dS(\mathbf{R})$$

$$\mathbf{B}(\mathbf{r}) = \frac{\mu}{4\pi}\int_S \frac{\mathbf{K}(\mathbf{R})\times(\mathbf{r}-\mathbf{R})}{|\mathbf{r}-\mathbf{R}|^3}dS(\mathbf{R})$$

For a line current $I(\mathbf{r})$ along a wire, these quantities are given respectively by

$$\mathbf{A}(\mathbf{r}) = \frac{\mu}{4\pi}\int_{\Gamma} \frac{I(\mathbf{R})}{|\mathbf{r}-\mathbf{R}|}d\mathbf{R}$$

$$\mathbf{B}(\mathbf{r}) = \frac{\mu}{4\pi}\int_{\Gamma} \frac{I(\mathbf{R})\times(\mathbf{r}-\mathbf{R})}{|\mathbf{r}-\mathbf{R}|^3}d\mathbf{R}$$

Here, V is the volume where the volume current distribution is present, S is the surface where the surface current is present and Γ is the wire along which the line current flows.

[17] Assume that a line current I flows along the $\hat{z}$ axis. Show that $\nabla \times \mathbf{H} = \mathbf{0}$ at all points except on the $\hat{z}$ axis. Hence, $\mathbf{H}$ can be derived from the magnetic scalar potential: $\mathbf{H} = -\nabla\Phi_m$. Assuming that at (ρ_1, ϕ_1, z_1), $\Phi_m = 0$, calculate Φ_m at (ρ_2, ϕ_2, z_2).

Hint: The difference between the magnetic scalar potential at two points is the negative line integral of $\mathbf{H}$ along any path starting from the first point and ending at the second.

[18] Derive an expression for the magnetic scalar potential along the axis of a circular loop of radius R carrying a current I.

[19] Derive from first principles an expression for the magnetic vector potential due to an infinite surface current sheet $J_S\hat{z}$ flowing on the surface $0 \le x \le a$, $y = 0$, $-\infty < z < \infty$. Derive also an expression for the magnetic vector potential due to the same current sheet but extending over the entire xz plane.

[20] Write down the explicit proof that $\nabla.\nabla \times \mathbf{A} = 0$ and $\nabla \times \nabla V = \mathbf{0}$ using an arbitrary curvilinear system of orthogonal co-ordinates.

[21] Consider two coaxial cylinders having their axis coinciding with the $\hat{z}$ axis. The first is a solid cylinder of radius a carrying current I_1 of constant volume density along the $\hat{z}$ axis. The second is a hollow cylinder with inner radius b and outer radius c. The region between $\rho = b$ and $\rho = c$ is solid and carries current I_2 of uniform density. Determine the magnetic field everywhere by using Ampere's law in integral form. Also determine expressions for the magnetic vector potential everywhere.

[22] Assume that the magnetic vector potential in space is given by $\mathbf{A}(\mathbf{r})$. Determine an integral expression for the magnetic flux through a loop Γ. Also determine an expression for the current density in space. When the vector potential is given by

$$\mathbf{A}(\mathbf{r}) = A_x(x)\hat{x} + A_y(\hat{y})\hat{y} + A_z(z)\hat{z}$$

and the equation of the closed loop is in parametric form given by $(x(s), y(s), z(s))$, $0 \le s \le 1$, then show that the flux crossing this loop is given by

$$\Phi_B = \int_0^1 A_x(x(s))x'(s)ds + \int_0^1 A_y(y(s))y'(s)ds + \int_0^1 A_z(z(s))z'(s)ds$$. Under what conditions on $A_x(x), A_y(y), A_z(z)$ will the current density in space be zero.

[23] A Transmission line consists of two infinite solid cylindrical conductors of radius R in space. The first conductor is described by the equation $x^2 + y^2 \le R^2$ and the second conductor by the equation $(x - a)^2 + y^2 \le R^2$. The first conductor carries a current I in the positive z direction with a uniform current density and the second conductor carries a current I in the negative z direction again with a uniform current density. Determine an expression for the magnetic field and vector potential in space.

[24] Consider a filamentary loop in the xy plane $ABCDA$, where AB is the segment $y = R$, $0 \le x \le L$, BC is the segment $x = L$, $-R \le y \le R$ with y decreasing from R to $-R$. CD is the segment $y = -R$, $0 \le x \le L$, with x decreasing from L to 0 and finally DA is the semicircle $\pi/2 \le \phi \le 3\pi/2$ with ϕ decreasing from $3\pi/2$ to $\pi/2$. The loop carries a current I. Determine the magnetic field due to the loop.

[25] Find an expression for the magnetic field due to a straight wire of length L carrying a current I in a plane perpendicular to the wire and passing through the centre of the wire. Use this expression to determine an expression for the magnetic field at the centre of a regular polygon of n sides having sidelength L when a wire is bent to occupy the position of the polygon and carries a current I. Also determine an expression for the magnetic field at an arbitrary point on the axis of the polygon.

Hint: Assume for the first part that the wire is located along the z axis with endpoints at $z = -L/2$ and $z = L/2$. Biot–Savart law gives

$$\mathbf{B}(x, y, 0) = \frac{\mu I}{4\pi} \int_{-L/2}^{L/2} \frac{(-y, x, 0)d\xi}{(\rho^2 + \xi^2)^{3/2}}$$

This has got only a $\hat{\phi}$ component which is given by

$$B_\phi = \frac{\mu I \rho}{4\pi} \int_{-L/2}^{L/2} \frac{d\xi}{(\rho^2 + \xi^2)^{3/2}}$$

$$= \frac{\mu I}{4\pi\rho} \int_{-\tan^{-1}(L/2\rho)}^{\tan^{-1}(L/2\rho)} \cos\theta d\theta$$

Use the principle of vector superposition to calculate the magnetic field due to the polygonal wire.

[26] A loop having the parametric equation $\mathbf{r} = \mathbf{r}(s)$, $0 \leq s \leq 1$ carries a current I. The magnetic field in space is given by $\mathbf{B}(\mathbf{r})$. Write down an integral expression for the torque of the magnetic force on the loop about the line passing through the point $\mathbf{r}_0$ and having direction defined by the unit vector $\mathbf{n}$.

[27] Determine using the result of problem [26] the torque on a circular loop of radius R having the equation $x^2 + y^2 = R^2$, $z = 0$ and carrying a current I due to a constant magnetic field $\mathbf{B} = B_0\, \mathbf{n}$ where B_0 is a constant and $\mathbf{n}$ a unit vector.

[28] Three infinite lines defined by the parametric equations $\mathbf{r}(s) = \mathbf{r}_i + s\mathbf{n}_i$, $i = 1,2,3$ carry currents I_1, I_2, I_3 respectively. Determine the force on the segment defined by $s_1 < s < s_2$ of the first wire due to the magnetic field produced by the latter two.

4

CHAPTER

FARADAY'S LAW OF INDUCTION

1. INTRODUCTION

We have seen that the equations of electrostatics are given by $\nabla.\mathbf{E} = \rho/\varepsilon$, $\nabla \times \mathbf{E} = \mathbf{0}$. The latter corresponds to the irrotational property of the electric field and implies that the electric field can be derived from a potential function, i.e. $\mathbf{E} = -\nabla V$. Plugging this into the former equation gave us Poisson's equation for the electrostatic potential $\nabla^2 V = -\rho/\varepsilon$. The solution to Poisson's equation was the Coulomb potential produced by a static charge distribution. When this discovery was made, it was thought that electric fields can only be produced by the presence of charges. However, Faraday devised an ingenious experiment in which he moved a magnet inside a coil and observed a flow of current through it. He concluded that electric fields can also be produced by changing magnetic fields, not only by charges. By performing more delicate experiments with coils and moving magnets, Faraday arrived at a quantitative relation between the magnetic field's rate of change and the electric field induced by this. This is known as the Faraday-Lenz law of induction. It states that if Γ is a loop and S the surface area enclosed by the loop, then

$$V = \int_{\Gamma} \mathbf{E}.\mathbf{dr} = -\frac{d\Phi_B}{dt} = -\frac{d}{dt}\int_S \mathbf{B}.\mathbf{n}dS$$

Here, V is the electromotive force or voltage induced around the loop and $\Phi_B = \int_S \mathbf{B}.\mathbf{n}dS$ is the flux of the magnetic field through the surface enclosed by the loop. So the Faraday-Lenz law states that the voltage induced around the loop equals the rate of change of the magnetic flux through the loop. In the above formula, **n** points in the direction such that looking opposite to it, the direction of Γ used in the contour integral is counter-clockwise, that is, **n** and the direction Γ of the contour are arranged according to the right hand rule. In electrostatics, the line integral $\int_\Gamma \mathbf{E}.d\mathbf{r}$ around a closed contour vanishes and this shows that the electric fields produced by changing magnetic fields are non-conservative in contrast to electrostatics. If the loop Γ had a finite resistance, then the voltage (or electromotive force) induced around it by the changing magnetic field would cause a current to flow through the loop which would, in turn produce a magnetic field (as we shall see later) in accordance with Ampere's right hand rule. This field will oppose the changing magnetic field owing to the negative sign appearing in the Faraday-Lenz law. Equivalently, if a changing current was initially flowing through the loop, it would in accordance with Ampere's law, produce a changing magnetic field which according to Lenz' law, will produce an electromotive force through the loop that opposes the direction of current flow. Faraday's law can be transformed into a differential law by the use of Stoke's theorem: If Γ is a closed contour bounding a surface S, then

$$\int_\Gamma \mathbf{E}.d\mathbf{r} = \int \nabla \times \mathbf{E}.\mathbf{n}dS = -\frac{d}{dt}\Phi_B = -\int \frac{\partial \mathbf{B}}{\partial t}.\mathbf{n}dS$$

or equivalently,

$$\int_S \left(\nabla \times \mathbf{E} + \frac{\partial \mathbf{B}}{\partial t}\right).\mathbf{n}dS = 0$$

If this is to hold for every surface element S, then

$$\nabla \times \mathbf{E} = -\frac{\partial \mathbf{B}}{\partial t}$$

must hold identically at all space-time points. This is the differential form of Faraday's law.

Solved Examples:

[1]. Consider a magnetic field $\mathbf{B} = B_z(x, y, z)\,\hat{z}$ and a loop in the x-y plane having the equation in plane polar co-ordinates $\rho = \rho_0(\phi)$, $0 < \phi < 2\pi$. Explain how you would calculate the voltage induced through the loop.

Solution: The flux of the magnetic field through the surface enclosed by the loop is given by

$$\Phi_B(t) = \int \mathbf{B}.\mathbf{n}dS = \int_0^{2\pi}\int_0^{\rho_0(\phi)} B_z(\rho.\cos\phi, \rho.\sin\phi, t)\rho.d\rho.d\phi$$

The induced emf through the loop is the negative time derivative of this:

$$V = -\frac{d\Phi_B}{dt} = -\int_0^{2\pi}\int_0^{\rho_0(\phi)} \frac{\partial \mathbf{B}}{\partial t}(\rho.\cos\phi, \rho.\sin\phi, t)\rho.d\rho.d\phi$$

[2] Consider a rectangular loop of wire having width a and length L falling with its plane perpendicular to the horizontal such that the width is parallel to the horizontal and the length parallel to the vertical. The plane of the loop remains on the x—z plane. After time t, if the height of the bottom width from the vertical is $\xi(t)$, then the vertices of the rectangular loop are $(0, 0, \xi(t) + L)$, $(a, 0, \xi(t) + L)$, $(a, 0, \xi(t))$, $(0, 0, \xi(t))$. A diagram showing this may be drawn. Assume that a magnetic field $B_y(x, z)$ exists on the x—z plane. Assume that the resistance of the loop is R. Using Faraday's law in conjunction with the Lorentz force equation, determine the equation of motion of the loop. The argument is shown in Fig. 4.1.

Solution: After time, t, the flux of the magnetic field through the loop is given by

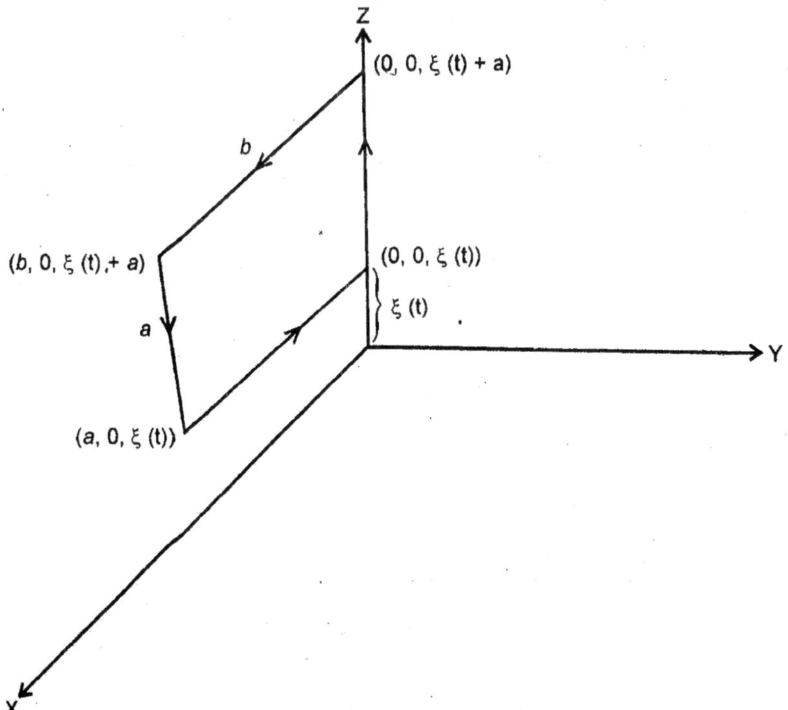

Fig. 4.1

$$\Phi_B(t) = \int_0^a \int_{\xi(t)}^{L+\xi(t)} B_y(x,z)dxdz$$

The voltage induced in the loop is thus

$$V = -\frac{d\Phi_B}{dt} = -\xi'(t) \int_0^a (B_y(x, L+\xi(t)) - B_y(x,\xi(t))dx$$

This voltage is induced counterclockwise in the loop looking along the $-\hat{y}$ direction. The current flowing clockwise is thus given by

$$i(t) = -V/R = \frac{\xi'(t)}{R} \int_0^a B_y(x, L+\xi(t)) - B_y(x,\xi(t))dx$$

The Lorentz force on this current due to the magnetic field is thus

$$F_z = -i(t) \int_0^b B_y(x, L+\xi(t))dx + i(t) \int_0^b B_y(x,\xi(t))dx$$

The first term here is the force on the top segment of the loop and the second term is the force on the bottom segment. The equation of motion of the loop is thus

$$m\frac{d^2\xi}{dt^2} = -mg - \frac{\xi'(t)}{R}\left(\int_0^a (B_y(x, L+\xi(t) - B_y(x, \xi(t))dx\right)^2$$

[3] Consider a rectangular loop of wire of length a and width b. One side of length a is hinged to the x axis as shown in Fig. 4.2. The hinge

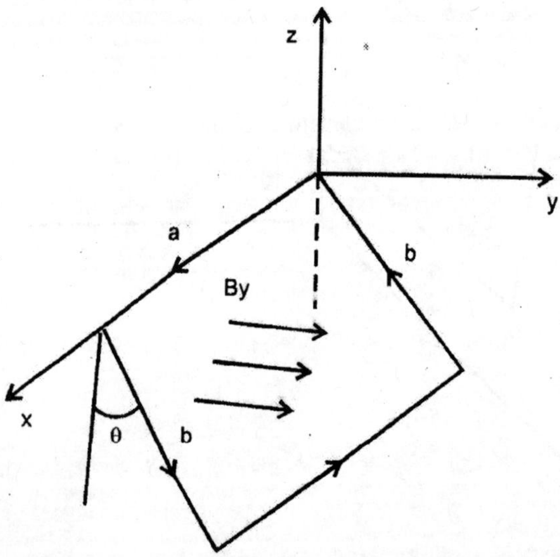

Fig. 4.2

extends from $x = 0$ up to $x = a$ along the x axis. The loop is free to oscillate about the hinge. A magnetic field $B_y(x, z)$ is present. The loop resistance is R. Determine the equation of motion of the loop taking into account Faraday's law of induction and the Lorentz force on a current element due to a magnetic field.

Ans. Assume that the loop makes an angle θ with the vertical at time t as shown in the figure. At this moment, the flux of the magnetic field through the loop is given by

$$\Phi_B(t) = \int_0^a \int_0^b B_y(x,-u\cos\theta(t)).\cos(\theta(t)).dxdu$$

The emf induced in the loop is the negative derivative of this:

$$V = -\frac{d\Phi_B}{dt} = -\frac{d\theta}{dt}.\sin(\theta).\int_0^a \int_0^b B_y(x,-u.\cos\theta)dxdu$$

$$-\frac{d\theta}{dt}.\sin(\theta).\cos\theta.\int_0^a \int_0^b \frac{\partial B_y(x,-u.\cos\theta)}{\partial z}dxdu$$

A current $i = V/R$ flows in the direction shown in the figure. The force on the current that contributes to the torque on the loop receives its contribution only from the lower segment of the loop. This force is

$$F_z = -i\int_0^a B_y(x,-b.\cos\theta)dx$$

The corresponding torque about the hinge is given by $\tau_c = F_z.b.\sin(\theta)$. In full form, the expression for the torque is

$$\tau_c = \frac{b}{R}.\sin^2\theta\frac{d\theta}{dt}.(\int_0^a B_y(x,-b.\cos\theta)dx)\int_0^a \int_0^b B_y(x,-u.\cos\theta)dxdu$$

$$-\frac{b}{R}\frac{d\theta}{dt}.\sin^2(\theta).\cos(\theta).(\int_0^a B_y(x,-b.\cos\theta)dx)\int_0^a \int_0^b \frac{\partial B_y(x,-u.\cos\theta)}{\partial z}dxdu$$

The subscript c below τ stands for the fact that the torque is due to the current. The torque due to gravity is receives its contribution from both the side legs and the bottom leg. It is given by

$$\tau_g = -2 = \frac{m_1 gb}{2}.\sin(\theta) - m_2.g.b.\sin(\theta)$$

where m_1 is the mass of one side leg and m_2 the mass of the bottom leg. In terms of the linear mass density ρ, we can write $m_1 = \rho.b$ and $m_2 = \rho.a$. Thus,

$$\tau_g = -(m_1 + m_2)gb\sin(\theta) = -\rho b(a+b)g.\sin\theta$$

With I denoting the moment of inertia of the loop about the hinge, the equation of angular motion of the loop is given by

$$I\frac{d^2\theta}{dt^2} = \tau_c + \tau_g$$

[4] A rectangular loop of wire lies on the X—Y plane with one side parallel to the y axis slideable. This arrangement is shown in the Fig. 4.3. A magnetic field $B_z(x, y)$ exists in space. Using a combination of Faraday's law of induction and the Lorentz force on a current loop due to a field, determine the equation of motion of the slideable side. Assume that the loop has a resistance R.

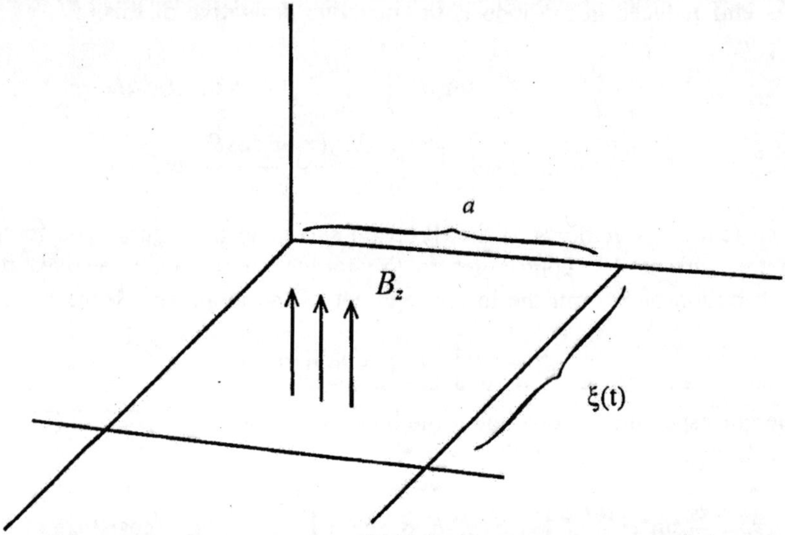

Fig. 4.3

Solution: Assume that after time t, the slideable side has traversed a distance $\xi(t)$ from the y axis. The flux of the magnetic field through the loop is then given by

$$\Phi_B(t) = \int_0^{\xi(t)} \int_0^a B_z(x, y)dxdy$$

The emf induced through the loop is given by

$$V = -\frac{d\Phi_B}{dt} = -\frac{d\xi(t)}{dt}\int_0^a B_z(\xi(t), y)dy$$

The current through the loop is $i = V/R$. The force of the magnetic field on the loop is then according to the Lorentz equation,

$$F_x = i\int_0^a B_z(\xi(t), y)dy = -\frac{1}{R}\frac{d\xi(t)}{dt}\left(\int_0^a B_z(\xi(t), y)dy\right)^2$$

Finally, the equation of motion of the slideable side which is assumed to have mass m is

$$m\frac{d^2\xi(t)}{dt^2} = F_x$$

[5] Explain how Faraday's law can be applied to a loop that moves in a static magnetic field. What happens when both the magnetic field varies as well as the loop moves ?

Solution: Consider a loop Γ moving in a magnetic field **B**(**r**) assumed to be static. A line element $d\mathbf{r}$ of the loop traces out an area $\mathbf{v}dt \times d\mathbf{r}$ in time dt where **v** is the velocity of the segment $d\mathbf{r}$ of the loop. This means that the change in the magnetic flux linkage due to the motion of this segment in time dt is given by $d(d\Phi_B) = \mathbf{B}.\mathbf{v}dt \times d\mathbf{r} = -d\mathbf{r}.\mathbf{v} \times \mathbf{B}dt$. The two *d's* appearing here before Φ_B correspond respectively to differential in time and differential in loop element. The change in flux linkage by this little segment per unit time is then $\dfrac{d(d\Phi_B)}{dt} = -d\mathbf{r}.\mathbf{v} \times \mathbf{B}$. Integrating this around the entire loop gives the rate of change of flux linkage due to motion of the entire loop:

$$\frac{d\Phi_B}{dt} = -\int_\Gamma \mathbf{v} \times \mathbf{B}.d\mathbf{r}$$

The induced emf in the loop is the negative of this quantity. We can give a different interpretation of the above expression. The segment $d\mathbf{r}$ of the loop moving with a velocity **B** has electrons in it. An electron e experiences a force $-e\mathbf{v} \times \mathbf{B}$. Relative to the loop, the electron is static and this force acting on the electron can be viewed as being produced by an electric field in the static reference frame of the loop segment of value $\mathbf{E} = \mathbf{v} \times \mathbf{B}$. The line integral of this field around the loop gives the induced emf and this agrees with the above expression.

6. The space between two concentric discs of radii a and b has a uniform magnetic field B_0 parallel to the axis. The two discs rotate with a uniform angular velocity of ω radians per second. Two wires are connected respectively to the two discs. Draw a diagram and determine the voltage induced in the wires.

Ans. Draw a radial line from the common centre of the two discs meeting the perimeter of the two discs. After time t, the area swept by this line between the two discs is given by $A(t) = (b^2 - a^2)\omega t/2$. The flux induced through this area is $BA(t)$ and its rate of change determines the voltage V. Thus, $V = (b^2 - a^2)\omega B/2$.

7. A conducting rod moves with a constant velocity $v\hat{z}$ parallel to a long straight wire carrying a current of I amperes. Assume that the motion of the rod is perpendicular to its length. Calculate the emf induced in the rod and state which end is at a higher potential.

Ans. Look at Fig. 4.4. Assume that the rod has a length L and the distance of its nearer end from the wire is d. After time t, the area awept by the rod equals vtL. The magnetic field due to the wire at a distance r is $B(r) = \mu I/2\pi r$ and this points into the paper, i.e. in the $\hat{y}$ direction. The total flux cut by the rod in time t is given by

$$\Phi = \int_d^{d+L} vtB(r)dr = \frac{vt\mu I}{2\pi}\log(1 + L/d)$$

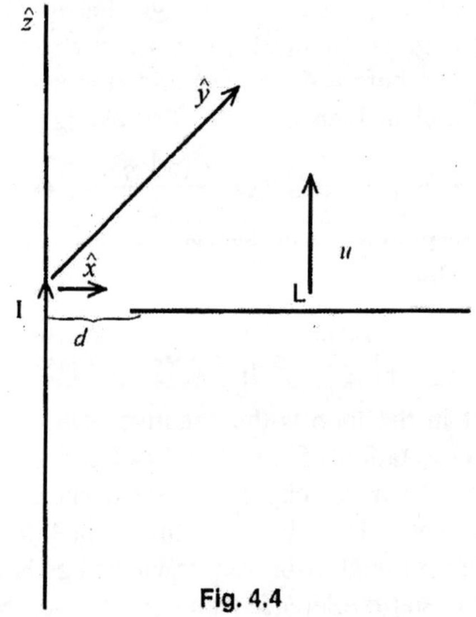

Fig. 4.4

The rate of change of this flux is the voltage induced in the rod. It is given by

$$V = \frac{d\Phi}{dt} = \frac{\mu vI}{2\pi}\log(1 + L/d)$$

The direction of this rate of change is $\hat{y}$. According to the left hand rule, the direction of this voltage is from the far end of the rod to the near end where far and near are with respect to the wire.

8. Consider Faraday's law in differential form

$$\nabla \times \mathbf{E} = -\frac{\partial \mathbf{B}}{\partial t}$$

By writing out the three components and integrating them, show explicitly how we can obtain the general solution for the electric field given the magnetic field **B** in space. Assume that $\nabla.\mathbf{B} = 0$ holds.

Ans. The three components of Faraday's law are

$$\frac{\partial E_z}{\partial y} - \frac{\partial E_y}{\partial z} = F_x$$

$$\frac{\partial E_x}{\partial z} - \frac{\partial E_z}{\partial x} = F_y$$

$$\frac{\partial E_y}{\partial x} - \frac{\partial E_x}{\partial y} = F_z$$

where

$$\mathbf{F} = -\frac{\partial \mathbf{B}}{\partial t}.$$

Integrating the first equation, we get

$$E_z = \frac{\partial}{\partial z}\int_0^y E_y dy + \int_0^y F_x dy + \psi(x,z)$$

Thus the second equation reduces to

$$\frac{\partial E_x}{\partial z} = \frac{\partial^2}{\partial z \partial x}\int_0^y E_y dy + \frac{\partial}{\partial x}\int_0^y F_x dy + \frac{\partial \psi}{\partial x} + F_y$$

Integrating this one, we get

$$E_x = \frac{\partial}{\partial x}\int_0^y E_y dy + \frac{\partial}{\partial x}\int_0^z\int_0^y F_x dy dz + \frac{\partial}{\partial x}\int_0^z \psi dz + \int_0^z F_y dz + \xi(x,y)$$

In view of this, the third equation can be expressed as

$$\frac{\partial E_y}{\partial x} = \frac{\partial E_y}{\partial x} + \frac{\partial}{\partial x}\int_0^z F_x dz + \frac{\partial}{\partial y}\int_0^z F_y dz + \frac{\partial \xi}{\partial y} + F_z$$

which is the same as

$$F_z + \int_0^z \left(\frac{\partial F_x}{\partial x} + \frac{\partial F_y}{\partial y}\right) dz + \frac{\partial \xi}{\partial y} = 0$$

or

$$\int_0^z \left(\frac{\partial F_x}{\partial x} + \frac{\partial F_y}{\partial y} + \frac{\partial F_z}{\partial z} \right) dz + F_z(x, y, 0) + \frac{\partial \xi}{\partial y} = 0$$

or since $\nabla.\mathbf{F} = 0$, this is the same as

$$F_z(x, y, 0) + \frac{\partial \xi}{\partial y} = 0$$

and hence we can take

$$\xi(x, y) = -\int_0^y F_z(x, y, 0)dy + \eta(x)$$

This means that the general solution is

$$E_x = \frac{\partial}{\partial x}\int_0^y E_y dy + \frac{\partial}{\partial x}\int_0^z \int_0^y F_x dydz + \eta(x) + \frac{\partial}{\partial x}\int_0^z \psi dz + \int_0^z F_y dz - \int_0^y F_x(x, y, 0)dy$$

$$E_y = E_y$$

$$E_z = \frac{\partial}{\partial z}\int_0^y E_y dy + \int_0^y F_x dy + \psi(x, z)$$

We may call $\int_0^x \eta(x)dx + \int_0^z \psi(x, z)dz$ as $v(x, z)$. Then the general solution can be expressed as

$$E_x = \frac{\partial}{\partial x}\int_0^y E_y dy + \frac{\partial}{\partial x}\int_0^z \int_0^y F_x dydz + \int_0^z F_y dz - \int_0^y F_x(x, y, 0)dy + \frac{\partial v}{\partial x}$$

$$E_y = E_y$$

$$E_z = \frac{\partial}{\partial z}\int_0^y E_y dy + \int_0^y F_x dy + \frac{\partial v}{\partial z}$$

In this general solution, the functions $E_y(x, y, z)$ and $v(x, z)$ can be chosen arbitrarily.

Remark: If **E** is a solution and g any scalar function, then $\mathbf{E} + \nabla g$ is also a solution since

$$\nabla \times (\mathbf{E} + \nabla g) = \nabla \times \mathbf{E} + \nabla \times \nabla g = \nabla \times \mathbf{E} = -\frac{\partial \mathbf{B}}{\partial t}$$

9. A square loop of side a recedes with a uniform velocity $u_0\hat{y}$ from an infinitely long filament carrying current I along $\hat{z}$ as shown in the Fig. 4.5. The plane of the loop is parallel to the yz plane and the lower edge is always along the $\hat{y}$ axis. Assuming that the distance of the near end of

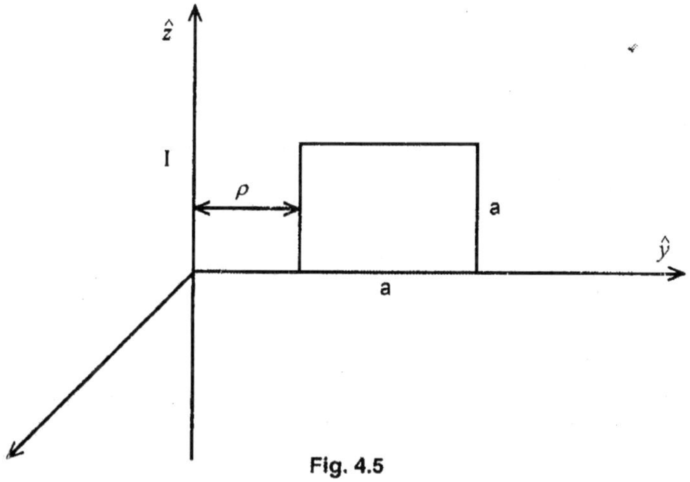

Fig. 4.5

the loop from the $\hat{z}$ axis is ρ_0 at time $t = 0$, find the voltage induced in the loop at time t when the near end is at a distance ρ from the $\hat{z}$ axis.

Ans. The magnetic field at $(0, \rho, z)$ which is on one vertical side of the loop is

$$B_x(\rho) = -\frac{\mu I}{2\pi\rho}$$

The magnetic field at $(0, \rho + a, z)$ which is on another vertical side of the loop is

$$B_x(\rho + a) = -\frac{\mu I}{2\pi(\rho + a)}$$

The voltage induced in the loop is due to motional emf. It is given by

$$V = \int_\Gamma \mathbf{u} \times \mathbf{B}.d\mathbf{r}$$

where Γ is the loop circuit. The contributions to the line integral come only from the two vertical sides. They evaluate to

$$V = -u_0 B_x(\rho)a + u_0 B_x(\rho + a)a = -u_0 a(B_x(\rho) - B_x(\rho + a))$$

$$= (u_0 a\mu I / 2\pi)(1/\rho - 1/(\rho + a)) = \frac{\mu I a^2 u_0}{\rho(\rho + a)}$$

10. The triangular region $x \geq y \geq 0$ has a magnetic field $B_z(x, y)$. We may assume this triangular region to be bounded by two straight wires, one along the $\hat{x}$ axis and another along the line $y = x$. Assume that a wire parallel to the $\hat{y}$ axis slides with a velocity $u_0\hat{x}$ so that it always makes

contact with the two straight wires. (See Fig. 4.6). Determine the emf induced in the sliding wire. Assume that at $t = 0$, the sliding wire passes through the origin.

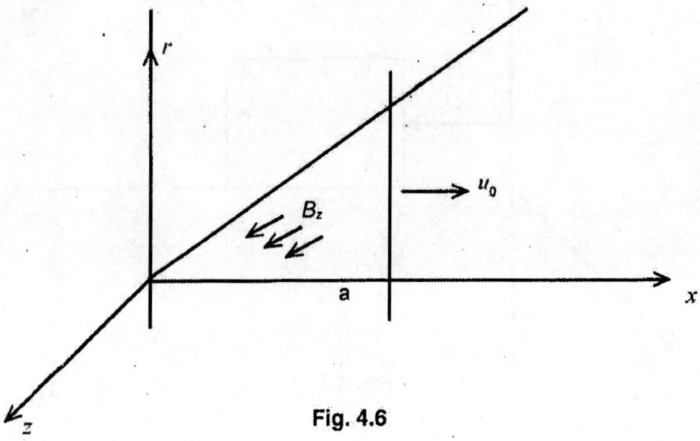

Fig. 4.6

Ans. After time t the region swept by the sliding wire is given by $\{(x, y) : 0 < x < u_0 t,\ 0 < y < x\}$. This is the triangular region shown. The flux of the magnetic field through this area is given by

$$\Phi = \int_0^{u_0 t} dx \int_0^x B_z(x, y) dy$$

and its rate of change gives the emf induced in the sliding wire:

$$V = \frac{d\Phi}{dt} = u_0 \psi(u_0 t)$$

where

$$\psi(x) = \int_0^x B_z(x, y) dy$$

Exercises

[1] Consider a coil that has a magnetic field varying through it. Consider the following two cases: (a) the coil is split at one point (b) the coil is not split. Explain why in both cases, an emf is induced in the coil but that in case (a), Joule heating does not occur while in case (b), Joule heating occurs.

Hint: A current will flow only if the coil is not split and Joule heating is proportional to the integral of the square of the current when the coil is not split.

(c) Show that if the coil is not split and $B(t)$ is the normal component of the magnetic field through the surface, then the Joule heating when the coil is not split equals $\frac{A^2}{R}\int_0^t \left(\frac{dB(\tau)}{d\tau}\right)^2 d\tau$, where A is the area of the loop. The magnetic field here is assumed to vary with time but remains constant across the loop area.

[2] A magnetic field $\mathbf{B} = B_0 \sin(\omega t)\hat{z}$ exists in space. A circular loop of radius R with centre at the origin rotates about the $\hat{x}$ axis with an angular velocity of ω_0. Show that the flux through the loop is given by $\Phi(t) = B_0 \pi R^2 \sin(\omega t).\cos(\omega_0 t)$ assuming that at $t = 0$, the plane of the coil is normal to the $\hat{z}$ axis.

[3] A rectangular loop of sidelengths a and b lies in the xy plane. The loop moves in an external magnetic field $B_0(x)\hat{z}$ in the xy plane in a direction parallel to the side a. Assume the resistance of the loop is R. Show that if i (t) is the current through the loop, then

$$m\frac{d^2x}{dt^2} = -i(t)bB_0(x) + i(t)bB_0(x+a)$$

$$Ri(t) = \frac{dx}{dt}B_0(x)b - \frac{dx}{dt}B_0(x+a)b$$

where m is the mass of the loop and x is the distance at time t of the nearer side of the loop from the $\hat{y}$ axis.

Hint: Consider the force acting on the sides of the loop perpendicular to the direction of motion. Consider also the flux of the magnetic field through the loop and its rate of change.

[4] Consider an electric field $\mathbf{E} = E(x, t)\hat{y}$ and a magnetic field $\mathbf{B} = B(x, t)\hat{z}$. Show that Faraday's law $\nabla \times \mathbf{E} = -\frac{\partial B}{\partial t}$ for this pair of fields simplifies to

$$\frac{\partial E(x,t)}{\partial x} = \frac{\partial B(x,t)}{\partial t}$$

[5] Consider a parallel plate capacitor with the plates circular of radius R and being charged with a current $i(t)$ from the centre. Assume the $\hat{z}$ direction to be directed along the axis joining the centre of the plates. Assume that the magnetic field inside the capacitor is $H_\phi(t, \rho, z)$ and that the electric field inside the capacitor is $E_z(t, \rho, z)$ where ρ is the distance from the central axis joining the plates. Assume that Ampere's law has been modified to the equation

$$\nabla \times \mathbf{H} = \mathbf{J} + \varepsilon \frac{\partial \mathbf{E}}{\partial t}$$

This modification will be carried out in the next chapter. Show by taking a rectangular loop with two sides parallel to the plates and two sides perpendicular to the plates with one of the latter being along the axis and the other at a distance ρ from the axis that Faraday's law implies

$$\int_0^z E_z(t,0,z)dz - \int_0^z E_z(t,\rho,z)dz = -\frac{\partial}{\partial t}\mu \int_0^{\rho}\int_0^z H_\phi(t,\rho,z)d\rho dz$$

Also show that the modified Ampere's law implies by taking a circular loop of radius ρ that

$$H_\phi(t,\rho,z).2\pi\rho = \frac{\partial}{\partial t}\varepsilon \int_0^{\rho} E_z(t,\rho,z)2\pi\rho d\rho$$

Show that these two equations imply

$$\frac{\partial E_z(t,\rho,z)}{\partial \rho} = \mu \frac{\partial H_\phi(t,\rho,z)}{\partial t}$$

$$\frac{1}{\rho}\frac{\partial \rho H_\phi(t,\rho,z)}{\partial \rho} = \varepsilon \frac{\partial E_z(t,\rho,z)}{\partial t}$$

Deduce from these two that E_z satisfies the equation

$$\frac{1}{\rho}\frac{\partial}{\partial \rho}\left(\rho \frac{\partial E_z}{\partial \rho}\right) = \varepsilon\mu \frac{\partial^2 E_z}{\partial t^2}$$

Show that Gauss' law, $\nabla.\mathbf{E} = 0$ implies that E_z cannot depend on z and hence H_ϕ also cannot depend upon z. Show finally, that the boundary condition for E_z is

$$\frac{d}{dt}\varepsilon \int_0^R E_z(t,\rho)2\pi\rho d\rho = i(t)$$

Explain how you would use these equations to derive a complete solution for E_z (t, ρ).

5

CHAPTER

MAXWELL'S EQUATIONS AND ELECTROMAGNETIC WAVES

1. The equation of charge conservation
2. Defect with Ampere's law and the displacement current correction term
3. Summary of the Maxwell equations
4. Maxwell equations with sinusoidally time varying sources and fields
5. Derivation of the wave propagation equations for fields in free space
6. The equations satisfied by the scalar and vector potentials
7. Complete solution to the Maxwell equations in terms of retarded potentials
8. Maxwell's equations in a conducting medium
9. Boundary conditions for Maxwell's equations
10. Relativistic transformation of the fields
11. Poynting's theorem, power flow and energy density of the electromagnetic field
12. Reflection and Refraction of electromagnetic waves
13. Propagation of electromagnetic waves in a waveguide
14. Transmission line theory
15. Some nonlinear problems associated with electromagnetic fields and wave propagation

INTRODUCTION

Ampere's law turns out to be incomplete in that it violates the principle of conservation of charge for non-steady currents. Therefore, a correction term in the form of displacement current needs to be added to this equation to bring it into agreement with charge conservation. This fact was first discovered by James Clerk Maxwell. The complete set of equations that describe all the phenomena of classical electromagnetism are thus (1) Gauss' law, (2) Ampere's law with Maxwell's displacement current correction, (3) Faraday's law (4) the fact that the magnetic lines of force are closed or equivalently, that there do not exist any magnetic monopoles. These four equations in differential form are commonly referred to as the Maxwell equations. In the absence of charge and current densities in space, one can derive from these equations as Maxwell first did, the wave equations for the electric and magnetic field. This shows that any electromagnetic field in space can be decomposed into a superposition of plane waves of definite frequencies, each component propagating at the speed of light. Further, the Maxwell equations also tell us about the nature of the polarization of any plane wave. They tell us that the directions of the electric and the magnetic field vectors are polarized perpendicular to the direction of propagation. The Maxwell equations for sinusoidally time varying sources are also important in applications for example in determining the field propagating inside a waveguide with sinusoidal excitation or the field produced by an antenna with sinusoidal feed in current. In order to effect the solution to the Maxwell equations, we introduce the electric scalar potential and the magnetic vector potential which are consequences of Faraday's law and the no magnetic monopole condition. The electric and magnetic fields can be expressed in a convenient form as linear combinations of partial derivatives of these four potentials. Further by introducing a Lorentz gauge condition, it can be shown that the scalar and vector potentials satisfy the wave equation with sources defined by the charge and current densities. A Fourier transformation in time transforms this wave equation with source into Helmholtz equations with source. The Green's function for this problem is shown to be $\exp(ikr)/r$ and this leads us to the retarded potential solution for the scalar and vector potentials. They correspond to the situation in which the electromagnetic wave travels from the source to the destination at the speed of light. Maxwell's equations in a conducting medium involve introducing a linear relationship between the current density and the electric field which is known as Ohm's law. Using this, we can derive a damped wave equation for the electromagnetic field in a conducting medium in the absence of external charges and currents. Then we discuss boundary conditions for the electric and magnetic fields. It is shown that the boundary conditions remain the same as for the corresponding electrostatic and magnetostatic cases. Specifically, the normal component of the electric displacement vector has a discontinuity at a surface equal to the surface charge density and the tangential component of the electric field is continuous. The normal component of the magnetic field is continuous and the tangential component has a disontinuity dictated by the surface current density. The topic of relativistic transformation of the fields involves computing the electromagnetic field in a frame moving with a constant velocity relative to another. The transformation of space and time co-ordinates that relates the two frames is the Lorentz transformation which was discovered by Einstein and Lorentz. The electromagnetic four potential is regarded as four vector and so is the charge and current density pair. This leads us to the conclusion that the Maxwell equations remain invariant under a Lorentz transformation and form the foundations of the special theory of relativity. The Lorentz transformation of the space-time co-ordinates and the charge-current density pair can be used to derive the transformation rule for the electromagnetic field. On can in fact, regard the electromagnetic field as the components of a rank four antisymmetric tensor and obtain the appropriate transformation rule. Poynting's theorem for the power transferred per unit area and the energy density of the electromagnetic field is derived as an elementary consequence of the Maxwell equations, the Lorentz force equation and the principle of energy conservation. It is useful in calculating the energy stored in a capacitor, the energy stored in a system of charges, in the magnetic field of an inductor or a transformer, in the power transferred by a waveguide or the power radiated into space by an antenna. The topic of reflection and refraction of electromagnetic waves at the interface of two dielectric media is based on assuming an incident, reflected and transmitted wavefield and applying the boundary conditions for the tangential components of the fields. This leads us to the Snell's laws of reflection and refraction and expressions for the ratios of the reflected and transmitted power density to the incident power density. These expressions are particularly useful in optics. Propagation of an electromagnetic wave through a waveguide is an important topic for engineers. The solution to the waveguide problem involves assuming harmonic dependence of the fields on frequency and exponential dependence upon the co-ordinate along which the wave propagates and application of the boundary conditions to the fields at the waveguide walls. This leads us to expressions for the components of the field perpendicular to the propagation direction in terms of the components, parallel to the direction of propagation, i.e. the longitudinal components. Two cases arise here, one in which the electric field does not have a longitudinal component, this is the TE or transverse electric mode, two in which the magnetic field does not have a longtitudinal mode, this is the TM or transverse magnetic mode. The general propagation is a linear combination of these two modes. Each mode has a certain cutoff frequency above which the wave gets attenuated as it propagates through the guide. The mode is defined by a pair of integers for rectangular waveguides and the cutoff frequency depends upon this pair of numbers. Propagation in a circular waveguide is also considered. Finally, we look at the theory of the transmission line in which there is a distribution of inductance, resistance, conductance and capacitance along the line and we are interested in the propagation of voltage and current as waves along the line. The line may be terminated with an impedance. The study of transmission lines is particularly important in power engineering where power is transmitted to different centres from a central substation. Finally, we take a look at certain non-linear problems in electromagnetic theory arising from the dependence of the permittivity and permeability on the fields.

The equation of charge conservation

Consider a charged fluid having a density $\rho(t,\mathbf{r})$ Coul/m^3 in space and moving with a velocity field $\mathbf{v}(t, \mathbf{r})$. The current density due to this charged fluid is given by $\mathbf{J} = \rho\mathbf{v}$. We take a volume V bounded by a closed surface S and compute the net rate at which charge enters this volume. It is given by $-\int_S \mathbf{J}.\mathbf{n}dS = -\int_S \rho\mathbf{v}.\mathbf{n}dS$. This rate of entry of charge into V must equal the rate at which charge inside V increases with time, i.e. $\frac{d}{dt}\int_V \rho(t,\mathbf{r})dV = \int \frac{\partial\rho(t,r)}{\rho t}\int_V dV$. From Gauss' divergence theorem we have $\int_S \mathbf{J}.\mathbf{n}dS = \int_V div\mathbf{J}dV$. We thus obtain the charge conservation equation

$$\int_V (\text{div}\mathbf{J} + \frac{\partial\rho}{\partial t})dV = 0$$

Since this must be valid for every volume V, it must follow that

$$\text{div}\mathbf{J} + \frac{\partial\rho}{\partial t}dV = 0$$

and this is the local charge conservation equation. This equation tells us that the rate at which charge enters a unit volume centred at a given point must equal the rate at which charge within this unit volume increases with time. Expressed in full in terms of Cartesian co-ordinates, it is

$$\frac{\partial\rho(t,x,y,z)}{\partial t} + \frac{\partial J_x(t,x,y,z)}{\partial x} + \frac{\partial J_y(t,x,y,z)}{\partial y} + \frac{\partial J_z(t,x,y,z)}{\partial z} = 0$$

In case there are generators of charge with the rate of charge generation being $g(t, \mathbf{r})$ at the location $\mathbf{r}$ at time t per unit volume, then the equation gets modified to

$$\text{div}\mathbf{J} + \frac{\partial\rho}{\partial t} = g(t,\mathbf{r})$$

For most of our problems unless otherwise mentioned, $g = 0$.

Defect with Ampere's law and its correction

Ampere's law states that $\nabla \times \mathbf{H} = \mathbf{J}$. Taking divergence on both sides and using the fact that the divergence of a curl is zero, we get $\nabla.\mathbf{J} = 0$. This equation is the correct charge conservation equation only if $\frac{\partial \rho}{\partial t} = 0$, i.e. for steady currents. For unsteady currents, it is an incorrect equation. This means that Ampere's law is not valid in time varying situations. This flaw was observed and corrected by the famous British physicist James Clerk Maxwell. Suppose we add a correction term on the right side of Ampere's law, say modifying it to $\nabla \times \mathbf{H} = \mathbf{J}_c + \mathbf{J}_d$, where $\mathbf{J}_c$ is the usual current (due to conduction) and $\mathbf{J}_d$ a correction term. Then, taking divergence gives $\nabla.\mathbf{J}_c + \nabla.\mathbf{J}_d = 0$. For this to agree with the equation of charge conservation, we must have $\nabla.\mathbf{J}_d = \frac{\partial \rho}{\partial t}$. Noting that from Gauss law, $\nabla.\mathbf{D} = \rho$, we can express this as $\nabla.\mathbf{J}_d = \nabla.\frac{\partial \mathbf{D}}{\partial t}$ or

$$\nabla.\left(\mathbf{J}_d - \frac{\partial \mathbf{D}}{\partial t}\right) = 0$$

Maxwell suggested that to satisfy this equation, we take $\mathbf{J}_d = \frac{\partial \mathbf{D}}{\partial t}$. This lead us to the following modification of Ampere's law:

$$\nabla \times \mathbf{H} = \mathbf{J}_c + \frac{\partial \mathbf{D}}{\partial t}$$

The term $\mathbf{J}_d = \frac{\partial \mathbf{D}}{\partial t}$ is called the Maxwell displacement current correction term. It is not a current, only a fictitious current. Its effect can be observed during the charging of a capacitor. Consider a parallel plate capacitor with a current I entering into a plate and a current I leaving the other plate. The charge on the top plate is $Q(t) = It$ and the charge on the bottom plate is $-Q(t) = -It$. An electric field is set up between the plates. This electric field is time varying because the charge on the plates is time varying. Consider now a closed surface S bounding a part of the top wire, the top capacitor plate but not the bottom plate of the capacitor. In other words, the one side of the surface covers the top plate and the other side passes through the space between the plates. A current I enters through the top surface via the top wire but no current leaves the bottom surface of S. Thus, from Ampere's law, $\nabla.\mathbf{J}_c = 0$ when integrated over the volume bounded by the surface would give $\int_S \mathbf{J}_c.\mathbf{n}dS = 0$ and this would imply that $I = 0$. This is false. (The surface integral equals $-I$.) However, if we use the Maxwell displacement current correction term in Ampere's law, then we would get $\nabla.(\mathbf{J}_c + \mathbf{J}_d) = 0$, where $\mathbf{J}_d = \frac{\partial \mathbf{D}}{\partial t} = \varepsilon.\frac{\partial \mathbf{E}}{\partial t}$ is the displacement current correction term. When integrated over the volume would give $\int_S(\mathbf{J}_c + \mathbf{J}_d).\mathbf{n}dS = 0$ and this condition could be met provided that the electric field in the region between the plates is such that $\int_S \frac{\partial \mathbf{D}}{\partial t}.\mathbf{n}dS = I$. Note that although no current flows in the region between the plates, there is a fictitious displacement current in this region caused by the changing electric field whose flux out of the surface matches the conduction current entering into the surface.

Summary of the Maxwell Equations

After incorporating the correction term, we can summarize the complete equations of electromagnetism. These are known as the Maxwell equations who was the first to express these in differential form as given below along with the displacement current correction term incorporated into Ampere's law.

Gauss' Law:

$$\nabla.\mathbf{D} = \rho$$

Faraday's Law:

$$\nabla \times \mathbf{E} = -\frac{\partial \mathbf{B}}{\partial t}$$

Ampere's law:

$$\nabla \times \mathbf{H} = \mathbf{J} + \frac{\partial \mathbf{D}}{\partial t}$$

The no-magnetic monopole condition:

$$\nabla.\mathbf{B} = 0$$

These along with constitutive relations for the medium relating **D** to **E** and **B** to **H** are the complete equations for classical electromagnetism. For a linear medium, the constitutive relations are $\mathbf{D} = \varepsilon\mathbf{E}, \mathbf{B} = \mu\mathbf{H}$. It should be noted that although Gauss' law has been derived from

Coulomb's law, Gauss' law is more general. It is valid even for time varying problems and charges in motion whereas Coulomb's law is not valid for moving charges. The four Maxwell equations describe a wide variety of physical phenomena ranging from fields in materials produced by charges, propagation of electromagnetic waves in different kinds of media, reflection, refraction, diffraction, interference, absorption and radiation of light in various media like dielectrics and conductors, motion of charges in semiconductors, and motion of charges in the higher layers of the atmosphere. It should also be mentioned that the Maxwell equations proved to be the starting point for the development of the special theory of relativity. These four equations, after Newton's laws of motion and the universal law of gravitation, form the next cornerstone of physics since its origin. The whole of twentieth century physics is essentially the Maxwell equations. Max Planck also used Maxwell's theory of electromagnetism to arrive at his quantum theory of radiation by deriving using classical arguments, the expression for the spectrum of black body radiation.

Maxwell Equations with Sinusoidally Time Varying Sources

Assuming the medium is linear, i.e. ε and μ are constants, it is clear that the Maxwell equations are linear in the field variables. This means that if (ρ,**J**) is a charge density and current density pair that produces fields (**E**,**B**) and (ρ',**J**$'$) is another charge density and current density pair that produces fields (**E**$'$,**B**$'$), then the charge density current density pair ($c\rho + c'\rho'$, $c\mathbf{J} + c'\mathbf{J}'$) will produce fields ($c\mathbf{E} + c'\mathbf{E}'$, $c\mathbf{B} + c'\mathbf{B}'$). This is the principle of superposition. Writing Maxwell equations for a linear medium in the form

$$\nabla.\mathbf{E} = \rho / \varepsilon$$

$$\nabla \times \mathbf{E} = -\frac{\partial \mathbf{B}}{\partial t}$$

$$\nabla \times \mathbf{B} = \mu\mathbf{J} + \varepsilon\mu\frac{\partial \mathbf{E}}{\partial t}$$

$$\nabla.\mathbf{B} = 0$$

we find likewise for the sources $\rho' / \mathbf{J}'$, and their response fields $\mathbf{E}'.\mathbf{B}'$,

$$\nabla.\mathbf{E}' = \rho' / \varepsilon$$

$$\nabla \times \mathbf{E}' = -\frac{\partial \mathbf{B}'}{\partial t}$$

$$\nabla \times \mathbf{B}' = \mu \mathbf{J}' + \varepsilon\mu \frac{\partial \mathbf{E}'}{\partial t}$$

$$\nabla . \mathbf{B}' = 0$$

Multiplying the first set of four equations by c, the second set by c' and adding gives

$$\nabla.(c\mathbf{E} + c'\mathbf{E}') = c\nabla.\mathbf{E} + c'\nabla.\mathbf{E}' = c\rho/\varepsilon + c'\rho'/\varepsilon = (c\rho + c'\rho')/\varepsilon$$

$$\nabla \times (c\mathbf{E} + c'\mathbf{E}') = c\nabla \times \mathbf{E} + c'\nabla \times \mathbf{E}' = -c\frac{\partial \mathbf{B}}{\partial t} - c'\frac{\partial \mathbf{B}'}{\partial t}$$

$$= -\frac{\partial}{\partial t}(c\mathbf{B} + c'\mathbf{B}')$$

$$\nabla \times (c\mathbf{B} + c'\mathbf{B}') = c\nabla \times \mathbf{B} + c'\nabla \times \mathbf{B}' = c\left(\mu \mathbf{J} + \varepsilon \frac{\partial \mathbf{E}}{\partial t}\right) + c'\left(\mu \mathbf{J}' + \varepsilon \frac{\partial \mathbf{E}'}{\partial t}\right)$$

$$= \mu(c\mathbf{J} + c'\mathbf{J}') + \mu\varepsilon \frac{\partial}{\partial t}(c\mathbf{E} + c'\mathbf{E}')$$

$$\nabla.(c\mathbf{B} + c'\mathbf{B}') = c\nabla.\mathbf{B} + c'\nabla.\mathbf{B}' = 0$$

This proves the linearity of the Maxwell equations. It stems essentially from the fact that the differential operators of divergence and curl are linear operators. Fourier analysis tells us that any function of time can be represented as a superposition of complex sinusoids. Thus, linearity of the Maxwell equations implies that the field produced by any source can be represented as the superposition of the fields corresponding to the different frequency components of the sources. This means the following: If $\mathbf{E}(\omega, t, \mathbf{r})$ and $\mathbf{B}(\omega, t, \mathbf{r})$ are respectively the electric and magnetic fields produced by charge density $\rho(\omega, \mathbf{r}) \exp(i\omega t)$ and current density $\mathbf{J}(\omega, \mathbf{r}) \exp(i\omega t)$, then the fields produced by the charge density $\int \rho(\omega, \mathbf{r}) \exp(i\omega t) d\omega$ and current density $\int \mathbf{J}(\omega, \mathbf{r}) \exp(i\omega t) dw$ are respectively given by $\int \mathbf{E}(\omega, t, \mathbf{r}) d\omega$ and $\int \mathbf{B}(\omega, t, r) d\omega$. This means that to determine the fields corresponding to any charge and current densities, it suffices to determine the fields corresponding to charge and current densities having only sinusoidal time variation. Further, the Maxwell equations are time invariant assuming that ε, μ are constants. This means that if the fields corresponding to charge and current densities $\rho(t,\mathbf{r})$ and $\mathbf{J}(t,\mathbf{r})$ are $\mathbf{E}(t,\mathbf{r})$ and $\mathbf{B}(t,\mathbf{r})$, then the fields corresponding to the charge and current densities $\rho(t - \tau, \mathbf{r})$ and $\mathbf{J}(t - \tau, \mathbf{r})$ are $\mathbf{E}(t - \tau, \mathbf{r})$ and $\mathbf{B}(t - \tau, \mathbf{r})$ respectively. This follows by direct substituting into the Maxwell equations noting that none of the coefficients in the system depends upon time. A consequence of linearity and time invariance is that the response to sinusoidally time varying charge and current densities will once again be sinusoidally time varying fields at the same frequency. Specifically, assuming the response to $\rho(\mathbf{r})\exp(i\omega t)$, $\mathbf{J}(\mathbf{r})\exp(i\omega t)$ to be $\mathbf{E}(\mathbf{r})\exp(i\omega t)$, $\mathbf{B}(\mathbf{r})\exp(i\omega t)$, we obtain on plugging these into the system of Maxwell equations,

$$\nabla.\mathbf{E} = 0,\ \nabla \times \mathbf{E} = -i\omega\mathbf{B},\ \nabla \times \mathbf{B} = \mu\mathbf{J} + i\omega\varepsilon\mathbf{E},\ \nabla.\mathbf{B} = 0$$

In these equations, $\mathbf{E}$, $\mathbf{B}$, ρ, $\mathbf{J}$ are all functions of position $\mathbf{r}$ only.

Wave propagation equations in free space

Maxwell equations in free space are

$$\nabla.\mathbf{E} = 0, \nabla \times \mathbf{E} = -\frac{\partial \mathbf{B}}{\partial t}, \nabla \times \mathbf{B} = \mu\varepsilon\frac{\partial \mathbf{E}}{\partial t}, \nabla.\mathbf{B} = 0$$

Taking the curl of the second equation gives

$$\nabla(\nabla.\mathbf{E}) - \nabla^2\mathbf{E} = \nabla \times (\nabla \times \mathbf{E}) = -\nabla \times \frac{\partial \mathbf{B}}{\partial t}$$

$$= -\frac{\partial}{\partial t}\nabla \times \mathbf{B}$$

However from the third Maxwell equation, we have

$$\frac{\partial}{\partial t}\nabla \times \mathbf{B} = \mu\varepsilon\frac{\partial^2 \mathbf{E}}{\partial t^2}$$

Also the first Maxwell equation is $\nabla.\mathbf{E} = 0$. We thus arrive at the wave equation for the electric field:

$$\nabla^2\mathbf{E} - \mu\varepsilon\frac{\partial^2 \mathbf{E}}{\partial t^2} = 0$$

Likewise, taking the curl of the third Maxwell equation gives

$$\nabla(\nabla.\mathbf{B}) - \nabla^2\mathbf{B} = \mu\varepsilon\frac{\partial}{\partial t}\nabla \times \mathbf{E}$$

Substituting for $\nabla \times \mathbf{E}$ into this equation from the second one and for $\nabla.\mathbf{B} = 0$, we get once again the wave equation for the magnetic field:

$$\nabla^2\mathbf{B} - \mu\varepsilon\frac{\partial^2 \mathbf{B}}{\partial t^2} = 0$$

These equations show that the electric and magnetic fields in a medium having constant permittivity ε and constant permeability μ propagate as waves at the speed of $c = \frac{1}{\sqrt{\varepsilon\mu}}$. The free space value of c comes out to be $\frac{1}{\sqrt{\varepsilon_0\mu_0}} = 3 \times 10^8$m/s. The wave equation in the special case of sinusoidally varying fields reduces to the Helmholtz equation. Defining $k = \omega\sqrt{\varepsilon\mu}$, the equations are

$$\nabla^2\mathbf{E}(\mathbf{r}) + k^2\mathbf{E}(\mathbf{r}) = 0,\ \nabla^2\mathbf{B}(\mathbf{r}) + k^2\mathbf{B}(\mathbf{r}) = 0$$

Solution to the Wave Equation

We look at the wave equation for a particular component of say the electric field. It can be expressed in full notation as

$$\frac{\partial^2 E(t,x,y,z)}{\partial x^2} + \frac{\partial^2 E(t,x,y,z)}{\partial y^2} + \frac{\partial^2 E(t,x,y,z)}{\partial z^2} - \frac{1}{c^2}\frac{\partial^2 E(t,x,y,z)}{\partial t^2} = 0$$

To solve this equation, we expand E in terms of its spatial Fourier transform:

$$E(t,x,y,z) = \int \hat{E}(t,k_x,k_y,k_z)\exp(-i(k_x x + k_y y + k_z z))dk_x dk_y dk_z$$

Substituting this into the wave equation and equating the coefficients of $\exp(-i(k_x x + k_y y + k_z z))$ to zero gives

$$\frac{\partial \hat{E}(t,\mathbf{k})}{\partial t^2} = -c^2k^2\hat{E}(t,\mathbf{k})$$

where $\mathbf{k} = (k_x, k_y, k_z)$. The general solution is

$$E(t,\mathbf{k}) = A(\mathbf{k})\exp(ikct) + B(\mathbf{k})\exp(-ikct),\ k = |\mathbf{k}|$$

This means that the general solution to the wave equation can be expressed as a superposition of the plane waves $\exp(i(kct - \mathbf{k}.\mathbf{r}))$ and $\exp(-i(kct + \mathbf{k}.\mathbf{r}))$. These plane waves are normally expressed in the form

$$\exp(i(\omega t - \mathbf{k}.\mathbf{r})),\ \exp(-i(\omega t + \mathbf{k}.\mathbf{r}))$$

where $\omega = |\mathbf{k}|c$. The relation between the frequency ω and the wave vector $\mathbf{k}$ is called the dispersion relation. As a special case, we can set $\mathbf{k} = \omega\mathbf{n}/c$ where $\mathbf{n}$ is a fixed unit vector and then we get the plane waves $\exp(i\omega(t - \mathbf{n}.\mathbf{r}/c))$ and $\exp(-i\omega(t + \mathbf{n}.\mathbf{r}/c))$. Superposition leads to a solution of the form $f(t - \mathbf{n}.\mathbf{r}/c) + g(t + \mathbf{n}.\mathbf{r}/c)$. The solution $f(t - \mathbf{n}.\mathbf{r}/c)$ corresponds to the plane wave produced by a signal $f(t)$ at $\mathbf{r} = \mathbf{0}$ and propagating in the direction $\mathbf{n}$ while the solution $g(t + \mathbf{n}.\mathbf{r}/c)$ corresponds to the plane wave produced by a signal $g(t)$ and propagating in the direction $-\mathbf{n}$. In particular, if the wave propagates along the positive and negative x directions, the general solution for it is given by $E(t,x) = f(t - x/c) + g(t + x/c)$. It is easily verified that E is the general solution to the one-dimensional wave equation

$$\frac{\partial^2 E(t,x)}{\partial t^2} = c^2 \frac{\partial^2 E(t,x)}{\partial x^2}$$

There is another solution to the three-dimensional wave equation of importance especially in the theory of antennas and radiation. It is called the spherical wave solution. To obtain it, we assume that the solution E depends only on t and the radial distance r from the origin, i.e. $E(t, r)$. The wave equation for such a function is obtained by replacing ∇^2 with just the radial part of the Laplacian. The relevant equation is

$$\frac{1}{r}\frac{\partial^2 rE(t,r)}{\partial r^2} - \frac{1}{c^2}\frac{\partial^2 E(t,r)}{\partial t^2} = 0$$

Defining $F(t, r) = rE(t, r)$, we get the ordinary one-dimensional wave equation for F:

$$\frac{\partial^2 F(t,r)}{\partial r^2} - \frac{1}{c^2}\frac{\partial^2 F(t,r)}{\partial t^2} = 0$$

The solution to this has already been obtained: $F(t,r) = f(t - r/c) + g(t + r/c)$. Thus,

$$E(t,r) = \frac{f(t - r/c)}{r} + \frac{g(t + r/c)}{r}$$

The first term here corresponds to an outgoing spherical wave from the origin and the second term is a spherical wave converging to the origin. Note that unlike the plane wave solution, the spherical wave solutions decay reciprocally with radial distance r. Before closing this section, we should mention that the Maxwell equations for sinusoidally time varying sources and fields can be manipulated directly to yield the Helmholtz equations for the fields. The derivation is routine. Start with the equations

$$\nabla.\mathbf{E} = 0,\ \nabla \times \mathbf{E} = -i\omega\mathbf{B},\ \nabla \times \mathbf{B} = i\omega\mu\varepsilon\,\mathbf{E},\ \nabla.\mathbf{B} = 0$$

Taking curl of the second equation and using the first and the third gives

$$\nabla(\nabla.\mathbf{E}) - \nabla^2\mathbf{E} = -\nabla^2\mathbf{E} = -i\omega\nabla \times \mathbf{B} = \mu\varepsilon\omega^2\mathbf{E}$$

so that

$$\nabla^2 \mathbf{E} + \omega^2 \mu \varepsilon \mathbf{E} = 0, \text{ and likewise } \nabla^2 \mathbf{B} + \omega^2 \mu \varepsilon \mathbf{B} = 0$$

which are the Helmholtz equations. It is important to see the nature of polarization of the waves as they propagate as waves. Take for example an electromagnetic field travelling in the form of plane waves at a definite frequency ω. The electric and magnetic vectors can be written as

$$\mathbf{E}(t, \mathbf{r}) = \text{Re}(\mathbf{E}_0 \exp(i(\omega t - \mathbf{k}.\mathbf{r})))$$
$$\mathbf{H}(t, \mathbf{r}) = \text{Re}(\mathbf{H}_0 \exp(i(\omega t - \mathbf{k}.\mathbf{r})))$$

An easy computation shows that

$$\nabla.\mathbf{E} = \text{Re}(-i\mathbf{k}.\mathbf{E}_0 \exp(i(\omega t - \mathbf{k}.\mathbf{r})))$$
$$\nabla.\mathbf{H} = \text{Re}(-i\mathbf{k}.\mathbf{H}_0 \exp(i(\omega t - \mathbf{k}.\mathbf{r})))$$

Since both of these are zero, we must have

$$\mathbf{k}.\mathbf{E}_0 = 0, \ \mathbf{k}.\mathbf{H}_0 = 0$$

and this means that the electric and magnetic vectors are polarized perpendicular to the direction of propagation. We also find by substituting these expressions for the electric and magnetic fields into the equation

$$\nabla \times \mathbf{E} = -\mu \frac{\partial \mathbf{H}}{\partial t}$$

that

$$\mathbf{k} \times \mathbf{E}_0 = \mu \omega \mathbf{H}_0$$

or since $\omega/k = 1/\sqrt{\varepsilon\mu}$, we have $\frac{\omega\mu}{k} = \sqrt{\varepsilon/\mu} = \frac{1}{\eta}$ and we have with $\mathbf{n}$ denoting the direction of the propagation vector $\mathbf{k}$, the equation

$$\mathbf{H}_0 = \frac{\mathbf{n} \times \mathbf{E}_0}{\eta}$$

This means that the vectors $(\mathbf{n}, \mathbf{E}_0, \mathbf{H}_0)$ from a right handed orthogonal co-ordinate system. This fact is usually depicted by a picture (see Fig. 5.1)

Fig. 5.1

which shows an electromagnetic wave propagating in the $\hat{z}$ direction, the electric field oscillating along the $\hat{x}$ direction and the magnetic field along the $\hat{y}$ direction. The ratio of the electric field to the magnetic field intensity is η, the impedance of the medium through which the wave travels. This situation is true even for nonsinusoidal waves provided that the propagation direction is definite. Consider for example an electromagnetic wave described by the vector amplitudes

$$\mathbf{E}(t, \mathbf{r}) = \mathbf{F}(t - \mathbf{n}.\mathbf{r}/c), \quad \mathbf{H}(t, \mathbf{r}) = \mathbf{G}(t - \mathbf{n}.\mathbf{r}/c)$$

where **F** and **G** have three components each, say $\mathbf{F} = (F_x, F_y, F_z)$ and $\mathbf{G} = (G_x, G_y, G_z)$. This means that

$$E_x(t, \mathbf{r}) = F_x(t - \mathbf{n}.\mathbf{r}/c), \ E_y(t, \mathbf{r}) = F_y(t - \mathbf{n}.\mathbf{r}/c), \ E_z(t, \mathbf{r}) = F_z(t - \mathbf{n}.\mathbf{r}/c)$$

with similar expressions for **H**. We find that

$$\nabla.\mathbf{E} = -\frac{1}{c}\mathbf{n}.\mathbf{F}'(t - \mathbf{n}.\mathbf{r}/c) = 0$$

implying that the vector **n** is orthogonal to the vector $\mathbf{F}' = (F_x', F_y', F_z')$ at each value of its argument. This yields $\mathbf{n}.\mathbf{F}'(\xi) = 0$ for all values of the argument ξ. Integration of this equation gives $\mathbf{n}.\mathbf{F} =$ const. and the value of this constant may be taken as zero assuming that at distant parts, there is no electric field. This proves that **n** is orthogonal to **E**. The equation $\nabla \times \mathbf{E} = -\mu\frac{\partial \mathbf{H}}{\partial t}$ implies that

$$-\frac{1}{c}\mathbf{n} \times \mathbf{F}'(t - \mathbf{n}.\mathbf{r}/c) = -\mu\mathbf{G}'(t - \mathbf{n}.\mathbf{r}/c)$$

which on integration, gives

$$\mathbf{n} \times \mathbf{F} = \mu c\mathbf{G} = \sqrt{\frac{\mu}{\varepsilon}}\,\mathbf{G} = \eta\mathbf{G}$$

The constant of integration is zero assuming that at distant parts, no electric or magnetic field is present. Thus,

$$\mathbf{H} = \frac{1}{\eta}\mathbf{n} \times \mathbf{F}$$

proving that (**n**,**F**,**G**) form a right-handed orthogonal co-ordinate system.

Equations satisfied by the vector and scalar potentials

The fourth Maxwell equation implies that **B** must be the divergence of a vector field, i.e., $\mathbf{B} = \nabla \times \mathbf{A}$. **A** is called the magnetic vector potential. Substituting this into the second Maxwell equation gives

$$\nabla\times\mathbf{E} = -\frac{\partial}{\partial t}\nabla\times\mathbf{A}$$

from which we conclude that

$$\nabla\times\left(\mathbf{E}+\frac{\partial\mathbf{A}}{\partial t}\right)=0$$

and hence the quantity within brackets must be the gradient of a scalar field, say $-\nabla V$. Thus,

$$\mathbf{E} = -\nabla V - \frac{\partial\mathbf{A}}{\partial t}$$

We may identify V with the scalar potential. In case the magnetic field is zero we have $\mathbf{A} = \mathbf{0}$ and $\mathbf{E} = -\nabla V$, which agrees with the definition of the electric field in electrostatics. Next, substituting the expression for the magnetic field and the electric field in terms of the vector and scalar potentials $\mathbf{A}$, V into the third Maxwell equation gives

$$\nabla\times\nabla\times\mathbf{A} = \mu\mathbf{J} + \varepsilon\mu\frac{\partial}{\partial t}\left(-\nabla V - \frac{\partial\mathbf{A}}{\partial t}\right)$$

or

$$\nabla(\nabla.\mathbf{A}) - \nabla^2\mathbf{A} = \mu\mathbf{J} - \varepsilon\mu\nabla\frac{\partial V}{\partial t} - \varepsilon\mu\frac{\partial^2\mathbf{A}}{\partial t^2}$$

which can be rearranged to give

$$\nabla^2\mathbf{A} - \varepsilon\mu\frac{\partial^2\mathbf{A}}{\partial t^2} = -\mu\mathbf{J} + \nabla\left(\nabla.\mathbf{A} + \varepsilon\mu\frac{\partial V}{\partial t}\right)$$

Now we may note that if $\mathbf{A}$ and V are respectively changed to $\mathbf{A}' = \mathbf{A} + \nabla f$ and $V' = V - \frac{\partial f}{\partial t}$, where f is an arbitrary scalar field, then $\mathbf{E}$ and $\mathbf{B}$ do not undergo any change. This follows from the identities

$$\nabla\times A' = \nabla\times(A+\nabla f) = \nabla\times\mathbf{A}$$

$$-\nabla V' - \frac{\partial\mathbf{A}'}{\partial t} = -\nabla\left(V - \frac{\partial f}{\partial t}\right) - \frac{\partial}{\partial t}(\mathbf{A}+\nabla f)$$

$$= -\nabla V - \frac{\partial\mathbf{A}}{\partial t}$$

Thus $(\mathbf{A}', V')$ may be used in place of $(\mathbf{A}, V)$. Now, the scalar function f

can be chosen so that $\nabla.\mathbf{A}' + \mu\varepsilon\frac{\partial V'}{\partial t} = 0$. This amounts to choosing f so that

$$\nabla.(\mathbf{A} + \nabla f) + \varepsilon\mu\frac{\partial}{\partial t}(V - \frac{\partial f}{\partial t}) = 0$$

or equivalently,

$$\nabla^f - \varepsilon\mu\frac{\partial^2 f}{\partial t^2} = -\left(\nabla.\mathbf{A} + \varepsilon\mu\frac{\partial V}{\partial t}\right)$$

Thus, without loss of generality, we may assume that the pair **A**, V has been obtained after performing a Gauge transformation and hence this pair satisfies the Lorentz Gauge condition $\nabla.\mathbf{A} + \varepsilon\mu\frac{\partial V}{\partial t} = 0$. It then follows that **A** satisfies the wave equation with source:

$$\nabla^2\mathbf{A} - \varepsilon\mu\frac{\partial^2\mathbf{A}}{\partial t^2} = -\mu\mathbf{J}$$

Similarly, we can derive a wave equation with source for the electric scalar potential. To do this, we substitute for the electric field the expression $\mathbf{E} = -\nabla V - \frac{\partial A}{\partial t}$ into the first Maxwell equation to get

$$\nabla.(\nabla V + \frac{\partial\mathbf{A}}{\partial t}) = -\rho/\varepsilon$$

This can be expressed as

$$\nabla^2 V - \varepsilon\mu\frac{\partial^2 V}{\partial t^2} + \frac{\partial}{\partial t}\left(\nabla.\mathbf{A} + \varepsilon\mu\frac{\partial V}{\partial t}\right) = -\rho/\varepsilon$$

Using the Lorentz Gauge condition, this reduces to the wave equation with source for the potential:

$$\nabla^2 V - \varepsilon\mu\frac{\partial^2 V}{\partial t^2} = -\rho/\varepsilon$$

Complete Solution to the Maxwell Equations

It is clear from the discussion of the preceding section that to solve Maxwell's equations, we need only solve the wave equation with source i.e.

$$\nabla^2\psi(t,\mathbf{r})-\frac{1}{c^2}\frac{\partial^2\psi(t,\mathbf{r})}{\partial t^2}=s(t,\mathbf{r})$$

To solve this, we Fourier transform with respect to time. Letting

$$\hat{\psi}(\omega,\mathbf{r})=\int_{-\infty}^{\infty}\psi(t,\mathbf{r})\exp(-i\omega t)dt$$

$$\hat{s}(\omega,\mathbf{r})=\int_{-\infty}^{\infty}s(t,\mathbf{r})\exp(-i\omega t)dt$$

we get on integration by parts,

$$\int_{-\infty}^{\infty}\frac{\partial^2\psi}{\partial t^2}\exp(-i\omega t)dt$$

$$=\int_{-\infty}^{\infty}\psi\frac{\partial^2}{\partial t^2}\exp(-i\omega t)dt=-\omega^2\int_{-\infty}^{\infty}\psi\exp(-i\omega t)dt$$

$$=-\omega^2\hat{\psi}$$

Thus, the wave equation transforms to

$$\nabla^2\hat{\psi}(\omega,\mathbf{r})+k^2\hat{\psi}(\omega,\mathbf{r})=\hat{s}(\omega,\mathbf{r})$$

This is the Helmholtz equation with a source term. The solution to this equation is accomplished by noting that

$$(\nabla^2+k^2)\exp(-ikr)/r=-4\pi\delta(\mathbf{r})$$

To see this, we use Green's theorem in the form

$$\int_V(f(r)\nabla^2(\exp(-ikr)/r)-(\exp(-ikr)/r)\nabla^2 f)dV$$

$$=(f(R)\frac{d}{dR}(\exp(-ikR)/R)-(\exp(-ikR)/R)f'(R))4\pi R^2$$

The volume to which Green's theorem has been applied is a sphere of radius R with $R\to\infty$. The right side tends to zero as $R\to\infty$ assuming that f decays sufficiently rapidly. We next note that

$$\int_V(\exp(-ikr)/r)\nabla^2 f(r)dV=\int_0^{\infty}(\exp(-ikr)/r^2)\frac{d^2rf(r)}{dr^2}4\pi r^2dr$$

$$=4\pi\int_0^{\infty}\exp(-ikr)\frac{d^2rf}{dr^2}dr=4\pi\int\exp(-ikr)(rf''(r)+2f'(r))dr$$

Integration by parts gives

$$\int_0^{\infty} \exp(-ikr)rf''(r)dr = \exp(-ikr)rf'(r)|_0^{\infty} - \int_0^{\infty} (\exp(-ikr)r)'f'(r)dr$$

$$= -(\exp(-ikr)r)'f(r)|_0^{\infty} + \int_0^{\infty} (\exp(-ikr)r)''f(r)dr$$

$$= f(0) - \int_0^{\infty} k^2 r\exp(-ikr)f(r)dr - 2ik \int_0^{\infty} \exp(-ikr)f(r)dr$$

Again, integration by parts gives

$$\int_0^{\infty} \exp(-ikr)f'(r)dr = \exp(-ikr)f(r)|_0^{\infty} + ik \int_0^{\infty} \exp(-ikr)f(r)dr$$

$$= -f(0) + ik \int_0^{\infty} \exp(-ikr)f(r)dr$$

Combining the above two identities derived using integration by parts, we get

$$\int_V (\exp(-ikr)/r)\nabla^2 f(r)dV = 4\pi \int_0^{\infty} \exp(-ikr)(rf''(r) + 2f'(r))dr$$

$$= -4\pi f(0) - 4\pi k^2 \int_0^{\infty} r\exp(-ikr)f(r)dr = -4\pi f(0) - k^2 \int_V f(r)(\exp(-ikr)/r)dV$$

Thus Green's identity results in the equation

$$\int_V f(\nabla^2(\exp(-ikr)/r) + k^2\exp(-ikr)/r)dV = -4\pi f(0)$$

and hence

$$(\nabla^2 + k^2)\,(\exp(-ikr)/r) = -4\pi\delta(\mathbf{r})$$

which is the desired identity. Consider now the equation

$$\nabla^2\hat{\psi}(\omega,\mathbf{r}) + k^2\hat{\psi}(\omega,\mathbf{r}) = \hat{s}(\omega,\mathbf{r})$$

We obtain by multiplying the equation

$$(\nabla^2 + k^2)\exp(-ik|\mathbf{r} - \mathbf{r}'|)/|\mathbf{r} - \mathbf{r}'| = -4\pi\delta(\mathbf{r} - \mathbf{r}')$$

by $\hat{s}(\omega,\mathbf{r}')$ and integrating overall of $\mathbf{r}'$, the identity

$$(\nabla^2 + k^2) \int \frac{\exp(-ik|\mathbf{r} - \mathbf{r}'|)}{|\mathbf{r} - \mathbf{r}'|}\hat{s}(\omega,\mathbf{r}')dV' = -4\pi\hat{s}(\omega,\mathbf{r})$$

This implies that the solution to the Hehnhdty equation with source is given by

$$\psi(\omega,\mathbf{r}) = -\frac{1}{4\pi}\int \hat{s}(\omega,\mathbf{r}')\frac{\exp(-ik\,|\,\mathbf{r}-\mathbf{r}'\,|)}{|\mathbf{r}-\mathbf{r}'|}dV'$$

the integral being over entire space. The solution to the wave equation with source

$$\nabla^2\psi(t,\mathbf{r}) - \frac{1}{c^2}\frac{\partial^2\psi(t,\mathbf{r})}{\partial t^2} = s(t,\mathbf{r})$$

is given by the inverse Fourier transform of $\hat{\psi}\,(\omega,\mathbf{r})$. Noting that $k = \omega/c$, we get

$$\psi(t,\mathbf{r}) = -\frac{1}{4\pi}\int \frac{dV'}{|\mathbf{r}-\mathbf{r}'|}\int s(\omega,\mathbf{r}')\exp(i\omega(t-|\mathbf{r}-\mathbf{r}'|/c))d\omega/2\pi$$

$$= -\frac{1}{4\pi}\int \frac{s(t-|\mathbf{r}-\mathbf{r}'|/c,\mathbf{r}')}{|\mathbf{r}-\mathbf{r}'|}dV'$$

This is called the retarded potential solution to the wave equation. From the wave equations with source satisfied by the vector and scalar potentials, we conclude that

$$V(t,\mathbf{r}) = \frac{1}{4\pi\varepsilon}\int \frac{\rho(t-|\mathbf{r}-\mathbf{r}'|/c,\mathbf{r}')}{|\mathbf{r}-\mathbf{r}'|}dV'$$

$$\mathbf{A}(t,\mathbf{r}) = \frac{\mu}{4\pi}\int \frac{\mathbf{J}(t-|\mathbf{r}-\mathbf{r}'|/c,\mathbf{r}')}{|\mathbf{r}-\mathbf{r}'|}dV'$$

where $c = 1/\sqrt{\varepsilon\mu}$ is the velocity of propagation of the waves in the medium. The meaning of the retarded potential solution is clear. To evaluate the contribution to the potential at time t and at the location $\mathbf{r}$ due to the charge density at $\mathbf{r}'$, we need to consider the charge density at the retarded time $t' = t - |\mathbf{r} - \mathbf{r}'|/c$ and then evaluate the potential as though it were a Coulomb potential. This is due to the fact that the electromagnetic wave from any point travels at the speed of light in that medium. It should be noted that formally, the wave equation with source also admits an advanced solution owing to the fact that

$$(\nabla^2 + k^2)\exp(+ikr)/r = -4\pi\delta(\mathbf{r})$$

The advanced solution is given by

$$\psi(t,r) = -\frac{1}{4\pi}\int \frac{s(t+|\mathbf{r}-\mathbf{r}'|/c,\mathbf{r}')}{|\mathbf{r}-\mathbf{r}'|}dV'$$

However, although this solution is mathematically correct, it cannot be used in the solution of the electromagnetic field because it is a noncausal solution. The effect of the source at $\mathbf{r}'$ is taken at a future time $t + |\mathbf{r} - \mathbf{r}'|/c$ to evaluate its effect at time t at the location $\mathbf{r}$. This means that the future values of the source affect the potential, a solution that is not physically reasonable. The general solution to the wave equation with source is a times the retarded potential solution plus b times the advanced potential solution where a, b are constants adding to unity. However from physical considerations, we must take $a = 1$, $b = 0$.

Maxwell's equations in a conducting medium

In a conducting medium, Ohm's law states that the current density is proportional to the electric field. This relation is expressed as $\mathbf{J} = \sigma\mathbf{E}$. σ is called the conductivity of the medium. Take a conducting bar of length L and cross- sectional area A and apply a potential difference of V volts across the length of the bar. The electric field in the bar is then $E = V/L$ and this is directed longitudinally along the bar. Ohm's law just described then yields $J = \sigma E = \sigma V/L$ and hence the current through the bar is given by $I = JA = \sigma AV/L$. This gives $V/I = L/\sigma A = R$. It is called the macroscopic form of Ohm's law and states that the voltage across the bar is proportional to the current through it. The proportionality constant is called the resistance R of the medium. $\rho = 1/\sigma$ is called the resistivity of the medium and we can write $R = \rho L/A$. The resistance of the bar is thus inversely proportional to the cross-sectional area of the bar and directly proportional to the length of the bar. Longer the bar, more the resistance, more the cross-sectional area, lesser the resistance. Now, assuming the medium is conducting with a permittivity of ε, a magnetic permeability of μ and a conductivity of σ, we have in the absence of external charges, the Maxwell equations

$$\nabla.\mathbf{E} = 0, \nabla \times \mathbf{E} = -\mu \frac{\partial \mathbf{H}}{\partial t}$$

$$\nabla \times \mathbf{H} = \sigma\mathbf{E} + \varepsilon \frac{\partial \mathbf{E}}{\partial t}, \nabla.\mathbf{H} = 0$$

Taking the curl of the second equation and using the first and the third equations gives

$$\nabla^2 \mathbf{E} - \sigma\mu \frac{\partial \mathbf{E}}{\partial t} - \varepsilon\mu \frac{\partial^2 \mathbf{E}}{\partial t^2} = 0$$

Likewise, taking the curl of the third Maxwell equation here and using the fourth and the second one gives

$$\nabla^2\mathbf{H} - \sigma\mu\frac{\partial \mathbf{H}}{\partial t} - \varepsilon\mu\frac{\partial^2 \mathbf{H}}{\partial t^2} = \mathbf{0}$$

These two equations show that the electric and magnetic fields inside a conducting medium in the absence of external charges propagate as damped waves. Assume that we are dealing with waves of a definite frequency ω and propagating along the z axis. Then, the electromagnetic phasor can be expressed as

$$\mathbf{E}(t, \mathbf{r}) = \mathbf{E}_0\exp(i\omega t - \gamma z), \ \mathbf{H}(t,\mathbf{r}) = \mathbf{H}_0\exp(i\omega t - \gamma z)$$

Substituting these into the damped wave equation gives

$$\gamma^2 - i\omega\sigma\mu + \varepsilon\mu\omega^2 = 0$$

or

$$\gamma = \sqrt{i\omega\mu(\sigma + i\omega\varepsilon)} = \alpha + i\beta$$

say. For example, this solution would be valid if we consider a wave impinging upon an infinite conducting slab with one face as the xy plane and extending in the positive negative z direction. The wave transmitted into the conducting medium then has γ as the propagation constant. For good conductors at that frequency, we have $\sigma >> \omega\varepsilon$ and hence

$$\gamma \approx \sqrt{i\omega\mu\sigma} = \sqrt{\omega\mu\sigma/2}(1+i)$$

Thus, for good conductors, $\alpha = \beta = \sqrt{\omega\mu\sigma/2} = 1/\delta$ and it follows that the phasor of the wave amplitude varies with z the distance of propagation as

$$\exp(-\gamma z) = \exp(-\alpha z)\exp(-i\beta z) = \exp(-z/\delta)\exp(-iz/\delta)$$

This means that the wave amplitude decays to a value of e^{-1} times its maximum amplitude after propagating a distance $\delta = \sqrt{2/\omega\mu\sigma}$ through the conductor. This distance is called the skin depth. A beam of electromagnetic radiation incident upon a conductor, penetrates a distance equal to the skin depth. We can also evaluate the power dissipated as heat in a square metre of the conductor. Assume that the beam of radiation of frequency ω is incident upon a square metre of the conductor which has infinite depth extending from $z = 0$ to $z = \infty$. The complex electric field amplitude at the surface is $\mathbf{E}_0$. Of course, $\mathbf{E}_0$ can have only x and y

components since the propagation direction is the z direction. The current density phasor inside the conductor is given by $\mathbf{J}(z) = \sigma\mathbf{E}(z) = \sigma\mathbf{E}_0\exp(-\gamma z)$. The time averaged power dissipated per unit square metre in a thickness of dz at a distance z is thus $\frac{1}{2}\,\mathrm{Re}(\mathbf{J}(z)\mathbf{E}(z)^*)dz = \frac{\sigma}{2}|\mathrm{E}_0|^2\exp(-2\alpha z)\,dz$. The total average power dissipated in a unit square metre of the conductor is then the integral of this from $z = 0$ to $z = \infty$. Evaluating the integral gives for the power dissipated per unit area

$$P_s = \sigma|\mathbf{E}_0|^2/4\alpha$$

For a good conductor, $\alpha = 1/\delta = \sqrt{\omega\mu\sigma/2}$ and this shows that P_s is proportional to $\sigma^{1/2}$ and also to $\omega^{-1/2}$. At high frequencies, lesser power is dissipated and higher the conductivity, more the power dissipated for a given electric field amplitude at the surface. It is also useful here to derive the relations between the electric and magnetic field phasors for a plane wave propagating in a conducting medium. Maxwell's equations for a definite frequency reduce to

$$\nabla.\mathbf{E} = 0,\ \nabla \times \mathbf{E} = -i\omega\mu\mathbf{H},\ \nabla \times \mathbf{H} = (\sigma + i\omega\varepsilon)\mathbf{E},\ \nabla.\mathbf{H} = 0$$

Letting

$$\mathbf{E} = \mathbf{E}_0\exp(-\gamma\,\mathbf{n}.\mathbf{r}),\ \mathbf{H} = \mathbf{H}_0\exp(-\gamma\,\mathbf{n}.\mathbf{r})$$

(Note that these are phasors, which means that the actual electric and magnetic fields are obtained by multiplying these amplitudes with $\exp(i\omega t)$ and taking the real part. The first Maxwell equation gives $\mathbf{n}.\mathbf{E} = 0$, so that the $\mathbf{E}$ vector is orthogonal to the direction of propagation. The last Maxwell equation likewise gives $\mathbf{n}.\mathbf{H} = 0$ implying that the $\mathbf{H}$ vector is also orthogonal to the direction of propagation. The second one gives

$$-\gamma\,\mathbf{n} \times \mathbf{E} = -i\omega\mu\mathbf{H}$$

or

$$\mathbf{H} = \frac{\gamma}{i\omega\mu}\mathbf{n} \times \mathbf{E}$$

Since $\gamma = \sqrt{i\omega\mu(\sigma + i\omega\varepsilon)}$, it follows that

$$\frac{\gamma}{i\omega\mu} = \sqrt{\frac{\sigma + i\omega\varepsilon}{i\omega\mu}}$$

We define the complex impedance of the medium as

$$\eta = \sqrt{\frac{i\omega\mu}{\sigma + i\omega\varepsilon}}$$

Then the relation between the electric and magnetic fields becomes

$$\mathbf{H} = \frac{1}{\eta}\mathbf{n} \times \mathbf{E}$$

Note that in all these relations, **E** and **H** may be replaced respectively by $\mathbf{E}_0$ and $\mathbf{H}_0$. When $\sigma = 0$, η reduces to the value $\sqrt{\mu/\varepsilon}$ for a dielectric. The other relation between **E**, **H** may be derived from the third Maxwell equation. It is

$$-\gamma \mathbf{n} \times \mathbf{H} = (\sigma + i\omega\varepsilon)\, \mathbf{E}$$

so that

$$\mathbf{E} = -\frac{\gamma}{\sigma + i\omega\varepsilon}\mathbf{n} \times \mathbf{H}$$

Taking the cross product on both sides of this equation with **n** and using $\mathbf{n}.\mathbf{H} = 0$ gives

$$\mathbf{n} \times \mathbf{E} = \frac{\gamma}{\sigma + i\omega\varepsilon}\mathbf{H}$$

This formula along with the equation $\mathbf{n} \times \mathbf{E} = \frac{i\omega\mu}{\gamma}\mathbf{H}$ obtained from the second Maxwell equation may be used to provide another derivation for the expression of the propagation constant. Indeed these two imply

$$\frac{\gamma}{\sigma + i\omega\varepsilon} = \frac{i\mu\omega}{\gamma}$$

so that

$$\gamma^2 = i\mu\omega\, (\sigma + i\omega\varepsilon)$$

which is the desired expression.

Boundary conditions for Maxwell's equations

We want to discuss the most general form for the boundary conditions satisfied by the electric and magnetic fields at the interface of two media. First consider two media separated by an interface. The first medium has parameters (ε_1, μ_1, σ_1) and the second (ε_2, μ_2, σ_2) respectively for the permittivity, magnetic permeability and conductivity. We are assuming the conductivities to be finite for both the media. Gauss' law $\nabla.\mathbf{D} = \rho$, where ρ is the external charge density, implies on applying to a small pill box with cross-sectional faces on both the sides of the interface, the boundary condition.

$$\mathbf{D}_2.\mathbf{n} - \mathbf{D}_1.\mathbf{n} = \sigma$$

where σ is the surface charge density of external charges. In particular, when there is no surface charge density, we have continuity of the normal component of **D**. The boundary condition may equivalently be expressed as

$$\varepsilon_2\mathbf{E}_2.\mathbf{n} - \varepsilon_1\mathbf{E}_1.\mathbf{n} = \sigma$$

Next, look at Faraday's law $\nabla \times \mathbf{E} = -\frac{\partial \mathbf{B}}{\partial t}$. Applying this to a loop of length L and inifinitesimal breadth with the length parallel to the interface gives

$$\int_\Gamma \mathbf{E}_1.d\mathbf{r} - \int_\Gamma \mathbf{E}_2.d\mathbf{r} = -\frac{\partial \Phi_B}{\partial t}$$

where Φ_B, the flux of the magnetic field through the area of the loop is zero since the loop has infinitesimal area. Here, Γ is the line along the length of the loop. Thus,

$$\int_\Gamma \mathbf{E}_1.d\mathbf{r} - \int_\Gamma \mathbf{E}_2.d\mathbf{r} = 0$$

Letting now the length Γ of the loop shrink to zero gives continuity of the tangential compnent of the electric field. That is

$$\mathbf{E}_{t,2} = \mathbf{E}_{t,1}$$

This may be expressed equivalently as

$$\mathbf{n} \times (\mathbf{E}_2 - \mathbf{E}_1) = \mathbf{0}$$

The third Maxwell equation in integral form is

$$\int_\Gamma \mathbf{H}.d\mathbf{r} = I + \frac{\partial}{\partial t}\int_S \mathbf{D}.\mathbf{n}dS$$

where Γ is a closed contour enclosing the surface S. Take the boundary separating the two media and a rectangle of length L with lengths on the two sides of the boundary and breadth negligible. The flux **D** through the surface is zero since the area enclosed by the loop is negligible. The current through the loop is KL where K is the surface current density, i.e. current per unit length. For definiteness, we assume the sides of the rectangle to be parallel to the x axis and the boundary to be the xy plane, so that the two media are respectively on the sides $z > 0$ and $z < 0$. The current is then along the y direction and we get on letting $L \to 0$ the condition

$$H_{x,2} - H_{x,1} = K_y$$

where the medium 2 is $z > 0$ and the medium 1 is $z < 0$. Generalizing this, we could write

$$\mathbf{H}_{t,2} - \mathbf{H}_{t,1} = \mathbf{K} \times \mathbf{n}$$

where $\mathbf{H}_{t,i}$ is the tangential component of the magnetic field on the medium i, $i = 1, 2$, or equivalently,

$$\hat{\mathbf{n}} \times (\vec{\mathbf{H}}_2 - \vec{\mathbf{H}}_1) = \vec{\mathbf{K}}$$

where **n** is the unit normal to the interface directed from medium 1 to medium 2. In the special case when one side, say medium 1 is a perfect conductor, $\sigma_1 = \infty$, this implies that there cannot be any electric and magnetic fields in this medium. The boundary conditions then reduce to

$$\mathbf{D}_{n,2} = \sigma,\ \mathbf{E}_{t,2} = 0,\ \mathbf{H}_{t,2} = \mathbf{K} \times \mathbf{n}$$

or equivalently,

$$\mathbf{D}_2.\mathbf{n} = \sigma, \mathbf{n} \times \mathbf{E}_2 = 0,\ \mathbf{n} \times \mathbf{H}_2 = \mathbf{K}$$

When there is no surface charge on the boundary, it follows that the normal component of **D** at the boundary is zero. If there is no surface current, the tangential component of **H** at the boundary is zero. Whether or not there is any surface charge density or surface current, the tangential component of **E** at the boundary of a perfect conductor is always zero.

Relativistic Transformation of the fields:

The wave equation is invariant under Lorentz transformations. This can be seen via the following arguments. Consider a scalar field $\phi(t, x)$ that satisfies the one-dimensional wave equation

$$\frac{\partial^2 \phi}{\partial x^2} - \frac{1}{c^2}\frac{\partial^2 \phi}{\partial t^2} = 0$$

We perform a Lorentz transformation that changes the time and spatial co-ordinates to

$$t' = \gamma\,(t - vx/c^2),\ x' = \gamma\,(x - vt)$$

where $\gamma = (1 - v^2/c^2)^{-1/2}$. This transformation satisfies the condition of linearity in the space-time co-ordinates and it preserves the speed of light, i.e.

$$x'^2 - c^2t'^2 = x^2 - c^2t^2$$

which means that a ray propagating at the speed of light in the unprimed system will also propagate at the speed of light in the primed system. We compute the transformation in the partial derivatives under this Lorentz transformation:

$$\frac{\partial}{\partial t} = \frac{\partial t'}{\partial t}\frac{\partial}{\partial t'} + \frac{\partial x'}{\partial t}\frac{\partial}{\partial x'}$$

$$= \gamma \frac{\partial}{\partial t'} - \gamma v \frac{\partial}{\partial x'}$$

$$\frac{\partial}{\partial x} = \frac{\partial t'}{\partial x}\frac{\partial}{\partial t'} + \frac{\partial x'}{\partial x}\frac{\partial}{\partial x'}$$

$$= -(\gamma v / c^2)\frac{\partial}{\partial t'} + \gamma \frac{\partial}{\partial x'}$$

Applying the operators twice, we get

$$\frac{\partial^2}{\partial t^2} = (\gamma \frac{\partial}{\partial t'} - \gamma v \frac{\partial}{\partial x'})^2$$

$$= \gamma^2 \frac{\partial^2}{\partial t^2} + \gamma^2 v^2 \frac{\partial^2}{\partial x'^2} - 2\gamma^2 v \frac{\partial^2}{\partial t' \partial x'}$$

and likewise

$$\frac{\partial^2}{\partial x^2} = (-(\gamma v / c^2)\frac{\partial}{\partial t'} + \gamma \frac{\partial}{\partial x'})^2$$

$$= (\gamma^2 v^2 / c^4)\frac{\partial^2}{\partial t'^2} + \gamma^2 \frac{\partial^2}{\partial x'^2} - 2(\gamma^2 v / c^2)\frac{\partial^2}{\partial t' \partial x'}$$

From these two equations, we get

$$\frac{\partial^2}{\partial t^2} - c^2 \frac{\partial^2}{\partial x^2} =$$

$$(\gamma^2 - \gamma^2 v^2 / c^2)\frac{\partial^2}{\partial t'^2} + (\gamma^2 v^2 - \gamma^2 c^2)\frac{\partial^2}{\partial x'^2}$$

$$= \frac{\partial^2}{\partial t'^2} - c^2 \frac{\partial^2}{\partial x'^2}$$

This analysis shows that the wave operator remains invariant under Lorentz transformations. Since the electromagnetic fields in free space satisfy the wave equation and in the presence of charges and currents, the vector and scalar potentials satisfy the wave equation with source terms, it is natural to expect a close link to exist between the Maxwell theory of electromagnetism and Einstein's special theory of relativity. In fact, Einstein first discovered the Lorentz transformations by analyzing the Maxwell theory and his first paper containing the germ of the special theory of relativity was termed as "On the electrodynamics of moving bodies".

More generally, we can consider the class of all linear transformations of $\Re^4$ consisting of the space-time points (t, x, y, z) which preserve the Lorentz metric $t^2 - x^2 - y^2 - z^2$. An easy way to represent the Lorentz metric is to introduce the metric tensor $\mathbf{G}$ which is a 4×4 matrix

$$\mathbf{G} = \begin{pmatrix} 1 & 0 & 0 & 0 \\ 0 & -1 & 0 & 0 \\ 0 & 0 & -1 & 0 \\ 0 & 0 & 0 & -1 \end{pmatrix}$$

The metric is then defined for

$$\mathbf{x} = [t, x, y, z]^T$$

by

$$t^2 - x^2 - y^2 - z^2 = \mathbf{x}^T\,\mathbf{G}\mathbf{x}$$

A Lorentz transformation $\mathbf{L}$ is a 4×4 real matrix that preserves this metric, i.e. if $\mathbf{x}' = \mathbf{L}\mathbf{x}$, then $\mathbf{x}'^T\,\mathbf{G}\mathbf{x}' = \mathbf{x}^T\,\mathbf{G}\mathbf{x}$ for all $\mathbf{x} \in \Re^4$. This is the same as

$$\mathbf{x}^T\mathbf{L}^T\mathbf{G}\mathbf{L}\mathbf{x} = \mathbf{x}^T\mathbf{G}\mathbf{x}, \ \forall \mathbf{x} \in \Re^4$$

or the same as

$$\mathbf{L}^T\mathbf{G}\mathbf{L} = \mathbf{G}$$

We can divide the Lorentz transformations into four classes. These classes are denoted by $L_+^{\uparrow}, L_+^{\downarrow}, L_-^{\uparrow}, L_-^{\downarrow}$. The transformations $\mathbf{L}$ in the first class preserve the direction of time, i.e. $t > 0$ implies $t' > 0$ and have positive determinant. These are the proper Lorentz transformations. The transformations $\mathbf{L}$ in the second class reverse the direction of time, i.e. $t > 0$ implies $t' < 0$ and have positive determinant. The transformations in the third class preserve the direction of time and have negative determinant and finally the transformations in the fourth class reverse the direction of time and have negative determinant. A Lorentz transformation that preserves the direction of time is also called orthochronous. For our purpose we shall require only proper Lorentz transformations. They consist of combinations of Lorentz boosts and rotations of space. A Lorentz boost is a transformation corresponding to primed frame moving with a uniform velocity relative to the unprimed frame along a given direction. Suppose for example, we want to compute the transformation for a boost with speed v along the **n** direction. Then, we have the obvious relations

$$\mathbf{r}'_{\parallel} = \gamma(\mathbf{r}_{\parallel} - \mathbf{v}t)$$

$$\mathbf{r}'_{\perp} = \mathbf{r}_{\perp}$$

where $\mathbf{r}_{\parallel} = (\mathbf{r}.\mathbf{n})\mathbf{n}$ is the component of **r** parallel to **n** and similarly for $\mathbf{r}'_{\parallel}$. $\mathbf{r}_{\perp} = \mathbf{r} - (\mathbf{r}.\mathbf{n})\mathbf{n}$ is the component of **r** perpendicular to **n**. We can thus express the overall transformation from the unprimed frame to the primed frame moving with a uniform velocity $\mathbf{v} = v\mathbf{n}$ relative to the unprimed frame as

$$\mathbf{r}' = \mathbf{r}'_{\parallel} + \mathbf{r}'_{\perp} = \gamma((\mathbf{r}.\mathbf{n}) - vt)\mathbf{n} + \mathbf{r} - (\mathbf{r}.\mathbf{n})\mathbf{n}$$
$$= ((\gamma - 1)(\mathbf{r}.\mathbf{n}) - \gamma vt)\mathbf{n} + \mathbf{r}$$

The transformation law for time is

$$t' = \gamma(t - v\mathbf{n}.\mathbf{r})$$

where $\gamma = (1 - v^2/c^2)^{-1/2}$. We leave it as an exercise to express these four equations in matrix form, i.e. write down explicitly the formula for the Lorentz transformation matrix **L**.

We can in general show that the three-dimensional wave operator and hence equation is preserved by any Lorentz transformation. Let $\mathbf{r}' = \mathbf{L}\mathbf{r}$ be a Lorentz transformation. In terms of components, we express this as

$$x'_i = \sum_{j=0}^{3} L_{ij}x_j, i = 0, 1, 2, 3$$

where for simplicity of notation, $x_0 = t$, $x_1 = x$, $x_2 = y$, $x_3 = z$. The wave operator in the unprimed system is

$$\frac{\partial^2}{\partial x_0^2} - \frac{\partial^2}{\partial x_1^2} - \frac{\partial^2}{\partial x_2^2} - \frac{\partial^2}{\partial x_3^2}$$
$$= \sum_{i,j=0}^{3} G_{ij} \frac{\partial^2}{\partial x_i \partial x_j}$$

with

$$G_{00} = 1,\ G_{0r} = G_{r0} = 0,\ r = 1, 2, 3,\ G_{rs} = -\delta_{rs},\ r, s = 1, 2, 3$$

Likewise, the wave operator in the primed system is

$$\sum_{i,j=0}^{3} G_{ij} \frac{\partial^2}{\partial x'_i \partial x'_j}$$

Now according to the chain rule for partial derivatives,

$$\frac{\partial}{\partial x_i} = \sum_{j=0}^{3} \frac{\partial x'_j}{\partial x_i} \frac{\partial}{\partial x'_j}$$

$$= \sum_{j=0}^{3} L_{ji} \frac{\partial}{\partial x'_j}$$

This gives on double application,

$$\sum_{i,k=0}^{3} G_{ik} \frac{\partial^2}{\partial x_i \partial x_k} = \sum_{i,j,k,m} G_{ik} L_{ji} L_{mk} \frac{\partial^2}{\partial x'_j \partial x'_m}$$

and noting that the equation $\mathbf{L}^T\mathbf{G}\mathbf{L} = \mathbf{G}$ in component from implies that

$$\sum L_{mk} G_{ik} L_{ji} = G_{mj}$$

we get

$$\sum_{i,k=0}^{3} G_{ik} \frac{\partial^2}{\partial x_i \partial x_k} = \sum G_{mj} \frac{\partial^2}{\partial x'_j \partial x'_m}$$

proving the equality of the wave operators in both the frames. The basic postulate that connects electromagnetic theory with the special theory of relativity is that the charge density and current density vector (ρ, J_x, J_y, J_z) forms a four vector and so does the scalar electric potential and magnetic vector potential (V/c^2, A_x, A_y, A_z). By a four vector, we mean a four tuple that under the Lorentz transformation, transforms like (t, x, y, z). An example of a four vector is the four velocity defined by

$$\left(\frac{dt}{d\tau}, \frac{dx}{d\tau}, \frac{dy}{d\tau}, \frac{dz}{d\tau}\right)$$

where

$$d\tau = \sqrt{dt^2 - \frac{1}{c^2}(dx^2 + dy^2 + dz^2)}$$

Note that $d\tau$ is an invariant under Lorentz transformations. We can write

$$d\tau = \gamma dt$$

where

$$\gamma = \sqrt{1 - \frac{1}{c^2}((dx/dt)^2 + (dy/dt)^2 + (dz/dt)^2)} = \sqrt{1 - v^2/c^2}$$

This means that the four velocity can be expressed as

$$(\gamma,\ \gamma.v_x,\ \gamma.v_y,\ \gamma.v_z) = (\gamma,\ \gamma\mathbf{v})$$

We have seen that the entire set of the four Maxwell equations is contained in the equivalent set

$$\nabla^2 V - \frac{1}{c^2}\frac{\partial^2 V}{\partial t^2} = -\rho/\varepsilon_0$$

$$\nabla^2 \mathbf{A} - \frac{1}{c^2}\frac{\partial^2 \mathbf{A}}{\partial t^2} = -\mu_0 \mathbf{J}$$

along with the equations that define the fields in terms of the potentials:

$$\mathbf{E} = -\nabla V - \frac{\partial \mathbf{A}}{\partial t}, \quad \mathbf{B} = \nabla \times \mathbf{A}$$

Consider for simplicity a Lorentz transformation corresponding to motion along the $\hat{x}$ direction with a velocity v. The wave operator remains unchanged as we have seen. Thus, we get

$$\nabla'^2 V' - \frac{1}{c^2}\frac{\partial^2 V'}{\partial t'^2} = \nabla^2 V' - \frac{1}{c^2}\frac{\partial^2 V'}{\partial t^2}$$

$$= \nabla^2 \gamma(V - vA_x) - \frac{1}{c^2}\frac{\partial^2}{\partial t^2}\gamma(V - vA_x)$$

$$= \gamma(\nabla^2 V - \frac{1}{c^2}\frac{\partial^2 V}{\partial t^2}) - \gamma v(\nabla^2 A_x - \frac{\partial^2 A_x}{\partial t^2})$$

$$= -\gamma\rho/\varepsilon_0 + \gamma v J_x/\varepsilon_0 c^2 = -(\gamma/\varepsilon_0)(\rho - vJ_x/c^2) = -\rho'/\varepsilon_0$$

Here, we have used the Lorentz transformation for the potential in the form

$$V'/c^2 = \gamma\,(V/c^2 - vA_x/c^2)$$

or

$$V' = \gamma\,(V - vA_x)$$

as well as the transformation rule for the charge density in the form

$$\rho' = \gamma(\rho - vJ_x/c^2)$$

Likewise, the transformation for the components of the vector potential is

$$A'_x = \gamma(A_x - vV/c^2),\ A'_y = A_y,\ A'_z = A_z$$

and the corresponding transformation rule for the components of the current density is

$$J'_x = \gamma(J_x - v\rho),\ J'_y = J_y,\ J'_z = J_z$$

This gives

$$\nabla'^2 A'_x - \frac{1}{c^2}\frac{\partial^2 A'_x}{\partial t'^2} = \nabla^2 A'_x - \frac{1}{c^2}\frac{\partial^2 A'_x}{\partial t^2}$$

$$= \nabla^2 \gamma(A_x - vV/c^2) - \frac{1}{c^2}\frac{\partial^2}{\partial t^2}\gamma(A_x - vV/c^2)$$

$$= \gamma\left(\nabla^2 A_x - \frac{1}{c^2}\frac{\partial^2 A_x}{\partial t^2}\right) - (\gamma v/c^2)\left(\nabla^2 V - \frac{1}{c^2}\frac{\partial^2 V}{\partial t^2}\right)$$

$$= -\gamma J_x/\varepsilon_0 c^2 + \gamma v\rho/\varepsilon_0 c^2 = -\frac{\gamma}{\varepsilon_0 c^2}(J_x - v\rho) = -J'_x/\varepsilon_0 c^2$$

The equations for A_y and A_z remain unaltered. These computations show that Maxwell equations remain valid even in the primed frame if they are valid in the unprimed frame. We need only derive the expressions for the fields in the primed frame in terms of the fields in the original frame. Note that since

$$x = \gamma(x' + vt'),\ t = \gamma(t' + vx'/c^2)$$

it follows that

$$\frac{\partial}{\partial x'} = \gamma\frac{\partial}{\partial x} + (\gamma v/c^2)\frac{\partial}{\partial t}$$

$$\frac{\partial}{\partial t'} = \gamma v\frac{\partial}{\partial x} + \gamma\frac{\partial}{\partial t}$$

$$E'_x = -\frac{\partial V'}{\partial x'} - \frac{\partial A'_x}{\partial t'}$$

$$= -\left(\gamma \frac{\partial}{\partial x} + (\gamma v / c^2) \frac{\partial}{\partial t}\right) \gamma (V - v A_x)$$

$$- \left(\gamma v \frac{\partial}{\partial x} + \gamma \frac{\partial}{\partial t}\right) \gamma (A_x - v V / c^2)$$

$$= -\gamma^2 \frac{\partial V}{\partial x} - (\gamma^2 v / c^2) \frac{\partial V}{\partial t} + (\gamma^2 v^2 / c^2) \frac{\partial A_x}{\partial t} + (\gamma^2 v^2 / c^2) \frac{\partial V}{\partial x} - \gamma^2 \frac{\partial A_x}{\partial t}$$

$$+ (\gamma^2 v / c^2) \frac{\partial V}{\partial t} = -\frac{\partial V}{\partial x} - \frac{\partial A_x}{\partial t} = E_x$$

This means that the x component of the electric field remains unchanged.

$$E'_y = -\frac{\partial V'}{\partial y'} - \frac{\partial A'_y}{\partial t'}$$

$$= -\frac{\partial}{\partial y} (\gamma (V - v A_x) - \left(\gamma v \frac{\partial}{\partial x} + \gamma \frac{\partial}{\partial t}\right) A_y$$

$$= -\gamma \frac{\partial V}{\partial y} + \gamma v \frac{\partial A_x}{\partial y} - \gamma v \frac{\partial A_y}{\partial x} - \gamma \frac{\partial A_y}{\partial t}$$

$$-\gamma \left(\frac{\partial V}{\partial y} + \frac{\partial A_y}{\partial t}\right) - \gamma v \left(\frac{\partial A_y}{\partial x} - \frac{\partial A_x}{\partial y}\right)$$

$$= \gamma E_y - \gamma v B_z = \gamma (E_y - v B_z)$$

$$E'_z = -\frac{\partial V'}{\partial z'} - \frac{\partial A'_z}{\partial t'}$$

$$= -\frac{\partial}{\partial z} \gamma (V - v A_x) - \left(\gamma v \frac{\partial}{\partial x} + \gamma \frac{\partial}{\partial t}\right) A_z$$

$$= -\gamma \frac{\partial V}{\partial z} + \gamma v \frac{\partial A_x}{\partial z} - \gamma v \frac{\partial A_z}{\partial x} - \gamma \frac{\partial A_z}{\partial t}$$

$$= -\gamma \left(\frac{\partial V}{\partial z} + \frac{\partial A_z}{\partial t}\right) + \gamma v \left(\frac{\partial A_x}{\partial z} - \frac{\partial A_z}{\partial x}\right)$$

$$= \gamma E_z + \gamma v B_y = \gamma (E_z + v B_y)$$

This completes the transformation rule for the three components of the electric field. For the magnetic field, we have

$$B'_x = \frac{\partial A'_z}{\partial y'} - \frac{\partial A'_y}{\partial z'} = \frac{\partial A_z}{\partial y} - \frac{\partial A_y}{\partial z} = B_x$$

$$B'_y = \frac{\partial A'_x}{\partial z'} - \frac{\partial A'_z}{\partial x'}$$

$$= \frac{\partial}{\partial z}\gamma(A_x - vV/c^2) - \left(\gamma\frac{\partial}{\partial x} + (\gamma v/c^2)\frac{\partial}{\partial t}\right)A_z$$

$$= \gamma\left(\frac{\partial A_x}{\partial z} - \frac{\partial A_z}{\partial x}\right) - (\gamma v/c^2)\left(\frac{\partial V}{\partial z} + \frac{\partial A_z}{\partial t}\right)$$

$$= \gamma B_y + (\gamma v/c^2)E_z = \gamma(B_y + vE_z/c^2)$$

$$B'_z = \frac{\partial A'_y}{\partial x'} - \frac{\partial A'_x}{\partial y'}$$

$$= \left(\gamma\frac{\partial}{\partial x} + (\gamma v/c^2)\frac{\partial}{\partial t}\right)A_y - \frac{\partial}{\partial y}\gamma(A_x - vV/c^2)$$

$$= \gamma\left(\frac{\partial A_y}{\partial x} - \frac{\partial A_x}{\partial y}\right) + (\gamma v/c^2)\left(\frac{\partial A_y}{\partial t} + \frac{\partial V}{\partial y}\right)$$

$$= \gamma B_z - (\gamma v/c^2)E_y = \gamma(B_z - vE_y/c^2)$$

This completes the transformation rules for the magnetic field.

Poynting's theorem

The objective of this section is to prove that the power flowing per unit area in an electromagnetic field is given by $\mathbf{P} = \mathbf{E} \times \mathbf{H}$ and the energy density of the electromagnetic field is given by $U = \frac{\varepsilon}{2}E^2 + \frac{\mu}{2}H^2$. Consider to this end a volume V in space containing charges and fields. The total force exerted by the fields on charge in a unit volume is given by the Lorentz force equation

$$\mathbf{F} = \rho(\mathbf{E} + \mathbf{v} \times \mathbf{B}) = \rho\mathbf{E} + \mathbf{J} \times \mathbf{B}$$

This force does work on charges in the unit volume at the rate of

$$\mathbf{F}.\mathbf{v} = \rho\mathbf{v}.\mathbf{E} = \mathbf{E}.\mathbf{J}$$

This must equal the rate at which energy is getting dissipated in the field per unit volume. Thus, if $\mathbf{P}$ denotes the power flowing per unit area and U the energy density of the field, we must have the energy balance equation

$$\frac{\partial}{\partial t}\int_V U dV = -\int_S \mathbf{P}.\mathbf{n}dS - \int_V \mathbf{E}.\mathbf{J}dV$$

This equation is read as "The rate at which energy in the electromagnetic field in the volume V increases with time must equal the rate at which energy flows into V from the boundary surface S of V minus the rate at which energy in V is getting dissipated in the form of heat caused by work done by the field on the charges. We now use the Maxwell equation

$$\mathbf{J} = \nabla \times \mathbf{H} - \frac{\partial \mathbf{D}}{\partial t}$$

to get

$$\mathbf{E}.\mathbf{J} = \mathbf{E}.\nabla \times \mathbf{H} - \mathbf{E}.\frac{\partial \mathbf{D}}{\partial t}$$

Also assuming constant permittivity ε, we have

$$\mathbf{E}.\frac{\partial \mathbf{D}}{\partial t} = \frac{\varepsilon}{2}\frac{\partial}{\partial t}E^2$$

Further, we make use of the vector identity

$$\nabla\cdot(\mathbf{E}\times\mathbf{H}) = \mathbf{H}\cdot\nabla\times\mathbf{E} - \mathbf{E}\cdot\nabla\times\mathbf{H}$$

Combining this with the Maxwell equation

$$\nabla \times \mathbf{E} = -\frac{\partial \mathbf{B}}{\partial t}$$

and the fact that constant permeability assumption implies that

$$\mathbf{H}.\frac{\partial \mathbf{B}}{\partial t} = \frac{\mu}{2}\frac{\partial}{\partial t}H^2$$

we get

$$\mathbf{E}.\nabla \times \mathbf{H} = -\frac{\mu}{2}\frac{\partial}{\partial t}H^2 - \nabla.(\mathbf{E}\times\mathbf{H})$$

Thus,

$$\mathbf{E}.\mathbf{J} = -\frac{\partial}{\partial t}\left(\frac{\varepsilon}{2}E^2 + \frac{\mu}{2}H^2\right) - \nabla.(\mathbf{E}\times\mathbf{H})$$

Thus making use of Gauss' theorem for evaluating the volume integral $\int_V \nabla.(\mathbf{E}\times\mathbf{H})dV$, we get

$$-\int_V \mathbf{E}.\mathbf{J}dV = \frac{\partial}{\partial t}\int_V \left(\frac{\varepsilon}{2}E^2 + \frac{\mu}{2}H^2\right)dV + \int_S \mathbf{E}\times\mathbf{H}.\mathbf{n}dS$$

Comparing this with the energy conservation equation, we deduce that

$$U = \frac{\varepsilon}{2}E^2 + \frac{\mu}{2}H^2, \mathbf{P} = \mathbf{E} \times \mathbf{H}$$

Reflection and Refraction of Electromagnetic Waves at a boundary

Consider a plane interface say $z = 0$ plane separating the regions $z < 0$ which has parameters (ε_1, μ_1, σ_1) and $z > 0$ having parameters (ε_2, μ_2, σ_2) for the permittivity, magnetic permeability and conductivity respectively. A plane wave in medium 1 impinges onto the boundary, part of which gets reflected and part transmitted into the region 2. We want to calculate the angles of reflection refraction as well as the ratios of the reflected to incident and transmitted to incident electric and magnetic fields. Two cases are possible. The first is normal incidence and the second is oblique incidence.

Normal Incidence:

This situation is shown in Fig. 5.2. An electromagnetic wave in medium

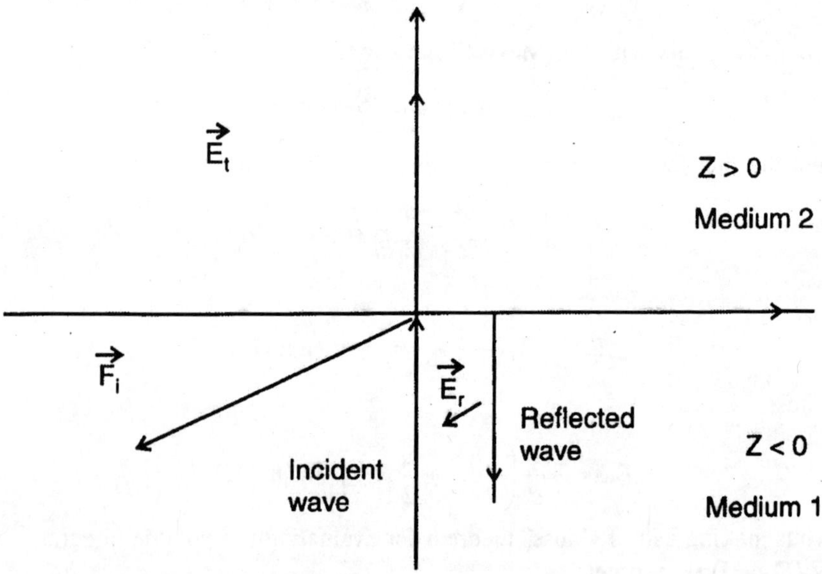

Fig. 5.2

1 impinges normally on the plane interface. From the symmetry of the problem, the directions of propagation of the reflected and transmitted waves will also be normal to the plane, i.e. along the z axis. The electric field phasor in the incident wave can be expressed as

$$\mathbf{E}_i(\mathbf{r}) = E_{i0}\hat{x}\exp(-\gamma_1 z)$$

Note that the interface plane is the xy plane and we are assuming without loss of generality that the electric fields are polarized along the $\hat{x}$ direction. The electric field in the reflected wave is

$$\mathbf{E}_r(\mathbf{r}) = E_{r0}\hat{x}\exp(\gamma_1 z)$$

and finally, the electric field in the transmitted wave is

$$\mathbf{E}_t(\mathbf{r}) = E_{t0}\hat{x}\exp(-\gamma_2 z)$$

Here, γ_1, γ_2 are respectively the propagation constants in medium 1 and medium 2. We have derived expressions for these:

$$\gamma_1 = \sqrt{i\omega\mu_1(\sigma_1 + i\omega\varepsilon_1)} = \alpha_1 + i\beta_1$$

$$\gamma_2 = \sqrt{i\omega\mu_2(\sigma_2 + i\omega\varepsilon_2)} = \alpha_2 + i\beta_2$$

In the case when medium 1 is a perfect dielectric, $\sigma_1 = 0$ and when 2 is a perfect dielectric, $\sigma_2 = 0$. In the former case, $\gamma_1 = i\omega \sqrt{\varepsilon_1\mu_1}$ and in the latter, $\gamma_2 = i\omega\sqrt{\varepsilon_2\mu_2}$. For a perfect dielectric, the waves propagate without any attenuation. The magnetic fields phasors in the incident, reflected and transmitted waves can be obtained from the Maxwell equations:

$$\nabla \times \mathbf{E}_i = -i\omega\mu_1\mathbf{H}_i$$

This gives

$$-\gamma_1\hat{z} \times \hat{x}.E_{i_0}\exp(-\gamma_1 z) = -i\omega\mu_1\mathbf{H}_i$$

and hence,

$$\mathbf{H}_i = H_{i0}\hat{y}\exp(-\gamma_1 z)$$

where

$$H_{i0} = \frac{E_{i0}}{\eta_1}$$

with

$$\eta_1 = \frac{i\omega\mu_1}{\gamma_1} = \sqrt{\frac{i\omega\mu_1}{\sigma_1 + i\omega\varepsilon_1}}$$

For the magnetic field $\mathbf{H}_r$ in the reflected wave, we find that

$$\nabla \times \mathbf{E}_r = -i\omega\mu_2\mathbf{H}_r$$

which gives

$$\mathbf{H}_r = H_{r0}\hat{y}\exp(\gamma_1 z)$$

with

$$H_{r0} = -\frac{E_{r0}}{\eta_1}$$

and finally, for the magnetic field in the transmitted field,

$$\mathbf{H}_t = H_{t0}y.\exp(-\gamma_2 z)$$

with

$$H_{to} = \frac{E_{to}}{\eta_2}$$

Continuity of the tangential component of the electric field at the boundary gives

$$E_{i0} + E_{r0} = H_{t0}$$

Continuity of the tangential component of the magnetic field at the boundary gives

$$H_{i0} + H_{r0} = H_{t0}$$

or

$$\frac{E_{i0}}{\eta_1} - \frac{E_{r0}}{\eta_1} = \frac{E_{t0}}{\eta_2}$$

Note that the tangential component of the magnetic field is continuous only when no surface current flows at the boundary. But no surface current can flow because that would imply an infinite current density which would in turn imply infinite electric field because of the finite conductivity and the relation $E = \frac{\mathbf{J}}{\sigma}$. Solving the above equations, we get the ratios

$$R = \frac{E_{r0}}{E_{i0}} = \frac{\eta_2 - \eta_1}{\eta_1 + \eta_2}$$

$$T = \frac{E_{t0}}{E_{i0}} = \frac{2\eta_2}{\eta_1 + \eta_2}$$

For perfect dielectrics, $\eta_1 = \sqrt{\frac{\mu_1}{\varepsilon_1}}$ and $\eta_2 = \sqrt{\frac{\mu_2}{\varepsilon_2}}$ and we get

$$R = \frac{\sqrt{\mu_2\varepsilon_1} - \sqrt{\mu_1\varepsilon_2}}{\sqrt{\mu_2\varepsilon_1} + \sqrt{\mu_1\varepsilon_2}}$$

$$T = \frac{2\sqrt{\mu_2\varepsilon_1}}{\sqrt{\mu_1\varepsilon_2} + \sqrt{\mu_2\varepsilon_1}}$$

If both the dielectrics, in addition, have the same magnetic permeability, $\mu_1 = \mu_2$ and we get

$$R = \frac{\sqrt{\varepsilon_1} - \sqrt{\varepsilon_2}}{\sqrt{\varepsilon_1} + \sqrt{\varepsilon_2}}$$

$$T = \frac{2\sqrt{\varepsilon_1}}{\sqrt{\varepsilon_1} + \sqrt{\varepsilon_2}}$$

The refractive index of medium 2 with respect to medium 1 is $n = \sqrt{\frac{\varepsilon_2}{\varepsilon_1}}$

since

$$n = \frac{c_1}{c_2} = \frac{1}{\sqrt{\mu\varepsilon_1}} \Big/ \frac{1}{\sqrt{\mu\varepsilon_2}}$$

where c_1, c_2 are respectively the speeds of light in medium 1 and medium 2. In terms of this, we can write

$$T = \frac{2}{1+n}, R = \frac{1-n}{1+n}$$

Note that we can also calculate the ratios of the reflected and transmitted magnetic fields to the incident field:

$$R' = \frac{H_{r0}}{H_{i0}} = -\frac{E_{r0}}{E_{i0}} = -R = \frac{\eta_1 - \eta_2}{\eta_1 + \eta_2}$$

$$T' = \frac{H_{t0}}{H_{i0}} = \frac{\eta_1}{\eta_2}\frac{E_{t0}}{E_{i0}} = \frac{\eta_1 T}{\eta_2} = \frac{2\eta_1}{\eta_1 + \eta_2}$$

The average power flux in the incident wave at $z = 0$ is given by

$$\mathbf{P}_\mathrm{i} = \frac{1}{2}\mathrm{Re}(E_{i0}\hat{x} \times \overline{H}_{i0}\hat{y}) = \frac{\hat{z}}{2}\mathrm{Re}(E_{i0}\overline{H}_{i0}) = \frac{\hat{z}}{2}|E_{i0}|^2 \mathrm{Re}(1/\overline{\eta}_1)$$

$$= \frac{\hat{z}}{2}|E_{i0}|^2 (\mathrm{Re}\,\eta_1)\big/|\eta_1|^2$$

Average power flux in the reflected wave is likewise

$$P_r = \frac{1}{2}\mathrm{Re}(E_{r0}\hat{x} \times \overline{H}_{r0}\hat{y}) = -\frac{\hat{z}}{2}|E_{i0}|^2 \mathrm{Re}(\eta_1)\big/|\eta_1|^2$$

Finally, the average power flow in the transmitted wave is

$$P_t = \frac{1}{2}\mathrm{Re}(E_{t0}\hat{x} \times \overline{H}_{t0}\hat{y}) = \frac{\hat{z}}{2}\mathrm{Re}(E_{t0}\hat{x} \times \overline{H}_{t0}\hat{y}) = \frac{\hat{z}}{2}|E_{t0}|^2 \mathrm{Re}(\eta_2)/|\eta_2|^2$$

We consider the case of ideal dielectrics, i.e. $\sigma_1 = \sigma_2 = 0$. Then, η_1, η_2 are real and the power flowing in the $\hat{z}$ direction in the incident wave is $|P_i| = |E_{i0}|^2/2\eta_1$. The power flowing in the $-\hat{z}$ direction in reflected wave is $|P_r| = |E_{r0}|^2\ 2\eta_2$ and the power flowing in the $\hat{z}$ direction in the transmitted wave is $|P_t| = |E_{t0}|^2/2\eta_2$. We can verify that $|P_i| = |P_r| + |P_t|$ in agreement with the energy conservation equation. Indeed,

$$|P_r| + |P_t| = |E_{i0}|^2 (R^2/2\eta_1 + T^2/2\eta_2) = \frac{|E_{i0}|^2}{2}$$

$$\left(\frac{(\eta_2 - \eta_1)^2}{\eta_1(\eta_1 + \eta_2)^2} + \frac{4\eta_2}{(\eta_1 + \eta_2)^2}\right) = \frac{|E_{i0}|^2}{2\eta_1} = |P_i|$$

This states that the incident power flux equals the reflected plus transmitted power flux. The case when medium 2 is a perfect conductor deserves special attention. In this case, the transmitted electric field is zero since otherwise there would be infinite power dissipation in medium 2. The boundary condition on the tangential electric field is $E_{i0} + E_{r0} = 0$. This gives $E_{r0} = -E_{i0}$ and hence,

$$\mathbf{E}_i = E_{i0}\hat{x}.\exp(-\gamma_1 z)$$

$$\mathbf{E}_r = -E_{i0}\hat{x}.\exp(-\gamma_1 z)$$

$$\mathbf{H}_i = \frac{E_{i0}}{\eta_1}\exp(-\gamma_1 z)\hat{y}$$

$$\mathbf{H}_r = \frac{E_{i0}}{\eta_1}\hat{y}\exp(\gamma_1 z)$$

For the transmitted magnetic field, we find

$$\mathbf{H}_t = H_{t0}\hat{y}.\exp(-\gamma_2 z)$$

where

$$H_{t0} = H_{i0} + H_{r0} = \frac{2E_{i0}}{\eta_1}$$

This boundary condition is subject to the condition that there is no surface current flowing. This equation makes it appear that there will be a magnetic field in medium 2, however that is not the case once we note that $\gamma_2 = \sqrt{j\omega\mu(\sigma_2 + j\omega\varepsilon)}$ is approximately $\sqrt{\omega\mu\sigma_2/2}\ (1+j)$ when $\sigma_2 \to \infty$. In the limit, the real part of γ_2 is ∞ and hence there will be infinite attenuation of the fields when $z > 0$. Thus, the magnetic field exists just at the interface of the two layers. We do not have any power flowing in region 2 since there is no electric field or magnetic field here. An alternate route to deriving the above expressions for E_{r0}, H_{r0}, H_{t0} for reflection at a perfect conductor is to take limits as $\sigma_2 \to \infty$ of the preceding expressions. Thus,

$$\lim_{\sigma 2\to\infty}\frac{E_{r0}}{E_{i0}} = \lim_{\sigma 2\to\infty}\frac{\eta_2-\eta_1}{\eta_2+\eta_1}$$

$$= -1$$

since $\eta_2 \to 0$ as $\sigma_2 \to \infty$. Thus for the limiting case, $E_{r0} = -E_{i0}$. Likewise,

$$\lim_{\sigma 2\to\infty}\frac{H_{r0}}{H_{i0}} = \lim_{\sigma 2\to\infty}\frac{\eta_1-\eta_2}{\eta_1+\eta_2} = 1$$

which means that for the limiting case, $H_{r0} = H_{i0}$. Next,

$$\lim_{\sigma 2\to\infty}\frac{E_{t0}}{E_{i0}} = \lim_{\sigma 2\to\infty}\frac{2\eta_2}{\eta_1+\eta_2} = 0$$

showing that there is no transmitted electric field. Finally,

$$\lim_{\sigma 2\to\infty}\frac{H_{t0}}{H_{i0}} = \lim_{\sigma 2\to\infty}\frac{2\eta_1}{\eta_1+\eta_2} = 2$$

in conformity with what we derived earlier.

Oblique Incidence

Consider the situation shown in Fig. 5.3. We have a wave incident at an angle θ_i to the normal of the interface in medium 1. The reflected wave makes an angle θ_r with the normal as shown and the transmitted wave an angle θ_t. Again, here we have two cases. The first case is called the perpendicular polarization case in which the electric field of the incident wave is polarized perpendicular to the plane of incidence. The plane of incidence is the plane containing the normal to the interface and the direction of the incident beam. This plane, from symmetry considerations, will also contain the directions of propagation of the reflected and transmitted/refracted waves. This plane is the *xz* plane and hence perpendicular polarization corresponds to the situation in which the electric fields of the incident, reflected and transmitted waves have only $\hat{y}$ components. The second case is that of parallel polarization in which the electric fields of the incident, reflected and transmitted waves are polarized parallel to the plane of incidence, i.e. they do not have any $\hat{y}$ component.

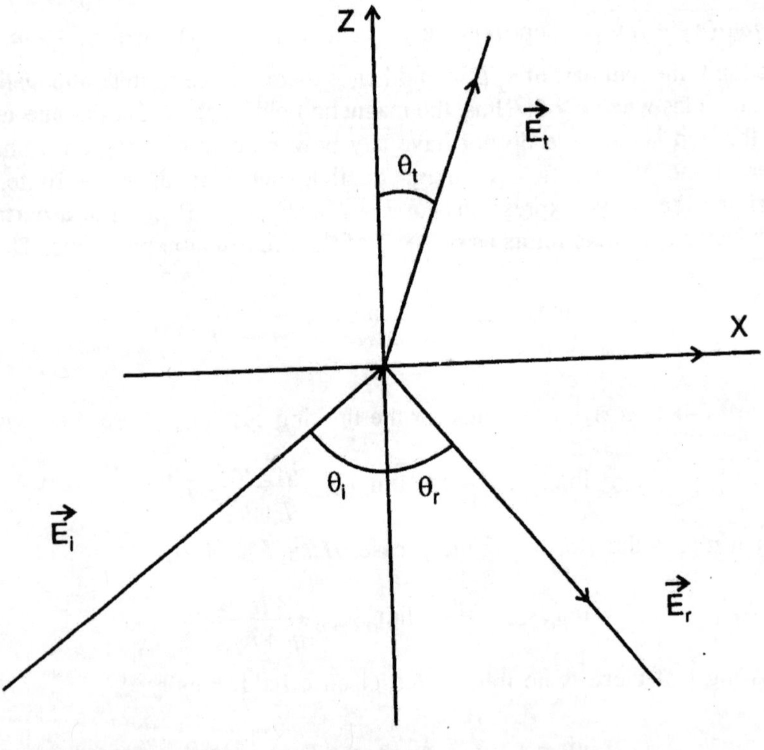

Fig. 5.3

Perpendicular polarization

The incident electric field is

$$\mathbf{E}_i = E_{i0}\,\hat{y}\exp(-\gamma_1(x.\sin\theta_i + z.\cos\theta_i))$$

The exponent here can be expressed as $-\gamma_1\mathbf{n}_i.\mathbf{r}$, where $\mathbf{n}_i = \hat{x}.\sin\theta_i + \hat{z}.\cos\theta_i$. The figure shows this fact clearly. The direction of propagation of the incident wave has projections $\sin\theta_i$, $\cos\theta_i$ respectively along the $\hat{x}$ and $\hat{z}$ directions. For the magnetic field in the incident wave, we use the equation

$$\nabla \times \mathbf{E}_i = -\,i\omega\mu_1\mathbf{H}_i$$

Now,

$$\nabla \times \mathbf{E}_i = -\gamma_1\mathbf{n}_i \times \hat{y}E_{i0}\exp(-\gamma_1\mathbf{n}_i.\mathbf{r}) = -\gamma_1\mathbf{n}_i \times \mathbf{E}_i$$

Thus,

$$\mathbf{H}_i = \frac{1}{\eta_1}\mathbf{n}_i \times \mathbf{E}_i$$

Note that

$$\mathbf{n}_i \times \hat{y} = (\hat{x}.\sin\theta_i + \hat{z}.\cos\theta_i) \times \hat{y} = \hat{z}.\sin\theta_i - \hat{x}.\cos\theta_i$$

Thus,

$$\mathbf{H}_i = \frac{E_{i0}}{\eta_1}(\hat{z}.\sin\theta_i - \hat{x}.\cos\theta_i).\exp(-\gamma_1\mathbf{n}_i.\mathbf{r})$$

For the reflected wave, we have

$$\mathbf{E}_r = E_{r0}\hat{y}\exp(-\gamma_1\mathbf{n}_r.\mathbf{r})$$

where

$$\mathbf{n}_r = -\hat{z}.\cos\theta_r + \hat{x}.\sin\theta_r$$

Thus,

$$\mathbf{n}_r.\mathbf{r} = x.\sin\theta_r - z.\cos\theta_r$$

We have

$$\mathbf{H}_r = \frac{1}{\eta_1}\mathbf{n}_r \times \mathbf{E}_r = \frac{E_{r0}}{\eta_1}\mathbf{n}_r \times \hat{y}.\exp(-\gamma_1\mathbf{n}_r.\mathbf{r})$$

where

$$\mathbf{n}_r \times \hat{y} = (\hat{x}.\sin\theta_r - \hat{z}.\cos\theta_r) \times \hat{y} = \hat{z}.\sin\theta_r + \hat{x}.\cos\theta_r$$

Thus,

$$\mathbf{H}_r = \frac{E_{r0}}{\eta_1}(\hat{z}.\sin\theta_r + \hat{x}.\cos\theta_r).\exp(-\gamma_1\mathbf{n}_r.\mathbf{r})$$

Finally,

$$\mathbf{E}_t = E_{t0}\hat{y}.\exp(-\gamma_2\mathbf{n}_t.\mathbf{r})$$

where

$$\mathbf{n}_t = \hat{x}.\sin\theta_t + \hat{z}.\cos\theta_t$$

The exponent is

$$\mathbf{n}_t.\mathbf{r} = x.\sin\theta_t + z.\cos\theta_t$$

$$\mathbf{H}_t = \frac{1}{\eta_2}\mathbf{n}_t \times \mathbf{E}_t = \frac{E_{t0}}{\eta_2}(\hat{z}.\sin\theta_t - \hat{x}.\cos\theta_t)\exp(-\gamma_2\mathbf{n}_t.\mathbf{r})$$

The boundary conditions are the usual ones: Continuity of the tangential components of the electric and magnetic fields at the boundary. Continuity of the tangential component of the electric field gives

$$E_{t0} = E_{i0} + E_{r0}$$

and

$$-\gamma_1.\sin\theta_i = -\gamma_1.\sin\theta_r = -\gamma_2.\sin\theta_t$$

Continuity of the tangential component of the magnetic field **H** at the boundary gives

$$-\frac{E_{i0}}{\eta_1}\cos\theta_i + \frac{E_{r0}}{\eta_1}\cos\theta_r = -\frac{E_{t0}}{\eta_2}\cos\theta_t$$

Note that continuity of the normal component of **B** implies that

$$\frac{\mu_1 E_{i0}}{\eta_1}\sin\theta_i + \frac{\mu_1 E_{r0}}{\eta_1}\sin\theta_r = \frac{\mu_2 E_{t0}}{\eta_2}\sin\theta_t$$

We shall first assume that both the media are perfect dielectrics so that $\gamma_1 = \omega\sqrt{\varepsilon_1\mu_1}$ and $\gamma_2 = \omega\sqrt{\varepsilon_2\mu_2}$ and $\eta_1 = \sqrt{\mu_1/\varepsilon_1}$, $\eta_2 = \sqrt{\mu_2/\varepsilon_2}$. The second equation gives

$$\sin\theta_i = \sin\theta_r$$

and hence, $\theta_i = \theta_r$. This is Snell's law of reflection. The second equation also gives

$$\frac{\sin\theta_i}{\sin\theta_t} = \frac{\sqrt{\varepsilon_2\mu_2}}{\sqrt{\varepsilon_1\mu_1}}$$

In the particular case, when $\mu_1 = \mu_2$, this ratio is $\sqrt{\varepsilon_2/\varepsilon_1}$. In general, we find that

$$n = \frac{c_1}{c_2} = \frac{\sqrt{\varepsilon_2\mu_2}}{\sqrt{\varepsilon_1\mu_1}}$$

is the refractive index of medium 2 with respect to medium 1. Thus,

$$\frac{\sin\theta_i}{\sin\theta_t} = n$$

This is in fact, Snell's law of refraction. Letting $R = \frac{E_{r0}}{E_{i0}}$, $T = \frac{E_{t0}}{E_{i0}}$, the first and the third equation can be expressed as

$$1 + R = T$$

$$\cos\theta_i - R.\cos\theta_r = \frac{\eta_1}{\eta_2} T.\cos\theta_t$$

The fourth equation is the same as

$$\sqrt{\varepsilon_1\mu_1}\sin\theta_i + \sqrt{\varepsilon_1\mu_1}\sin\theta_r.R = \sqrt{\varepsilon_2\mu_2}\sin\theta_t.T$$

and in view of Snell's laws, this is the same as

$$1 + R = T$$

So this last equation is in agreement with the first equation and conveys nothing new. It just shows that our equations are consistent. The first two equations give

$$\cos\theta_i - R.\cos\theta_r = \frac{\eta_1}{\eta_2}(1+R).\cos\theta_t$$

or

$$R\left(\frac{\eta_1}{\eta_2}\cos\theta_t + \cos\theta_r\right) = \cos\theta_i - \frac{\eta_1}{\eta_2}\cos\theta_t$$

and since $\theta_r = \theta_i$, it follows that

$$R = \frac{\eta_2\cos\theta_i - \eta_1\cos\theta_t}{\eta_1\cos\theta_t + \eta_2\cos\theta_i}$$

In the special case when $\mu_2 = \mu_1$, this reduces to

$$R = \frac{\sqrt{\varepsilon_1}\cos\theta_i - \sqrt{\varepsilon_2}\cos\theta_t}{\sqrt{\varepsilon_1}\cos\theta_i + \sqrt{\varepsilon_2}\cos\theta_t}$$

and

$$T = 1 + R = \frac{2\eta_2\cos\theta_i}{\eta_1\cos\theta_t + \eta_2\cos\theta_i}$$

which in the above special case, reduces to

$$T = \frac{2\sqrt{\varepsilon_1}\cos\theta_i}{\sqrt{\varepsilon_1}\cos\theta_i + \sqrt{\varepsilon_2}\cos\theta_t}$$

The power flux in the incident beam is

$$P_i = \frac{|E_{i0}|^2}{2\eta_1}$$

and this flows at an angle θ_i to the surface normal. Thus, the normal component of the incident power flux is given by

$$P_{i,n} = P_i\cos\theta_i = \frac{|E_{i0}|^2}{2\eta_1}\cos\theta_i$$

The power flux in the reflected beam is

$$P_r = \frac{|E_{r0}|^2}{2\eta_1}$$

and its normal component is

$$P_{r,n} = P_r\cos\theta_r = \frac{|E_{r0}|^2}{2\eta_1}\cos\theta_i$$

Finally, the normal component of the transmitted/refracted power flux is given by

$$P_{t,n} = \frac{|E_{t0}|^2}{2\eta_2}\cos\theta_t$$

We must verify the energy conservation principle:

$$P_{i,n} = P_{r,n} + P_{t,n}$$

We find that

$$P_{r,n} + P_{t,n} = \frac{|E_{i0}|^2}{2}(R^2\cos\theta_i/\eta_1 + T^2\cos\theta_t/\eta_2)$$

We find

$$R^2\cos\theta_i/\eta_1 + T^2\cos\theta_t/\eta_2 =$$

$$\left(\frac{\eta_2\cos\theta_i - \eta_1\cos\theta_t}{\eta_1\cos\theta_t + \eta_2\cos\theta_i}\right)^2\cos\theta_i/\eta_1 + \left(\frac{2\eta_2\cos\theta_i}{\eta_1\cos\theta_t + \eta_2\cos\theta_i}\right)^2\cos\theta_t/\eta_2$$

$$= \frac{(\eta_2\cos\theta_i - \eta_1\cos\theta_t)^2(\cos\theta_i)/\eta_1 + 4\eta_2\cos^2\theta_i\cos\theta_t}{(\eta_2\cos\theta_i + \eta_1\cos\theta_t)^2}$$

$$= (\cos\theta_i)/\eta_1$$

This proves the energy conservation principle. We leave it as an exercise to compute the ratios of the magnetic fields in the reflected and transmitted beams to that in the incident beam. Finally, consider the case when medium 2 is a perfect conductor. Then, $\mathbf{E}_t = 0$ (as follows from the expression for T and the fact that $\eta_2 \to 0$ as $\sigma_2 \to \infty$ and we have the boundary condition

$$\mathbf{E}_i + \mathbf{E}_r = \mathbf{0}$$

This gives $\theta_r = \theta_i$ and $E_{r0} = -E_{i0}$. The magnetic field in the incident beam is

$$\mathbf{H}_i = \frac{E_{i0}}{\eta_1}(\hat{z}.\sin\theta_i - \hat{x}.\cos\theta_i)\exp(-\gamma_1\mathbf{n}_i.\mathbf{r})$$

and that in the reflected beam is

$$\mathbf{H}_r = -\frac{E_{i0}}{\eta_1}(\hat{z}.\sin\theta_i + \hat{x}.\cos\theta_i)\exp(\gamma_1\mathbf{n}_r.\mathbf{r})$$

where

$$\mathbf{n}_i.\mathbf{r} = x.\sin\theta_i + z.\cos\theta_i$$
$$\mathbf{n}_r.\mathbf{r} = x.\sin\theta_i - z.\cos\theta_i$$

The magnetic field in the transmitted beam is polarized in the plane of incidence. It must therefore have the form

$$\mathbf{H}_t = A.(\hat{z}.\sin\theta_t - \hat{x}.\cos\theta_t).\exp(-\gamma_2\mathbf{n}_t.\mathbf{r})$$

where

$$\mathbf{n}_t.\mathbf{r} = x.\sin\theta_t + z.\cos\theta_t$$

Equating tangential components of **H** at the boundary (i.e. the $\hat{x}$ component), we get

$$-\frac{E_{i0}}{\eta_1}\cos\theta_i - \frac{E_{i0}}{\eta_1}\cos\theta_i = -A.\cos\theta_t$$

This gives

$$A = \frac{2E_{i0}\cos\theta_i}{\eta_1.\cos\theta_t}$$

This determines the amplitude of the boundary magnetic field at $x = 0$, $z = 0$, i.e.

$$\mathbf{H}_t|_{x=z=0} = \frac{2E_{i0}\cos\theta_i}{\eta_1.\cos\theta_t}(\hat{z}.\sin\theta_t - \hat{x}.\cos\theta_t)$$

When $x > 0$ or $z > 0$, the magnetic field in the transmitted wave is zero since γ_2 has infinite real part causing infinite attenuation.

Critical Angle of Incidence

Consider once again the case when both the media are perfect dielectrics with $\mu_1 = \mu_2$. We have seen that the coefficient of reflection is given by

$$R = \frac{\sqrt{\varepsilon_1}\cos\theta_i - \sqrt{\varepsilon_2}\cos\theta_t}{\sqrt{\varepsilon_1}\cos\theta_i + \sqrt{\varepsilon_2}\cos\theta_t}$$

and since

$$\sin\theta_t = \sqrt{\varepsilon_1/\varepsilon_2}.\sin\theta_i$$

and hence

$$\cos\theta_t = \sqrt{1 - \varepsilon_1.\sin^2\theta_i/\varepsilon_2}$$

Thus, the reflection coefficient can be expressed as

$$R = \frac{\sqrt{\varepsilon_1}\cos\theta_i - \sqrt{\varepsilon_2 - \varepsilon_1\sin^2\theta_i}}{\sqrt{\varepsilon_1}\cos\theta_i + \sqrt{\varepsilon_2 - \varepsilon_1\sin^2\theta_i}}$$

Suppose we assume that $\varepsilon_2 < \varepsilon_1$ and $\sin\theta_i > \sqrt{\varepsilon_2/\varepsilon_1}$. Then, we find that R acquires the form

$$R = \frac{a + ib}{a - ib}$$

where a, b are real numbers. This means that $|R| = 1$ and hence all the incident energy is transmitted to the reflected wave. This does not mean, however, that there does not exist any transmitted wave. There will be a transmitted wave since $T \neq 0$ but its amplitude will decay exponentially as it propagates and hence cannot propagate energy. This can be seen from the fact that when $\sin\theta_i > \sqrt{\varepsilon_2/\varepsilon_1}$, we have $\sin\theta_t > 1$ from Snell's law of refraction which means that θ_t is complex. Thus, $\cos\theta_t$ is pure imaginary and $\gamma_2 \mathbf{n}_t.\mathbf{r} = j\omega\sqrt{\mu_2\varepsilon_2}\,(x.\sin\theta_t + z.\cos\theta_t)$ has a real part of the form $-\alpha z$ where $\alpha > 0$. This corresponds to the fact that the transmitted wave decays exponentially with increasing z. This angle of incidence $\theta_{ic} = \sin^{-1}(\sqrt{\varepsilon_2/\varepsilon_1})$ at angles greater than which no energy is transmitted is called the critical angle of incidence. We say that at angles of incidence greater than this critical angle, total internal reflection occurs.

Parallel Polarization

In this case, the electric field is polarized parallel to the plane of incidence. This is equivalent to the magnetic field being polarized perpendicular to the plane of incidence. Thus, the incident magnetic field phasor is given by

$$\mathbf{H}_i = H_{i0}\hat{y}\exp(-\gamma_1 \mathbf{n}_i.\mathbf{r})$$

The reflected magnetic field is

$$\mathbf{H}_r = H_{r0}\hat{y}\exp(-\gamma_1 \mathbf{n}_r.\mathbf{r})$$

and finally, the transmitted magnetic field is

$$\mathbf{H}_t = H_{t0}\hat{y}\exp(-\gamma_2 \mathbf{n}_t.\mathbf{r})$$

Here, assuming θ_i, θ_r, θ_t to be the angles of incidence, reflection and refraction, we have

$$\mathbf{n}_i = \hat{x}.\sin\theta_i + \hat{z}.\cos\theta_i$$
$$\mathbf{n}_r = \hat{x}.\sin\theta_r - \hat{z}.\cos\theta_r$$
$$\mathbf{n}_t = \hat{x}.\sin\theta_t + \hat{z}.\cos\theta_t$$

We calculate

$$\nabla \times \mathbf{H}_i = -\gamma_1 \mathbf{n}_i \times \hat{y}H_{i0}\exp(-\gamma_1 \mathbf{n}_i.\mathbf{r})$$
$$\nabla \times \mathbf{H}_r = -\gamma_1 \mathbf{n}_r \times \hat{y}H_{i0}\exp(-\gamma_1 \mathbf{n}_r.\mathbf{r})$$
$$\nabla \times \mathbf{H}_t = -\gamma_2 \mathbf{n}_t \times \hat{y}H_{t0}\exp(-\gamma_2 \mathbf{n}_t.\mathbf{r})$$

Now,

$$\mathbf{n}_i \times \hat{y} = \hat{z}.\sin\theta_i - \hat{x}.\cos\theta_i$$
$$\mathbf{n}_r \times \hat{y} = \hat{z}.\sin\theta_r + \hat{x}.\cos\theta_r$$
$$\mathbf{n}_t \times \hat{y} = \hat{z}.\sin\theta_t - \hat{x}.\cos\theta_t$$

Now, the Maxwell equation $\nabla \times \mathbf{H}_i = (\sigma_1 + i\omega\varepsilon_1)\mathbf{E}_i$ implies

$$\mathbf{E}_i = -\frac{\gamma_1}{\sigma + i\omega\varepsilon_1}(\hat{z}\sin\theta_i - \hat{x}.\cos\theta_i)H_{i0}\exp(-\gamma_i \mathbf{n}_i.\mathbf{r})$$

Now,

$$\frac{\gamma_1}{\sigma_1 + i\omega\varepsilon_1} = \sqrt{\frac{i\omega\mu_1}{\sigma_1 + i\omega\varepsilon_1}} = \eta_1$$

Thus,

$$\mathbf{E}_i = \eta_1(\hat{x}.\cos\theta_i - \hat{z}.\sin\theta_i)\; H_{i0}\exp(-\gamma_1\mathbf{n}_i.\mathbf{r})$$

Similarly

$$\mathbf{E}_r = \eta_1(-\hat{x}.\cos\theta_r - \hat{z}.\sin\theta_r)\; H_{r0}\exp(-\gamma_1\mathbf{n}_r.\mathbf{r})$$

$$\mathbf{E}_t = \eta_2(\hat{x}.\cos\theta_t - \hat{z}.\sin\theta_t)\; H_{t0}\exp(-\gamma_2\mathbf{n}_t.\mathbf{r})$$

Equating the tangential component of **H** at the boundary gives

$$H_{i0}\exp(-\gamma_1.x\sin\theta_i) + H_{r0}\exp(-\gamma_1 x.\sin\theta_r) = H_{t0}\exp(-\gamma_2 x.\sin\theta_t)$$

We shall assume henceforth both the media to be perfect dielectrics, so that $\sigma_1 = \sigma_2 = 0$. It then follows from the above that

$$H_{i0} + H_{r0} = H_{t0}$$

$$\gamma_1\sin\theta_i = \gamma_1\sin\theta_r = \gamma_2\sin\theta_t$$

so that

$$\theta_i = \theta_r, \frac{\sin\theta_i}{\sin\theta_t} = \sqrt{\frac{\varepsilon_2\mu_1}{\varepsilon_1\mu_2}}$$

which means the validity of Snell's law once again. Equating tangential components of **E** at the boundary gives

$$\eta_1.(\cos\theta_i).H_{i0} - \eta_1(\cos\theta_r).H_{r0} = \eta_2.(\cos\theta_t).H_{t0}$$

Defining

$$R' = \frac{H_{r0}}{H_{i0}}, T' = \frac{H_{t0}}{H_{i0}}$$

we get

$$1 + R' = T',\; \eta_1\cos\theta_i - \eta_1(\cos\theta_r)R' = \eta_2.(\cos\theta_t)T'$$

Thus,

$$\eta_1\cos\theta_i - \eta_1(\cos\theta_r)R' = \eta_2(\cos\theta_t)(1 + R')$$

This gives on using $\theta_r = \theta_i$,

$$R' = \frac{\eta_1\cos\theta_i - \eta_2\cos\theta_t}{\eta_1\cos\theta_i + \eta_2\cos\theta_t}$$

$$T' = \frac{2\eta_1\cos\theta_i}{\eta_1\cos\theta_i + \eta_2\cos\theta_t}$$

The above expressions give

$$R = \frac{|E_r|}{|E_i|} = \frac{|H_r|}{|H_i|} = |R'|$$

$$T = \frac{|E_t|}{|E_i|} = \frac{\eta_2}{\eta_1} T' = \left| \frac{2\eta_2 \cos\theta_i}{\eta_1 \cos\theta_i + \eta_2 \cos\theta_t} \right|$$

Brewster angle

The expression for R can be expressed in the case when $\mu_2 = \mu_1$ as

$$R = \frac{\sqrt{\varepsilon_2}\cos\theta_i - \sqrt{\varepsilon_1}\cos\theta_t}{\sqrt{\varepsilon_2}\cos\theta_i + \sqrt{\varepsilon_1}\cos\theta_t}$$

Clearly $R = 0$ when

$$\varepsilon_2 \cos^2\theta_i = \varepsilon_1 \cos^2\theta_t$$

or in view of Snell's law of refraction, when

$$\varepsilon_2 \cos^2\theta_i = \varepsilon_1 (1 - \frac{\varepsilon_1}{\varepsilon_2} \sin^2\theta_i)$$

or when

$$\sin\theta_i = \sqrt{\frac{\varepsilon_2}{\varepsilon_1 + \varepsilon_2}}$$

This angle is called the Brewster angle. At this critical angle all the power in the incident beam is transmitted, nothing is reflected.

Propagation of waves in rectangular waveguides

Consider a rectangular waveguide as shown in Fig. 5.4. with axis parallel to the $\hat{z}$ axis and sides parallel respectively to the $\hat{x}$ and $\hat{y}$ axes. We assume that one side of the rectangle is along the x axis from $x = 0$ to $x = a$ and the other side is orthogonal to it falling along the $\hat{y}$ axis stretching from $y = 0$ to $y = b$. The region within the waveguide is a dielectric which may have a conductivity. The parameters are ε, μ, σ. The boundary conditions are that the tangential component of the electric field **E** and the normal component of the magnetic field **B** vanishes at the walls of the waveguide. This is assuming that there do not exist any fields outside the waveguide. We are assuming that the electromagnetic waves have a sinusoidal time dependence and propagate along the z axis. Thus, the field phasors have the following dependence on space and time:

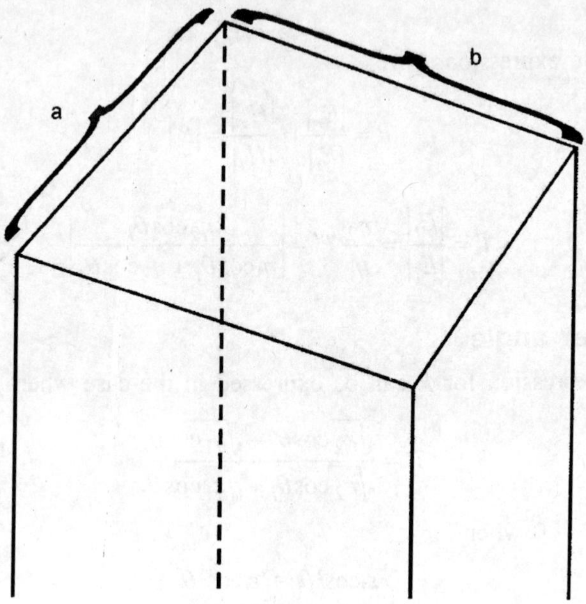

Fig. 5.4

$$\mathbf{E}(t, \mathbf{r}) = \mathbf{E}(x,y)\exp(j\omega t - \gamma z)$$

$$\mathbf{H}(t, \mathbf{r}) = \mathbf{H}(x,y)\exp(j\omega t - \gamma z)$$

The actual fields are the real parts of these phasors. Substituting these into the Maxwell equations and noting that the z derivatives are to be replaced by multiplication with $-\gamma$, the second Maxwell equations is in component form

$$\frac{\partial E_z}{\partial y} + \gamma E_y = -i\omega\mu H_x$$

$$\gamma E_x + \frac{\partial E_z}{\partial x} = i\omega\mu H_y$$

$$\frac{\partial E_y}{\partial x} - \frac{\partial E_x}{\partial y} = -i\omega\mu H_z$$

and the third Maxwell equation in component form is

$$\frac{\partial H_z}{\partial y} + \gamma H_y = (\sigma + i\omega\varepsilon)E_x$$

$$\gamma H_x + \frac{\partial H_z}{\partial x} = -(\sigma + i\omega\varepsilon)E_y$$

$$\frac{\partial H_y}{\partial x} - \frac{\partial H_x}{\partial y} = (\sigma + i\omega\varepsilon)E_z$$

Consider now the second equation in the set of first three and the first equation in the second set of three equations. The first one involves E_x, H_y and a derivative of E_z and the second one involves E_x, H_y and a derivative of H_z. We can thus solve these two equations to obtain E_x, H_y in terms of the partial derivatives of E_z and H_z. To do this elimination, we first cast the two equations in matrix form.

$$\begin{pmatrix} \gamma & -i\omega\mu \\ \sigma + i\omega\varepsilon & -\gamma \end{pmatrix}\begin{pmatrix} E_x \\ H_y \end{pmatrix} = \begin{pmatrix} -\dfrac{\partial E_z}{\partial x} \\ \dfrac{\partial H_z}{\partial y} \end{pmatrix}$$

The solution to this set is given by

$$\begin{pmatrix} E_x \\ H_y \end{pmatrix} = \frac{1}{\gamma^2 - i\mu\omega(\sigma + i\omega\varepsilon)}\begin{pmatrix} \gamma & -i\omega\mu \\ \sigma + i\omega\varepsilon & -\gamma \end{pmatrix}\begin{pmatrix} -\dfrac{\partial E_z}{\partial x} \\ \dfrac{\partial H_z}{\partial y} \end{pmatrix}$$

In full form

$$E_x = -\frac{1}{\gamma^2 - i\mu\omega(\sigma + i\omega\varepsilon)}\left(\gamma\frac{\partial E_z}{\partial x} + i\omega\mu\frac{\partial H_z}{\partial y}\right)$$

$$H_y = -\frac{1}{\gamma^2 - i\omega\mu(\sigma + i\omega\varepsilon)}\left((\sigma + i\omega\varepsilon)\frac{\partial E_z}{\partial x} + \gamma\frac{\partial H_z}{\partial y}\right)$$

Similarly we can solve the first equation of the first set and the second equation of the second set simultaneously to obtain E_y, H_x in terms of the partial derivatives of E_z and H_z. Cast in matrix form, these two equations read

$$\begin{pmatrix} \gamma & i\omega\mu \\ -(\sigma + i\omega\varepsilon) & -\gamma \end{pmatrix}\begin{pmatrix} E_y \\ H_x \end{pmatrix} = \begin{pmatrix} -\dfrac{\partial E_z}{\partial y} \\ \dfrac{\partial H_z}{\partial x} \end{pmatrix}$$

The solution to this pair is

$$E_y = \frac{1}{\gamma^2 - i\omega\mu(\sigma + i\omega\varepsilon)}\left(-\gamma\frac{\partial E_z}{\partial y} + i\omega\mu\frac{\partial H_z}{\partial x}\right)$$

$$H_x = \frac{1}{\gamma^2 - i\omega\mu(\sigma + i\omega\varepsilon)}\left((\sigma + i\omega\varepsilon)\frac{\partial E_z}{\partial y} - \gamma\frac{\partial H_z}{\partial x}\right)$$

This means that given E_z, H_z inside the guide, all the components of the electric and magnetic field are completely determined. We can consider two cases separately. The first is one in which $E_z \neq 0$, $H_z = 0$. Such waves are called transverse magnetic waves or TM modes. The second is one in which $E_z = 0$, $H_z \neq 0$. Such waves are called transverse electric waves or TE modes. The general case is clearly from the above formula obtained by superposing the solutions for the two modes. For the transverse magnetic mode:

$$E_x = -\frac{\gamma}{h^2}\frac{\partial E_z}{\partial x}$$

$$H_y = -\frac{\sigma + i\omega\varepsilon}{h^2}\frac{\partial E_z}{\partial x}$$

$$E_y = -\frac{\gamma}{h^2}\frac{\partial E_z}{\partial y}$$

$$H_x = -\frac{\sigma + i\omega\varepsilon}{h^2}\frac{\partial E_z}{\partial y}$$

and for the transverse electric mode,

$$E_x = -\frac{i\omega\mu}{h^2}\frac{\partial H_z}{\partial y}$$

$$H_y = -\frac{\gamma}{h^2}\frac{\partial H_z}{\partial y}$$

$$E_y = \frac{i\omega\mu}{h^2}\frac{\partial H_z}{\partial x}$$

$$H_x = -\frac{\gamma}{h^2}\frac{\partial H_z}{\partial x}$$

where

$$h^2 = \gamma^2 - i\omega\mu\,(\sigma + i\omega\varepsilon)$$

We consider the special case when $\sigma = 0$ that is the space filling the waveguide is a perfect dielectric. We then set $\gamma^2 = -\beta^2$ and the equation for h becomes

$$h^2 = -\beta^2 + \omega^2\mu\varepsilon$$

The Maxwell equations imply that E_z, H_z satisfy the wave equation and this translates to

$$\frac{\partial^2 E_z}{\partial x^2} + \frac{\partial^2 E_z}{\partial y^2} + (\gamma^2 + \omega^2\mu\varepsilon)E_z = 0$$

or

$$\frac{\partial^2 E_z}{\partial x^2} + \frac{\partial^2 E_z}{\partial y^2} = -h^2 E_z$$

and

$$\frac{\partial^2 H_z}{\partial x^2} + \frac{\partial^2 H_z}{\partial y^2} = -h^2 H_z$$

The solution of the first equation corresponding to the boundary condition $E_z = 0$ when $x = 0$, $x = a$, $y = 0$, $y = b$ is given by a superposition of the solutions

$$E_z(x, y) = A_{nm}\sin(n\pi x/a)\sin(m\pi y/b)$$

where

$$h^2 = (n\pi/a)^2 + (m\pi/b)^2$$

with n, m being positive integers. We do not have any transverse magnetic mode corresponding to $(0, m)$ or $(n, 0)$. The equation

$$\beta = \sqrt{\omega^2\mu\varepsilon - h^2}$$

means that if $\omega < \dfrac{h}{\sqrt{\mu\varepsilon}} = \omega_c$, β becomes imaginary implying that the wave gets attenuated as it propagates along the $\hat{z}$ direction. This means that for the (m,n)th mode of the wave to propagate without any attenuation along the $\hat{z}$ direction, the condition $\omega > \omega_c$ must be satisfied. ω_c is called the cutoff frequency of the (m,n)th mode. This is as far as the transverse magnetic mode is concerned. For the transverse electric mode, $E_z = 0$ but H_z is non-zero. The solution for H_z will be of the form

$$(A_1\cos(h_x x) + A_2\sin(h_x x)).\ (B_1\cos(h_y y) + B_2\sin(h_y y))$$

where $h^2 = h_x^2 + h_y^2$. E_x as seen above is proportional to $\frac{\partial H_z}{\partial y}$ which is in turn proportional to

$$(A_1\cos(h_x x) + A_2\sin(h_x x)).\ (B_1\sin(h_y y) + B_2\cos(h_y y))$$

This must vanish when $y = 0$ and $y = b$. This means that $B_2 = 0$. $H_y = m\pi/b$. Likewise, E_y is proportional to $\frac{\partial H_z}{\partial x}$ which in turn, is proportional to $(A_1\sin(h_x x) - A_2\cos(h_x x)).(B_1\cos(h_y y) + B_2\sin(h_y y))$. This must vanish when $x = 0$ and $x = a$. Thus, $A_2 = 0$, $h_x = n\pi/a$. Thus, the transverse electric modes have the H_z components proportional to $\cos(n\pi x/a)\cos(m\pi y/b)$. Once again, $\beta = \sqrt{\omega^2\mu\varepsilon - h^2}$ with $h^2 = (n\pi/a)^2 + (m\pi/b)^2$. Here, we can have $m = 0$ or $n = 0$. We leave it as an exercise to the student to compute the other components of the electric and magnetic field for the *TE* and *TM* modes.

Remark

When the medium filling the waveguides has a definite value for the conductivity σ, the expression for h remains unaltered, i.e. $h^2 = (n\pi/a)^2 + (m\pi/b)^2$ and so do the dependence of the solutions on x and y. However, the dependence of the solutions on z is of the form $\exp(-\gamma z)$ where

$$\gamma = \sqrt{h^2 + j\mu\omega(\sigma + j\omega\varepsilon)}$$

Propagation of waves in circular waveguides

Consider the propagation of electromagnetic waves through a cylindrical waveguide of radius R with perfectly conducting walls. The boundary conditions are that E_z, E_ϕ, H_θ should radius at $\rho = R$.

We first look at the equation $\nabla \times \mathbf{E} = -i\omega\mu\,\mathbf{H}$ in cylindrical co-ordinates. We have as in chapter 1,

$$\nabla \times \mathbf{E} = \det\begin{pmatrix} \frac{\hat{\rho}}{\rho} & \hat{\phi} & \frac{\hat{z}}{\rho} \\ \frac{\partial}{\partial \rho} & \frac{\partial}{\partial \phi} & \frac{\partial}{\partial z} \\ E_\rho & \rho E_\phi & E_z \end{pmatrix}$$

Equating this componentwise to $-j\omega\mu\mathbf{H}$, we get

$$\frac{1}{\rho}\frac{\partial E_z}{\partial \phi}-\frac{\partial E_\phi}{\partial z}=-i\omega\mu H_\rho$$

$$\frac{\partial E_\rho}{\partial z}-\frac{\partial E_z}{\partial \rho}=-i\omega\mu H_\phi$$

$$\frac{E_\phi}{\rho}+\frac{\partial E_\phi}{\partial \rho}-\frac{1}{\rho}\frac{\partial E_\rho}{\partial \phi}=-i\omega\mu H_z$$

Likewise the equation $\nabla \times \mathbf{H} = i\omega\varepsilon\,\mathbf{E}$ boils down to the system

$$\frac{1}{\rho}\frac{\partial H_z}{\partial \phi}-\frac{\partial H_\phi}{\partial z}=i\omega\varepsilon E_\rho$$

$$\frac{\partial H_\rho}{\partial z}-\frac{\partial H_z}{\partial \rho}=i\omega\varepsilon E_\phi$$

$$\frac{H_\phi}{\rho}+\frac{\partial H_\phi}{\partial \rho}-\frac{1}{\rho}\frac{\partial H_\rho}{\partial \phi}=i\omega\varepsilon E_z$$

We replace $\mathbf{E}(\rho, \phi, z)$ and $\mathbf{H}(\rho, \phi, z)$ respectively by $\mathbf{E}(\rho, \phi)\exp(-\gamma z)$ and $\mathbf{H}(\rho, \phi)\exp(-\gamma z)$ in the above equations to get

$$\frac{1}{\rho}\frac{\partial E_z}{\partial \phi}+\gamma E_\phi=-i\omega\mu H_\rho$$

$$\gamma E_\rho+\frac{\partial E_z}{\partial \rho}=i\omega\mu H_\phi$$

$$\frac{1}{\rho}E_\phi+\frac{\partial E_\phi}{\partial \rho}-\frac{1}{\rho}\frac{\partial E_\rho}{\partial \phi}=-i\omega\mu H_z$$

$$\frac{1}{\rho}\frac{\partial H_z}{\partial \phi}+\gamma H_\phi=i\omega\varepsilon E_\rho$$

$$\gamma H_\rho+\frac{\partial H_z}{\partial \rho}=-i\omega\varepsilon E_\phi$$

$$\frac{1}{\rho}H_\phi+\frac{\partial H_\phi}{\partial \rho}-\frac{1}{\rho}\frac{\partial H_\rho}{\partial \phi}=i\omega\varepsilon E_z$$

Solving the first equation of the first set and the second equation of the second set for E_ϕ, H_ρ, we get

$$E_\phi = \frac{j\omega\mu}{h^2}\frac{\partial H_z}{\partial \rho} - \frac{\gamma}{\rho h^2}\frac{\partial E_z}{\partial \phi}$$

$$H_\rho = \frac{j\omega\varepsilon}{\rho h^2}\frac{\partial E_z}{\partial \phi} - \frac{\gamma}{h^2}\frac{\partial H_z}{\partial \rho}$$

Solving the second equation of the first set with the first of the second set gives

$$E_\rho = -\frac{\gamma}{h^2}\frac{\partial E_z}{\partial \rho} - \frac{j\omega\mu}{\rho h^2}\frac{\partial H_z}{\partial \phi}$$

$$H_\phi = -\frac{j\omega\varepsilon}{h^2}\frac{\partial E_z}{\partial \rho} - \frac{\gamma}{\rho h^2}\frac{\partial H_z}{\partial \phi}$$

Once again, we can divide the class of solutions into two parts. The first is the transverse electric TE mode and the second is the transverse magnetic TM mode. The general solution is obtained as superposition of these two kinds of solution. For the TM mode, $H_z = 0$ and we have

$$E_\rho = -\frac{\gamma}{h^2}\frac{\partial E_z}{\partial \rho}, \quad E_\phi = -\frac{\gamma}{\rho h^2}\frac{\partial E_z}{\partial \phi}$$

$$H_\rho = \frac{j\omega\varepsilon}{\rho h^2}\frac{\partial E_z}{\partial \phi}, \quad H_\phi = -\frac{j\omega\varepsilon}{h^2}\frac{\partial E_z}{\partial \rho}$$

and for the TE mode, we have $E_z = 0$ and we have

$$E_\rho = -\frac{j\omega\mu}{\rho h^2}\frac{\partial H_z}{\partial \phi}, E_\phi = \frac{j\omega\mu}{h^2}\frac{\partial H_z}{\partial \rho}$$

$$H_\rho = -\frac{\gamma}{h^2}\frac{\partial H_z}{\partial \rho}, H_\phi = -\frac{\gamma}{\rho h^2}\frac{\partial H_z}{\partial \phi}$$

We first look at the TM solutions. H_z satisfies the wave equation

$$\frac{1}{\rho}\frac{\partial}{\partial \rho}\rho\frac{\partial H_z}{\partial \rho} + \frac{1}{\rho^2}\frac{\partial^2 H_z}{\partial \phi^2} + h^2 H_z = 0$$

where

$$h^2 = \omega^2\mu\varepsilon + \gamma^2 = \omega^2\mu\varepsilon - \beta^2$$

assuming $\gamma = j\beta$. We try a solution of the form $H_z(\rho, \phi) = R(\rho)\,\Phi(\phi)$. Plugging this into the wave equation gives

$$(\rho.R'(\rho))'/\rho.R(\rho) + \Phi(\phi)/\rho^2\,\Phi(\phi) + h^2 = 0$$

or

$$\rho(\rho.R'(\rho))'/R(\rho) + h^2\rho^2 = -\Phi(\phi)/\Phi(\phi)$$

The left side is a function of ρ only and the right side of ϕ only. Hence, both sides must equal a constant, say m^2. This gives $\Phi(\phi) = A.\cos(m\phi) + B.\sin(m\phi)$, $m = 1, 2, \ldots$ and the equation satisfied by $R(\rho)$ is

$$\rho^2R''(\rho) + \rho.R'(\rho) + (h^2\rho^2 - m^2)R(\rho) = 0$$

the solution to which having no singularity at $\rho = 0$ is

$$R(\rho) = C.J_m(h\rho)$$

h is determined by the condition that E_ϕ should vanish at the walls of the waveguide. Now,

$$H_z(\rho,\phi) = J_m(h\rho)(A.\cos(m\phi) + B.\sin(m\phi))$$

Plugging this into the expression for E_ϕ, we get

$$E_\phi = \frac{j\omega\mu}{h^2}\frac{\partial H_z}{\partial\rho}$$

which is proportional to $J'_m(h\rho)$. This must vanish when $\rho = R$ and hence, $h = \alpha_{m,n}/R$, $n = 1, 2, \ldots$, where $\alpha'_{m,n}s$ are the roots of $J'_m(x) = 0$. We leave the exercise of calculating the explicit expressions for the components of the electric and magnetic fields inside the waveguide. The cutoff frequency below which the (m,n)th mode does not propagate but rather gets attenuated is given by $\omega_c = \alpha_{m,n}/\sqrt{\mu\varepsilon}$. For $\omega > \omega_c$, the propagation constant is given by

$$\beta = \sqrt{\omega^2\mu\varepsilon - \alpha_{m,n}^2 / R^2}$$

We next look at the TE solutions. E_z satisfies the same wave equation as H_z and we can write its solution as

$$E_z = J_m(h\rho).(A.\cos(m\phi) + B.\sin(m\phi))$$

The boundary condition that E_z vanish when $\rho = R$ gives $h = \delta_{m,n}/R$, $n = 1, 2, \ldots$, where $\delta_{m,n}$, $n = 1, 2, \ldots$ are the roots of $J_m(x) = 0$. This completes the discussion of wave propagation in waveguides.

Transmission Line Theory

We consider a pair of long wires parallel to each other connecting a source at one end to a load at another end. We assume that per unit length the capacitance between the two wires is $C(z)$ where z is the distance along the wires measured from the source end. The inductance parallel to the wires is $L(z)$ per unit length and the resistance parallel to the wires per unit length is $R(z)$. The conductance (reciprocal resistance) between the wires per unit length is $G(z)$. Consider section of the line between z and $z + dz$. The circuit diagram of this section looks as shown in Fig. 5.5. It consists of a series connection of a resistance $R(z)dz$ and an inductor $L(z)dz$ and a parallel connection of a conductance $G(z)dz$ and a capacitance $C(z)dz$. The current entering into the section is $I(t, z)$ and the current leaving the section is $I(t, z + dz)$. The voltage at the node where the current enters is $V(t, z)$ and the voltage at the node where the current leaves is $V(t, z + dz)$. The drop across the section is

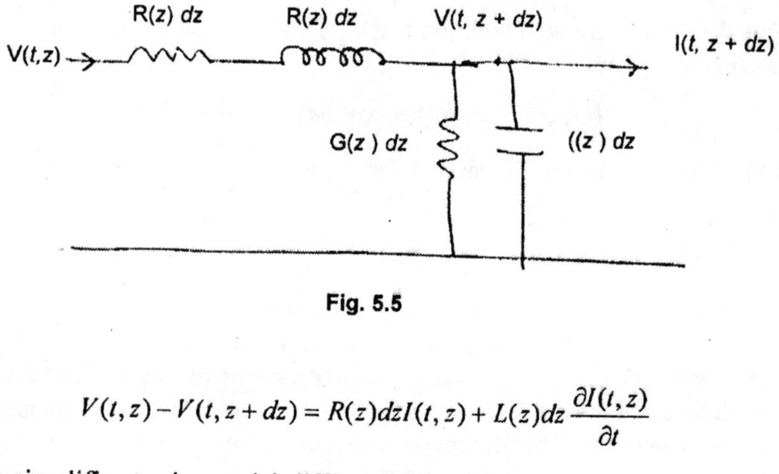

Fig. 5.5

$$V(t,z) - V(t,z+dz) = R(z)dzI(t,z) + L(z)dz\frac{\partial I(t,z)}{\partial t}$$

This simplifies to the partial differential equation

$$-\frac{\partial V}{\partial z} = RI + L\frac{\partial I}{\partial t}$$

The current entering into the parallel combination of the capacitance and the conductance is

$$I(t,z) - I(t,z+dz) = \left(C\frac{\partial V(t,z)}{\partial t} + G(z)V(t,z)\right)dz$$

which simplifies to the partial differential equation

$$-\frac{\partial I}{\partial z} = GV + C\frac{\partial V}{\partial t}$$

We can eliminate I between these two equations and derive a partial differential equation for V or equivalently, eliminate V between the two equations and derive a partial differential equation for I. For example letting $V(t, 0)$ denote the voltage at the source end, we have on integrating the first equation

$$V(t,z) = V(t,0) - \int_0^z R(\xi)I(t,\xi)d\xi - \int_0^z L(\xi)\frac{\partial I(t,\xi)}{\partial t}d\xi$$

Plugging this into the second one gives

$$-\frac{\partial I}{\partial z} = GV(t,0) - G(z)\int_0^z R(\xi)I(t,\xi)d\xi - G(z)\int_0^z L(\xi)\frac{\partial I(t,\xi)}{\partial t}d\xi$$
$$+C(z)\frac{\partial V(t,0)}{\partial t} - C(z)\int_0^z R(\xi)\frac{\partial I(t,\xi)}{\partial t}d\xi - C(z)\int_0^z L(\xi)\frac{\partial^2 I(t,\xi)}{\partial t^2}d\xi$$

We devide both sides by $C(z)$ and take the partial derivative with respect to z to get

$$-\frac{\partial}{\partial z}\frac{1}{C(z)}\frac{\partial I}{\partial z} = -R(z)\frac{\partial I}{\partial t} - L(z)\frac{\partial^2 I}{\partial t^2}$$
$$+\frac{\partial}{\partial z}\frac{G(z)}{C(z)}V(t,0) - \frac{\partial}{\partial z}\frac{G(z)}{C(z)}\int R(\xi)I(t,\xi)d\xi$$
$$-\frac{\partial}{\partial z}\frac{G(z)}{C(z)}\int_0^z L(\xi)\frac{\partial I(t,\xi)}{\partial t}d\xi$$

which can be rearranged as

$$L(z)\frac{\partial^2 I}{\partial t^2} + R(z)\frac{\partial I}{\partial t} - \frac{\partial}{\partial z}\frac{1}{C(z)}\frac{\partial I}{\partial z}$$
$$-\frac{\partial}{\partial z}\frac{G(z)}{C(z)}V(t,0) + \frac{\partial}{\partial z}\frac{G(z)}{C(z)}\int R(\xi)I(t,\xi)d\xi$$
$$+\frac{\partial}{\partial z}\frac{G(z)}{C(z)}\int_0^z L(\xi)\frac{\partial I(t,\xi)}{\partial t}d\xi = 0$$

This is an integro differential equation for the current. We can obtain the solution for the current and voltage waveforms in the special case when the parameters, R, L, C, G are constant as a function of the distance from source. The relevant differential equations are

$$-\frac{\partial I}{\partial z} = GV + C\frac{\partial V}{\partial t}$$

$$-\frac{\partial V}{\partial z} = RI + L\frac{\partial I}{\partial t}$$

with G, C, R, L constants. Assuming sinusoidal current and voltage variations, i.e. setting $I(t, z) = \text{Re}(I(z)\exp(i\omega t))$ and $V(t, z) = \text{Re}(V(z)\exp(i\omega t))$, we get

$$-\frac{dI(z)}{dz} = (G + i\omega C)V(z)$$

$$-\frac{dV(z)}{dz} = (R + i\omega L)I(z)$$

We differentiate the first equation here with respect to z, multiply the second equation by $(G + i\omega C)$ and add the two to get

$$\frac{d^2 I(z)}{dz^2} = \gamma^2 I(z)$$

Likewise, eliminating $V(z)$ between the two equations gives

$$\frac{d^2 V(z)}{dz^2} = \gamma^2 V(z)$$

where

$$\gamma^2 = (G + i\omega C)\,(R + i\omega L)$$

The solution for the voltage phasor along the line is given by

$$V(z) = V_+\exp(-\gamma z) + V_-\exp(\gamma z)$$

We are assuming that the source end $z = 0$ is excited by a sinusoidal waveform of frequency ω. The current is determined from the equation

$$(R + i\omega L)I = -\frac{dV(z)}{dz} = \gamma V_+\exp(-\gamma z) - \gamma V_-\exp(\gamma z)$$

Thus,

$$I(z) = \frac{V_+}{Z_0}\exp(-\gamma z) - \frac{V_-}{Z_0}\exp(\gamma z)$$

where

$$Z_0 = \frac{R + i\omega L}{\gamma} = \sqrt{\frac{R + i\omega L}{G + i\omega C}}$$

is called the characteristic impedance of the transmission line. Suppose that we have a load Z_L that terminates the end of the transmission line at $z = L$. Then,

$$Z_L = \frac{V(L)}{I(L)} = Z_0 \frac{V_+ \exp(-\gamma L) + V_- \exp(\gamma L)}{V_+ \exp(-\gamma L) - V_- \exp(\gamma L)}$$

and defining the ratio

$$r = \frac{V_-}{V_+}$$

which is the ratio of the amplitude of the reflected wave to the transmitted wave at the source end $z = 0$, we can express the above equation as

$$\frac{Z_L}{Z_0} = \frac{1 + r.\exp(2\gamma L)}{1 - r.\exp(2\gamma L)}$$

and hence

$$r.\exp(2\gamma L) = \frac{Z_L - Z_0}{Z_L + Z_0}$$

We can use this expression to calculate the impedance looking into the transmission line from the point z along the positive z direction. This impedance is given by

$$Z(z) = \frac{V(z)}{I(z)} = Z_0 \frac{V_+ \exp(-\gamma z) + V_- \exp(\gamma z)}{V_+ \exp(-\gamma z) - V_- \exp(\gamma z)}$$

$$= Z_0 \frac{1 + r.\exp(2\gamma z)}{1 - r.\exp(-2\gamma z)} = Z_0 . \frac{(Z_L + Z_0) + (Z_L - Z_0)\exp(-2\gamma d)}{(Z_L + Z_0) - (Z_L - Z_0)\exp(-2\gamma d)}$$

$$= Z_0 \frac{Z_L \cosh(\gamma d) + Z_0 \sinh(\gamma d)}{Z_L \sinh(\gamma d) + Z_0 \cosh(\gamma d)} = Z_0 \frac{Z_0 \tanh(\gamma d) + Z_L}{Z_L \tanh(\gamma d) + Z_0}$$

with $d = L - z$. In conclusion, it should be mentioned that the transmission line is lossless when $R = 0$, $G = 0$. The equations satisfied by the voltage and current phasors are then

$$-\frac{dI(z)}{dz} = i\omega C V(z)$$

$$-\frac{dV(z)}{dz} = i\omega LI(z)$$

Elimination of $I(z)$ leads to

$$\frac{d^2V(z)}{dz^2} + k^2V(z) = 0$$

and likewise elimination of $I(z)$ gives

$$\frac{d^2I(z)}{dz^2} + k^2I(z) = 0$$

where $k^2 = \omega^2 LC$. The characteristic impedance of the line is given by $Z_0 = \sqrt{\frac{L}{C}}$. The voltage and current propagate as waves in both the forward and reverse direction without attenuation:

$$V(z) = V_+\exp(-ikz) + V_-\exp(ikz),\ I(z) = \frac{V_+}{Z_0}\exp(-ikz) - \frac{V_-}{Z_0}\exp(ikz)$$

Consider for example, an open-ended transmission line at $z = L$ with a sinusoidal voltage excitation $V(t, 0) = A.\cos(\omega t)$. We then have $V(0) = A$ for the voltage phasor at $z = 0$. The boundary conditions are

$$V(0) = A = V_+ + V_-,\ I(L) = 0 = \frac{V_+}{Z_0}\exp(-ikL) - \frac{V_-}{Z_0}\exp(ikL)$$

This gives

$$V_+ = \frac{A.\exp(ikL)}{2.\cos(kL)},\ V_- = \frac{A.\exp(-ikL)}{2.\cos(kL)}$$

and hence

$$V(t,z) = \text{Re}(V(z)\exp(i\omega t)) = \frac{A}{2\cos(kL)}(\cos(\omega t + kd) + \cos(\omega t - kd))$$

$$= \frac{A.\cos(kd)}{\cos(kL)}\cos(\omega t)$$

where $d = L - z$. This shows that the voltage is a standing wave along an open-ended line.

Some nonlinear problems associated with electromagnetic fields and wave propagation

The first nonlinear problem we consider here is the electrostatic problem involving determining the potential and field in a medium whose permittivity depends on the field itself. The relevant equation is the Poisson equation with the field dependent permittivity introduced. That is

$$\nabla.(\varepsilon(\mathbf{E})\mathbf{E}) = 0, \ \nabla \times \mathbf{E} = \mathbf{0}$$

The second equation implies that $\mathbf{E} = -\nabla V$ where $\mathbf{E}$ is the electrostatic potential and the first one yields

$$\nabla.(\varepsilon(\nabla V)\nabla V) = 0$$

We may make a Taylor expansion of the permittivity retaining only terms up to second order in the fields. Thus,

$$\varepsilon(\mathbf{E}) \approx \varepsilon_0 + \sum_{r=1}^{3} \varepsilon_{1,r} E_r + \frac{1}{2} \sum_{r,s=1}^{3} \varepsilon_{2,r,s} E_r E_s$$

wherefor convenience of notation, E_1, E_2, E_3 are used in place of E_x, E_y, E_z respectively. When expressed in terms of the potential, these are

$$\varepsilon(\mathbf{E}) \approx \varepsilon_0 - \sum_{r=1}^{3} \varepsilon_{1,r} \frac{\partial V}{\partial x_r} + \frac{1}{2} \sum \varepsilon_{2,r,s} \frac{\partial V}{\partial x_r} \frac{\partial V}{\partial x_s}$$

The modified Poisson equation is

$$\varepsilon \nabla^2 V + (\nabla \varepsilon, \nabla V) = 0$$

or

$$\nabla^2 V + (\nabla \log(\varepsilon), \nabla V) = 0$$

Now,

$$\log(\varepsilon) \approx \log(\varepsilon_0) - \sum_{r=1}^{3} \frac{\varepsilon_{1,r}}{\varepsilon_0} E_{1,r} + \frac{1}{2} \sum_{r,s=1}^{3} \left(\frac{\varepsilon_{2,r,s}}{\varepsilon_0} - \frac{\varepsilon_{1,r}\varepsilon_{1,s}}{\varepsilon_0^2} \right) E_r E_s$$

For ease of notation we define

$$\chi_{1,r} = \frac{\varepsilon_{1,r}}{\varepsilon_0}, \chi_{2,r,s} = \frac{\varepsilon_{2,r,s}}{\varepsilon_0} - \frac{\varepsilon_{1,r}\varepsilon_{1,s}}{\varepsilon_0^2}$$

Then,

$$\frac{\partial \log(\varepsilon)}{\partial x_m} \approx -\sum_{r=1}^{3} \chi_{1,r} \frac{\partial^2 V}{\partial x_r x_m} + \frac{1}{2} \sum_{r,s=1}^{3} \chi_{2,r,s} \frac{\partial}{\partial x_m} \left(\frac{\partial V}{\partial x_r} \frac{\partial V}{\partial x_s} \right)$$

This gives up to terms of third degree in V, the equation

$$\nabla^2 V - \sum_{r,m=1}^{3} \chi_{1,r} \frac{\partial V}{\partial x_m} \frac{\partial^2 V}{\partial x_r x_m} + \frac{1}{2} \sum_{r,s,m=1}^{3} \chi_{2,r,s} \frac{\partial V}{\partial x_m} \frac{\partial}{\partial x_m} \left(\frac{\partial V}{\partial x_r} \frac{\partial V}{\partial x_s} \right) = \mathbf{0}$$

We leave it as an exercise to the student to formulate numerical methods for solving this equation with Dirichlet or Neumann boundary conditions and hence determine the relation connecting potential and charge for conductors of given shape. The other interesting nonlinear problem in propagation of electromagnetic waves is the situation when there is no charge or current but the permittivity, permeability and conductivity are field dependent. This happens for example while dealing with waves propagating in the ionosphere. For simplicity assume that only the permittivity ε depends on the electric field and that the permeability and conductivity are constant. Then, the relevant Maxwell equations are

$$\nabla . \varepsilon \mathbf{E} = 0, \qquad \nabla \times \mathbf{E} = -\mu \frac{\partial \mathbf{H}}{\partial t}$$

$$\nabla \times \mathbf{H} = \sigma \mathbf{E} + \frac{\partial \varepsilon \mathbf{E}}{\partial t}, \nabla . \mathbf{H} = 0$$

with $\varepsilon = \varepsilon(\mathbf{E})$. Taking the curl of the second equation and using the expression for $\nabla \times \mathbf{H}$ from the third and $\nabla.\mathbf{E} = -\nabla\varepsilon.\mathbf{E}/\varepsilon$ from the first, we get the nonlinear wave propagation equation

$$\nabla^2 \mathbf{E} - \mu\sigma \frac{\partial \mathbf{E}}{\partial t} - \nabla(\nabla\varepsilon / \varepsilon, \mathbf{E}) - \mu \frac{\partial^2 \varepsilon \mathbf{E}}{\partial t^2} = \mathbf{0}$$

We leave it as an exercise to derive the full expansion of this equation making a Taylor expansion of ε in powers of the electric field components and truncating up to two terms the expansion. This nonlinear wave propagation leads to frequency mixing observed in experiments of nonlinear optics. An incident electric field having many frequencies, after propagating through the channel produces frequencies at the output which are linear combinations of the input frequencies.

Solved Problems and Review Questions

1. Derive the expressions for the electric and magnetic fields produced by an infinitesimal current element when the current has sinusoidal variation.

Ans. Assume that the current element dl is oriented along the $\hat{z}$ axis and is located at the origin. Then, the magnetic vector potential is given by

$$\mathbf{A}(\mathbf{r}) = \frac{\mu I dl}{4\pi} \frac{\exp(-ikr)}{r} \hat{z}$$

where $k = \omega/c$ with $c = 1/\sqrt{\varepsilon\mu}$ and ω the alternating frequency of the current. We find for the polar components of the vector potential

$$A_r = A_z \cos\theta = \frac{\mu I dl}{4\pi} \frac{\exp(-ikr)}{r} \cos\theta$$

$$A_\theta = A_z \sin\theta = \frac{\mu I dl}{4\pi} \frac{\exp(-ikr)}{r} \sin\theta$$

$$A_\phi = 0$$

We can also calculate the divergence of **A**:

$$\nabla.\mathbf{A} = \frac{\partial A_z}{\partial_z} = \frac{z}{r} \frac{\mu I dl}{4\pi} \frac{d}{dr} \frac{\exp(-ikr)}{r}$$

$$= \frac{\mu I dl \cos\theta}{4\pi} \exp(-ikr)(-ik/r - 1/r^2)$$

The Lorentz Gauge condition translates for sinusoidally varying fields to $\nabla.\mathbf{A} = -i\omega\varepsilon\mu V$. This gives

$$V = \frac{i}{\omega\varepsilon\mu} \nabla.\mathbf{A} = -\frac{iI\cos\theta}{4\pi\varepsilon\omega} \exp(-ikr)(ik/r + 1/r^2)$$

The expression for the electric field phasor is

$$\mathbf{E} = -\nabla V - i\omega\mathbf{A}$$

∇V is computed using the expression for the gradient in polar co-ordinates:

$$\nabla V = \hat{r}\frac{\partial V}{\partial r} + \frac{\hat{\theta}}{r}\frac{\partial V}{\partial \theta}$$

since V does not depend upon ϕ. Now,

$$\frac{\partial V}{\partial r} = \frac{iI\cos\theta}{4\pi\varepsilon\omega} \exp(-ikr)(ik/r^2 + 2/r^3) - \frac{I\cos\theta}{4\pi\varepsilon\omega}(ik^2/r + k/r^2)$$

Noting that $\omega = k/\sqrt{\varepsilon\mu}$ and $\eta = \sqrt{\mu/\varepsilon}$ where η is the impedance of the medium, we can write

$$\frac{\partial V}{\partial r} = -\frac{I\eta\cos\theta}{4\pi r^2}\exp(-ikr)(1+2/ikr) - \frac{I\eta\cos\theta}{4\pi}\exp(-ikr)(ik/r+1/r^2)$$

$$= -\frac{I\eta\cos\theta}{4\pi}\exp(-ikr)(ik/r+2/r^2+2/ikr^3)$$

Also

$$\frac{1}{r}\frac{\partial V}{\partial \theta} = \frac{iI\sin\theta}{4\pi\varepsilon\omega}\exp(-ikr)(ik/r^2+1/r^3)$$

Thus,

$$E_r = -i\omega A_r - \frac{\partial V}{\partial r} = \frac{-i\omega\mu l}{4\pi}\frac{\exp(-ikr)}{r}\cos\theta + \frac{I\eta\cos\theta}{4\pi}\exp(-ikr)(ik/$$

$$r+2/r^2+2/ikr^3) = -\frac{ik\eta I\cos\theta}{4\pi r}\exp(-ikr) + \frac{I\eta\cos\theta}{4\pi}\exp(-ikr)(ik/$$

$$r+2/r^2+2/ikr^3) = -\frac{I\eta\cos\theta}{2\pi r^2}\exp(-ikr)(1+1/ikr)$$

Likewise, we find

$$E_\theta = -i\omega A_\theta - \frac{1}{r}\frac{\partial V}{\partial \theta} = -\frac{i\omega\mu I}{4\pi r}\exp(ikr)\sin\theta - \frac{iI\sin\theta}{4\pi\varepsilon\omega}\exp(ikr)$$

$$(ik/r=1/r^2) - \frac{i\eta kI\sin\theta}{4\pi r}\exp(-ikr) - \frac{I\eta I\sin\theta}{4\pi}\exp(ikr)\,(i/r^2+1/kr^3)$$

$$\frac{\eta I\sin\theta}{4\pi}\exp(-ikr)(-ik/r=1/r^2-i/kr^3)$$

We see that only the $\hat{\theta}$ component of the electric field has a $1/r$ dependence at large values of r. This contributes to radiation.

2. A plane defined by the equation $lx + my + nz = p$ separates two media one and two. Medium one is the side $lx + my + nz < p$ and medium two is the side $lx + my + nz > p$. The permittivity of medium 1 is ε_1 and the permittivity of medium 2 is ε_2. Let σ denote the charge density on the plane. Write down the relations between the components of the electric fields on both sides of the plane assuming these fields to be constant. Repeat the problem for the magnetic fields on both sides of the plane if the plane carries a constant surface current **K** in a direction tangential to the plane and the magnetic permeabilities on the two sides are respectively μ_1 and μ_2.

Ans. The condition $\varepsilon_2\mathbf{E}_2.\mathbf{n} - \varepsilon_1\mathbf{E}_1.\mathbf{n} = \sigma$ gives

$$\varepsilon_2(lE_{2x} + mE_{2y} + nE_{2z}) - \varepsilon_1(lE_{1x} + mE_{1y} + nE_{1z}) = \sigma$$

The condition $\mathbf{n} \times (\mathbf{E}_2 - \mathbf{E}_1) = \mathbf{0}$ gives three equations out of which any two are linearly independent. These three equations are

$$mE_{2z} - nE_{2y} = mE_{1z} - nE_{1y}$$

$$nE_{2x} - lE_{2z} = nE_{1x} - lE_{1z}$$

$$lE_{2y} - mE_{2x} = lE_{1y} - mE_{1x}$$

We leave it as an exercise to the student to arrange the three linearly independent equations so obtained into matrix form and solve for $\mathbf{E}_2$ in terms of $\mathbf{E}_1$. The continuity equation for the normal component of the magnetic field is $\mathbf{B}_2.\mathbf{n} - \mathbf{B}_1.\mathbf{n} = 0$, which can be expressed as

$$\mu_2(lH_{2x} + mH_{2y} + nH_{2z}) = \mu_1(lH_{1x} + mH_{1y} + nH_{1z})$$

The equation $\mathbf{n} \times (\mathbf{H}_1 - \mathbf{H}_2) = -\mathbf{K}$, is

$$(mH_{2z} + nH_{2y}) - (mH_{1z} - nH_{1y}) = -K_x$$

$$(nH_{2z} + lH_{2x}) - (nH_{1z} - lH_{1x}) = -K_y$$

$$(lH_{2y} + mH_{2x}) - (lH_{1y} - mH_{1x}) = -K_z$$

Note that $lK_x + mK_y + nK_z = 0$. Only two of the last three equations are linearly independent. We leave it as an exercise to solve for $\mathbf{H}_2$ in terms of $\mathbf{H}_1$ after arranging the equations in matrix form.

3. A uniform plane wave in air with $\mathbf{E}_i = E_{i0}.\cos(\omega t - k_x x - k_z z)\ \hat{y}$ V/m is incident on a dielectric slab $z \geq 0$ with relative permeability μ_r and dielectric constant ε_r, $\sigma = 0$. Find

(a) The polarization of the wave

(b) The angle of incidence

(c) The reflected **E** field

(d) The transmitted **H** field

Your answer should be in terms of E_{i0}, k_x, k_z, μ_r, ε_r only.

Ans. We have the propagation vector of the incident wave given by $(k_x, k_y, k_z) = (k_x, 0, k_z)$. We are assuming that k_x, k_z are positive. The incident electric field has a $\hat{y}$ component which is perpendicular to the plane of incidence. Thus, the polarization is perpendicular. The angle of incidence is given by

$$\cos(\theta_i) = (k_x, 0, k_z).(0,0,1)/\sqrt{k_x^2 + k_z^2} = \frac{k_z}{\sqrt{k_x^2 + k_z^2}} = \alpha$$

say. Thus, $\theta_i = \cos^{-1}(\alpha)$. We assume the reflected wave to be given by

$$\mathbf{E}_r = E_{r0}.\cos(\omega t - k'_x x - k'_z z + \phi')\,\hat{y}$$

and the transmitted wave by

$$\mathbf{E}_t = E_{t0}.\cos(\omega t - k''_x x - k''_z z + \phi'')\,\hat{y}$$

Equating the tangential components of **E** at the boundary $z = 0$ gives

$$E_{i0}\cos(\omega t - k_x x) + E_{r0}\cos(\omega t - k'_x x + \phi') = E_{t0}\cos(\omega t - k''_x x + \phi'')$$

from which we conclude that

$$k_x = k'_x = k''_x$$

and hence

$$(E_{i0} + E_{r0}\cos(\phi') - E_{t0}\cos(\phi'')).\cos(\omega t - k_x x) - (E_{r0}\sin\phi' - E_{t0}\sin\phi'').$$
$$\sin(\omega t - k_x x) = 0$$

so that

$$E_{i0} + E_{r0}\cos\phi' = E_{t0}\cos\phi''$$
$$E_{r0}\sin\phi' = E_{t0}\sin\phi''$$

We next calculate the incident, reflected and transmitted magnetic fields.

$$\nabla \times \mathbf{E}_i = E_{i0}(k_x\hat{x} + k_z\hat{z}) \times \hat{y}.\sin(\omega t - k_x x - k_z z)$$
$$= E_{i0}(k_x\hat{z} - k_z\hat{x}).\sin(\omega t - k_x x - k_z z)$$

This must equal $-\mu_0 \dfrac{\partial H_i}{\partial t}$. It follows that

$$\mathbf{H}_i = \frac{E_{i0}}{\mu_0 \omega}(k_x\hat{z} - k_z\hat{x}).\cos(\omega t - k_x x - k_z z)$$

Likewise,

$$\nabla \times \mathbf{E}_r = E_{r0}(k'_x\hat{z} - k'_z\hat{x}).\sin(\omega t - k'_x x - k'_z z + \phi')$$

and hence

$$\mathbf{H}_r = \frac{E_{r0}}{\mu_0 \omega}(k'_x\hat{z} - k'_z\hat{x}).\cos(\omega t - k'_x x - k'_z z + \phi')$$

Finally,

$$\mathbf{H}_t = \frac{E_{t0}}{\mu_0 \mu_r \omega}(k_x'' \hat{z} - k_z'' \hat{x}).\cos(\omega t - k_x'' x - k_z'' z + \phi'')$$

Equating tangential components of **H** at the boundary gives

$$\frac{E_{i0}}{\mu_0 \omega} k_z.\cos(\omega t - k_x x) + \frac{E_{r0}}{\mu_0 \omega} k_z'.\cos(\omega t - k_x' x + \phi') =$$

$$\frac{E_{t0}}{\mu_0 \mu_r \omega} k_z''.\cos(\omega t - k_x'' x + \phi'')$$

Recalling that $k_x = k_x' = k_x''$ and equating coefficients of cos $(\omega t - k_x x)$ and sin $(\omega t - k_x x)$ on both sides, we get

$$\frac{E_{i0} k_z}{\mu_0 \omega} + \frac{E_{r0} k_z'}{\mu_0 \omega}\cos\phi' = \frac{E_{t0} k_z''}{\mu_0 \mu_r \omega}\cos\phi''$$

$$\frac{E_{r0} k_z'}{\mu_0 \omega}\sin\phi' = \frac{E_{t0} k_z''}{\mu_0 \mu_r \omega}\sin\phi''$$

We also note that if c_1 denotes the velocity of light in medium 1 and c_2 that in medium 2, then $c_2 = c_1 / \sqrt{\mu_r \varepsilon_r}$. Also,

$$k_x^2 + k_z^2 = \omega^2 / c_1^2$$

so that

$$\omega = c_1 \sqrt{k_x^2 + k_z^2}$$

$$k_x'^2 + k_z'^2 = k_x^2 + k_z'^2 = \omega^2 / c_1^2$$

and hence $k_z' = \pm k_z$. Since the reflected wave cannot coincide with the incident wave and since $k_x' = k_x$, it must therefore follow that $k_z' = -k_z$. Finally,

$$k_x''^2 + k_z''^2 = k_x^2 + k_z''^2 = \omega^2 / c_2^2 == \omega^2 \mu_r \varepsilon_r / c_1^2 = (k_x^2 + k_z^2)\mu_r \varepsilon_r$$

and hence

$$k_z'' = \sqrt{(k_x^2 + k_z^2)\mu_r \varepsilon_r - k_x^2} = \sqrt{(\mu_r \varepsilon_r - 1)k_x^2 + \mu_r \varepsilon_r k_z^2}$$

We have derived two equations connecting $\sin\phi'$ and $\sin\phi''$. They are of the form

$$E_{r0}\sin\phi' = E_{t0}\sin\phi''$$

and

$$AE_{r0}\sin\phi' = BE_{t0}\sin\phi''$$

where

$$B/A = k''_z / k'_z \mu_0\mu_r$$

is negative since $k'_z < 0$. It must therefore follow that $\sin\phi' = \sin\phi'' = 0$, or $\phi' = \phi'' = 0$. The other two equations involving $\cos\phi'$, $\cos\phi''$ thus reduce to

$$E_{i0} + E_{r0} = E_{t0}$$
$$k_z E_{i0} + k'_z E_{r0} = k''_z E_{t0}$$

the second one here is the same as

$$E_{i0} - E_{r0} = \beta.E_{t0}$$

where

$$\beta = k''_z / k_z = \sqrt{(\mu_r\varepsilon_r - 1)(k_x / k_z)^2 + \mu_r\varepsilon_r}$$

Solving for E_{r0}, E_{t0}, we get

$$E_{t0} = \frac{2E_{i0}}{1+\beta}, E_{r0} = \left(\frac{1-\beta}{1+\beta}\right)E_{i0}$$

The reflected electric field is thus

$$\mathbf{E}_r = \frac{(1-\beta)E_{i0}}{1+\beta}.\cos\left(c_1\sqrt{k_x^2 + k_z^2}\,t - k_x x + k_z z\right)$$

and the transmitted magnetic field is

$$\mathbf{H}_t = \frac{E_{t0}}{\mu_0\mu_r\omega}(k_x\hat{z} - k''_z\hat{x}).\cos(\omega t - k_x x - k''_z z)$$

where the expressions for E_{t0}, ω, k''_z are as given above.

4. For static time independent fields, which of the following current densities are possible?

(a) $\mathbf{J} = 2x^3y\hat{x} + 4x^2z\hat{y} - 6x^2yz\hat{z}$

(b) $\mathbf{J} = xy\hat{x} + (z + 1)\hat{y} + 2y\hat{z}$

(c) $\mathbf{J} = \frac{z^2}{\rho}\hat{\rho} + z\cos(\phi)\hat{z}$

(d) $\mathbf{J} = \frac{\sin\theta}{r^2}\hat{r}$

Ans. A static current density should satisfy $\nabla.\mathbf{J} = 0$. For (a),

$$\nabla.\mathbf{J} = \frac{\partial}{\partial x}(2x^3y) + \frac{\partial}{\partial y}(4x^2z) + \frac{\partial}{\partial z}(-6x^2yz)$$
$$= 6x^2y + 0 - 6x^2y = 0$$

Thus, this current density is possible. For (b),

$$\nabla.\mathbf{J} = \frac{\partial}{\partial x}(xy) + \frac{\partial}{\partial y}(z+1) + \frac{\partial}{\partial z}(2y)$$
$$= y + 0 + 0 = y \neq 0$$

This current density is not possible. For (c),

$$\nabla.\mathbf{J} = \frac{1}{\rho}\frac{\partial}{\partial \rho}(\rho z^2 / \rho) + \frac{\partial}{\partial z}(z.\cos\phi)$$
$$= \cos\phi \neq 0$$

So this current density is not feasible. For (d),

$$\nabla.\mathbf{J} = \frac{1}{r^2}\frac{\partial}{\partial r}(r^2 \sin\theta / r^2) = 0$$

So, this current density is feasible.

5. The excess charge in a certain medium decreases to α fraction of its initial value in τ seconds. (a) If σ is the conductivity of the medium, what is the dielectric constant? (b) What is the relaxation time? (c) After τ' seconds, what fraction of the charge will remain?

Ans. The equation of charge decay is obtained from $\nabla.\mathbf{E} = \rho/\varepsilon$, $\nabla.\mathbf{J} + \frac{\partial \rho}{\partial t} = 0$ and $\mathbf{J} = \sigma\mathbf{E}$. These three result in the equation.

$$\sigma\rho / \varepsilon + \frac{\partial \rho}{\partial t} = 0$$

so that

$$\rho(t) = \rho(0)\exp(-\sigma t/\varepsilon)$$

The first condition gives $\exp(-\sigma\tau/\varepsilon) = \alpha$ so that $\varepsilon = -\sigma\tau/\log(\alpha)$. Relaxation time is clearly $\tau_r = \varepsilon/\sigma$. After r' seconds, $\sigma' = \exp(-\sigma\tau'/\varepsilon)$ fraction of the initial charge will remain.

6. A plane wave travelling in the $\hat{y}$ direction in a lossy medium having parameters ε_r, μ_r, σ for the relative permittivity, relative permeability and conductivity has $\mathbf{E} = A\cos(\omega t + \phi)\,\hat{z}$ at $y = 0$. Find

(a) **E** at $y = dm$ and $t = t_0$sec.

(b) The distance travelled by the wave to have a phase shift of ϕ_0 relative to its initial phase.

(c) The distance travelled by the wave to have its amplitude reduced by p per cent.

(d) **H** at $y = dm$, $t = t_0$ sec.

Ans. The equation of wave propagation in the y direction is

$$E_z(0, t) = \text{Re}(A\exp(j\phi)\exp(j\omega t))$$

$$E_z(y, t) = \text{Re}(A\exp(j\phi)\exp(j\omega t)\exp(-\gamma y))$$

where

$$\gamma = \sqrt{j\omega\mu(\sigma + j\omega\varepsilon} = \alpha + j\beta$$

say. where $\mu = \mu_0\mu_r$, $\varepsilon = \varepsilon_0\varepsilon_r$. Thus,

$$E_z(y, t) = A\exp(-\alpha y)\cos(\omega t - \beta y + \phi)$$

Setting $y = d$, $t = t_0$ gives

$$E_z(d, t_0) = A\exp(-\alpha d)\cos(\omega t_0 - \beta d + \phi)$$

The distance y travelled by the wave to acquire a phase shift of ϕ_0 relative to the initial phase is given by

$$\beta d = \phi_0$$

or

$$d = \phi_0/\beta$$

The distance y travelled by the wave to have its amplitude reduced by p per cent is given by

$$\exp(-\alpha y) = p/100$$

so that

$$y = \frac{1}{\alpha}\log_e(100/p)$$

The magnetic field will have only an x component. Assume that it is given by

$$H_x(y, t) = \text{Re}(B.\exp(j\omega t)\exp(-\gamma y))$$

Then, the Maxwell equation

$$\nabla \times \mathbf{H} = \varepsilon\frac{\partial \mathbf{E}}{\partial t}$$

gives

$$-\frac{\partial H_x}{\partial y} = \varepsilon\frac{\partial E_z}{\partial t}$$

or

$$\mathrm{Re}(\gamma.B.\exp(j\omega t)\exp(-\gamma y)) = \varepsilon.\mathrm{Re}(A\exp(j\phi)j\omega\exp(j\omega t)\exp(-\gamma y))$$

so that

$$\gamma.B = j\omega.A.\varepsilon\exp(j\phi)$$

$$B = \frac{j\omega A}{\gamma}\varepsilon\exp(j\phi)$$

We leave the details of expressing H_x in terms of a real cosine wave to the student.

7. In a certain region of space,

$$\mathbf{J}(t, x, y, z) = (J_x(x, y, z)\hat{x} + J_y(x, y, z)\hat{y} + J_z(x, y, z)\hat{z})f(t)$$

Find the charge density $\rho\,(t, x, y, z)$ if $\rho\,(0, x, y, z) = \rho_0\,(x, y, z)$.

Ans: We use the equation of charge conservation:

$$\nabla.\mathbf{J} + \frac{\partial\rho}{\partial t} = 0$$

This gives

$$f(t)\left(\frac{\partial J_x}{\partial x} + \frac{\partial J_y}{\partial y} + \frac{\partial J_z}{\partial z}\right) = -\frac{\partial\rho}{\partial t}$$

Integration with the initial condition gives

$$\rho(t,x,y,z) = \rho_0(x,y,z) - \left(\frac{\partial J_x}{\partial x} + \frac{\partial J_y}{\partial y} + \frac{\partial J_z}{\partial z}\right)\int_0^t f(\tau)d\tau$$

8. In a source free region $\mathbf{J} = 0$, $\rho = 0$, Maxwell's equations can be reduced to two. Identify the all embracing equations.

Ans. The two equations are

$$\nabla\times\mathbf{E} = -\frac{\partial\mathbf{B}}{\partial t},\quad \nabla\times\mathbf{B} = \varepsilon\mu\frac{\partial\mathbf{E}}{\partial t}$$

Taking the divergence of the first equation here gives

$$\frac{\partial}{\partial t}\nabla.\mathbf{B} = 0$$

which integrates to give $\nabla.\mathbf{B} = C$, where C is constant in time, i.e. dependent only upon the spatial variables. We can assume this to be zero. This is a necessary boundary condition. Hence, taking the divergence of Faraday's law leads us to the fourth Maxwell equation, namely, the no monopole condition. Taking the divergence of the second equation, namely Ampere's law with no current but Maxwell's displacement current correction term gives

$$\frac{\partial}{\partial t}\nabla.\mathbf{E} = 0$$

which integrates to $\nabla.\mathbf{E} = C'$ where again C' depends only upon the spatial variables. C'' can be taken as zero, a necessary boundary condition.

9. In a source free region, the electric field satisfies the diffusion equation under a certain approximation. Derive this equation stating the approximation and obtain the general solution to it using Fourier transforms. What equation is satisfied by **E** when there is a charge and a current density?

Ans. We start with the Maxwell equations

$$\nabla.\mathbf{E} = 0, \nabla\times\mathbf{E} = -\frac{\partial\mathbf{B}}{\partial t}, \nabla\times\mathbf{B} = \mu\sigma\mathbf{E} + \mu\varepsilon\frac{\partial\mathbf{E}}{\partial t}$$

Taking the curl of the second equation and using $\nabla\times(\nabla\times\mathbf{E}) = \nabla(\nabla.\mathbf{E}) - \nabla^2\mathbf{E}$ and substituting $\nabla.\mathbf{E} = 0$ from the first and for $\nabla\times\mathbf{B}$ from the third, we get

$$\nabla^2\mathbf{E} - \mu\sigma\frac{\partial\mathbf{E}}{\partial t} - \mu\varepsilon\frac{\partial^2\mathbf{E}}{\partial t^2} = 0$$

This equation we have derived earlier. Suppose the duration T over which the field changes significantly is such that

$$\mu\sigma/T >> \mu\varepsilon/T^2$$

that is $T > \varepsilon/\sigma$. This means that over the duration $[0,\varepsilon/\sigma]$ the field does not change appreciably. Then the above equation can be approximated by the diffusion equation

$$\nabla^2\mathbf{E} - \mu\sigma\frac{\partial\mathbf{E}}{\partial t} = 0$$

We shall first indicate how the general equation when no approximation is made is solved using Fourier transforms. The Fourier transform

$$\hat{\mathbf{E}}(\mathbf{r},\omega) = \int_{-\infty}^{\infty}\mathbf{E}(t,\mathbf{r})\exp(-i\omega t)dt$$

satisfies the Helmholtz equation

$$\nabla^2\hat{\mathbf{E}} - \gamma^2\hat{\mathbf{E}} = 0$$

where

$$\gamma^2 = i\mu\sigma\omega - \mu\varepsilon\omega^2 = i\mu\omega(\sigma + i\omega\varepsilon)$$

Note that γ is a function of ω. The solution to this equation is a superposition of the solutions

$$\mathbf{E}(\mathbf{r},\omega) = \mathbf{A}(\mathbf{n},\omega)\exp(\gamma\mathbf{n}.\mathbf{r})$$

where $\mathbf{n}$ is any unit vector. The vector $\mathbf{A}(\mathbf{n},\omega)$ must be chosen so that $\mathbf{n}.\mathbf{A}(\mathbf{n},\omega) = 0$ in view of the condition $\nabla.\mathbf{E} = 0$. This can be done by choosing two direction perpendicular to each other and to $\mathbf{n}$ calling these directions as $\mathbf{l}_1$, $\mathbf{l}_2$ and setting

$$\mathbf{A}(\mathbf{n},\omega) = A_1(\omega)\mathbf{l}_1 + A_2(\omega)\mathbf{l}_2$$

The general solution is then given by

$$\mathbf{E}(\mathbf{r},\omega) = \int \mathbf{A}(\mathbf{n},\omega)\exp(\gamma(\omega)\mathbf{n}.\mathbf{r})d\Omega(\mathbf{n})d\omega$$

Consider now the diffusion equation for a scalar function ψ:

$$\nabla^2\psi(t,\mathbf{r}) - \mu\sigma\frac{\partial\psi(t,\mathbf{r})}{\partial t} = 0$$

we shall solve this by a different method. Taking the Fourier transform with respect to the spatial variables

$$\hat{\psi}(t,\mathbf{k}) = \int \psi(t,\mathbf{r})\exp(i\mathbf{k}.\mathbf{r})d\mathbf{k}$$

we get

$$\mu\sigma\frac{\partial\hat{\psi}}{\partial t} = -k^2\hat{\psi}$$

and hence

$$\hat{\psi}(t,\mathbf{k}) = \hat{A}(\mathbf{k})\exp(-k^2t/\mu\sigma)$$

The inverse Fourier transform of $\exp(-k^2t/\mu\sigma)$ is given by

$$\frac{1}{(2\pi)^3}\int \exp(-k^2t/\mu\sigma)\exp(-i\mathbf{k}.\mathbf{r})d\mathbf{k} = (\mu\sigma/4\pi t)^{3/2}$$

$$\exp(-r^2\mu\sigma/4t) = K(t,\mathbf{r})$$

say. Thus, the general solution to the diffusion equation is

$$\psi(t,\mathbf{r}) = A(\mathbf{r}) * K(t,\mathbf{r}) = \int A(\mathbf{r}')K(t,\mathbf{r}-\mathbf{r}')d\mathbf{r}'$$

Each component of the electric field is of this form. Thus, in vectorial notation,

$$\mathbf{E}(t,\mathbf{r}) = (\mu\sigma/4\pi t)^{3/2} \int \mathbf{A}(\mathbf{r}')\exp(-\mu\sigma|\mathbf{r}-\mathbf{r}'|^2/4t)d\mathbf{r}'$$

The vector valued functions $\mathbf{A}(\mathbf{r})$ must be chosen so that its Fourier transform $\hat{\mathbf{A}}(\mathbf{k})$ is orthogonal to the vector $\mathbf{k}$ that is $\mathbf{k}.\hat{\mathbf{A}}(\mathbf{k}) = 0$. This is equivalent to requiring that $\nabla.\mathbf{A} = 0$, or equivalently, the initial electric field is divergenceless, i.e. $\nabla.\mathbf{E}(0,\mathbf{r}) = 0$.

Finally consider the case when there is a charge and a current density. Then we use the four Maxwell equations in the form

$$\nabla.\mathbf{E} = \rho/\varepsilon, \quad \nabla\times\mathbf{E} = -\frac{\partial\mathbf{B}}{\partial t}$$

$$\nabla\times\mathbf{B} = \mu\mathbf{J} + \mu\sigma\mathbf{E} + \mu\varepsilon\frac{\partial\mathbf{E}}{\partial t}, \quad \nabla.\mathbf{B} = 0$$

Taking the curl of the second equation and using the first and the third one gives

$$\nabla(\rho)/\varepsilon - \nabla^2\mathbf{E} = -\mu\sigma\frac{\partial\mathbf{E}}{\partial t} - \mu\varepsilon\frac{\partial^2\mathbf{E}}{\partial t^2} - \mu\frac{\partial\mathbf{J}}{\partial t}$$

which can be rearranged as

$$\nabla^2\mathbf{E} - \mu\varepsilon\frac{\partial^2\mathbf{E}}{\partial t^2} - \mu\sigma\frac{\partial\mathbf{E}}{\partial t} = \nabla\rho/\varepsilon + \mu\frac{\partial\mathbf{J}}{\partial t}$$

10. A square waveguide is filled with a dielectric having complex permittivity $\varepsilon = \varepsilon_0(A - jB)$, i.e. $\varepsilon = \varepsilon_0 + \sigma/j\omega\varepsilon_0$, where σ is the conductivity. It is excited with the $TM_{n,m}$ mode. If the waveguide operates at a frequency r fraction above the cutoff frequency, calculate the attenuation α and the distance travelled by the wave before its magnitude is reduced by a fraction s.

Ans. The propagation constant is given by

$$\gamma = \alpha + j\beta = \sqrt{(\pi/a)^2(n^2+m^2) - \omega^2\mu_0\varepsilon}$$

$$= \sqrt{(\pi/a)^2(n^2+m^2) - (\omega^2/c^2)(A - jB)}$$

since $\mu_0\varepsilon_0 = 1/c^2$. The cutoff frequency is defined as the frequency below which only attenuation is there when the conductivity is zero. For this case, it is obtained by setting $B = 0$ and we get

$$\omega_c = \frac{\pi c}{a\sqrt{A}}\sqrt{n^2+m^2}$$

The frequency of operation is $\omega = \omega_c(1 + r)$ and we get

$$\gamma = \alpha + j\beta = \sqrt{(\pi/a)^2(n^2+m^2) - \omega_c^2(1+r)(A - jB)}$$

$$= \sqrt{(\pi/a)^2(n^2+m^2) - (\pi/a)^2(n^2+m^2)(1 - jB/A)} =$$

$$(\pi/a)\sqrt{n^2+m^2}\ \sqrt{\frac{B}{A}}\left(\frac{1+j}{\sqrt{2}}\right)$$

The real part of this expression gives the attenuation constant α and the distance d travelled by the wave before its magnitude is reduced by a fraction s is given by $\exp(-\alpha d) = s$ or $d = -\log(s)/\alpha$.

11. Find the power transmitted in a rectangular waveguide corresponding to the $TE_{n,m}$ and $TM_{n,m}$ modes.

Ans. Consider first the $TE_{n,m}$ mode. The electric field phasors are given by

$$E_z(x, y) = E_0 \sin(n\pi x/a)\sin(m\pi y/b)\exp(-\gamma z)$$

where

$$\gamma = \sqrt{(n\pi/a)^2 + (m\pi/b)^2 + j\mu\omega(\sigma + j\omega\varepsilon)}$$

We derive from this expression, formulae for the transverse components of the electric and magnetic field phasors:

$$E_x = -\frac{\gamma}{h^2}\frac{\partial E_z}{\partial x} = -\frac{\gamma m\pi E_0}{ah^2}\cos(n\pi x/a)\sin(m\pi y/b)\exp(-\gamma z)$$

$$H_y = -\frac{\sigma + i\omega\varepsilon}{h^2}\frac{\partial E_z}{\partial x} = -\frac{(\sigma + i\omega\varepsilon)n\pi E_0}{ah^2}\cos(n\pi x/a)\sin(m\pi y/b)\exp(-\gamma z)$$

$$E_y = -\frac{\gamma}{h^2}\frac{\partial E_z}{\partial y} = -\frac{\gamma m\pi E_0}{bh^2}\sin(n\pi x/a)\cos(m\pi y/b)\exp(-\gamma z)$$

$$H_x = \frac{\sigma + i\omega\varepsilon}{h^2}\frac{\partial E_z}{\partial y} = \frac{(\sigma + i\omega\varepsilon)m\pi E_0}{bh^2}\sin(n\pi x/a)\cos(m\pi y/b)\exp(-\gamma z)$$

From this we calculate the Poynting vector

$$\mathbf{P} = \frac{1}{2}\mathrm{Re}(\mathbf{E} \times \mathbf{H}^*)$$

The only nonvanishing component of this is the z component. It is given by

$$(\mathbf{E}\times\mathbf{H}^*)_z = E_x H_y^* - E_y H_x^*$$

$$\frac{\gamma(\sigma - i\omega\varepsilon)n^2\pi^2|E_0|^2}{a^2h^4}\cos^2(n\pi x/a)\sin^2(m\pi y/b)\exp(-(\gamma+\gamma^*)z)$$

$$+\frac{\gamma(\sigma - i\omega\varepsilon)m^2\pi^2|E_0|^2}{b^2h^4}\sin^2(n\pi x/a)\cos^2(m\pi y/b)\exp(-(\gamma+\gamma^*)z)$$

Using the integrals

$$\int_0^a \cos^2(n\pi x/a)dx = a/2 = \int_0^a \sin^2(n\pi x/a)dx$$

$$\int_0^b \cos^2(m\pi y/b)dy = b/2 = \int \sin^2(m\pi y/b)dy$$

and the fact $\gamma = \alpha + i\beta$ so that $\gamma + \gamma^* = 2\alpha$, we find the total transmitted power at the point z is given by

$$W = \frac{1}{2}\int_0^a\int_0^b \mathrm{Re}(\mathbf{E}\times\mathbf{H}^*)dxdyW = \frac{1}{2}\int_0^a\int_0^b \mathrm{Re}(\mathbf{E}\times\mathbf{H}^*)dxdy$$

$$\left(\frac{(\alpha\sigma+\beta\omega\varepsilon)n^2\pi^2|E_0|^2 b}{8ah^4}+\frac{(\alpha\sigma+\beta\omega\varepsilon)m^2\pi^2|E_0|^2 a}{8bh^4}\right)\exp(-2\alpha z)$$

$$=\left(\alpha\sigma+\beta\omega\varepsilon)\pi^2(n^2b/a+m^2a/b)|E_0|^2/8h^4\right)\exp(-2\alpha z)$$

A similar computation holds for the transverse electric mode. We shall in any case repeat the computation since it involves deriving expressions for the electric and magnetic field components in the rectangular waveguide which are useful to keep as a record. We start with

$$H_z = H_0\cos(n\pi x/a)\cos(m\pi y/b)\exp(-\gamma z)$$

Then,

$$E_x = -\frac{i\omega\mu}{h^2}\frac{\partial H_z}{\partial y} = \frac{i\omega\mu m\pi}{bh^2}H_0\cos(n\pi x/a)\sin(m\pi y/b)\exp(-\gamma z)$$

$$H_y = -\frac{\gamma}{h^2}\frac{\partial H_z}{\partial y} = \frac{\gamma m\pi}{bh^2}H_0\cos(n\pi x/a)\sin(m\pi y/b)\exp(-\gamma z)$$

$$E_y = \frac{i\omega\mu}{h^2}\frac{\partial H_z}{\partial x} = -\frac{i\omega\mu n\pi}{ah^2} H_0 \sin(n\pi x/a)\cos(m\pi y/b)\exp(-\gamma z)$$

$$H_x = -\frac{\gamma}{h^2}\frac{\partial H_z}{\partial x} = \frac{\gamma n\pi}{ah^2} H_0 \sin(n\pi x/a)\cos(m\pi y/b)\exp(-\gamma z)$$

Thus,

$$E_x H_y^* - E_y H_x^* = \frac{i\omega\mu m^2\pi^2\gamma^*}{b^2h^4}|H_0|^2\cos^2(n\pi x/a)\sin^2(m\pi y/b)\exp(-2\alpha z)$$

$$+\frac{i\omega\mu n^2\pi^2\gamma^*}{a^2h^4}|H_0|^2\sin^2(n\pi x/a)\cos^2(m\pi y/b)\exp(-2\alpha z)$$

Hence,

$$\mathrm{Re}(E_x H_y^* - E_y H_x^* = \frac{\omega\mu m^2\pi^2\beta}{b^2h^4}|H_0|^2\cos^2(n\pi x/a)\sin^2(m\pi y/b)$$

$$\exp(-2\alpha z) + \frac{\omega\mu n^2\pi^2\beta}{a^2h^4}|H_0|^2\sin^2(n\pi x/a)\cos^2(m\pi y/b)$$

and for the power transmitted, we get the expression

$$W = \frac{1}{2}\int_0^a\int_0^b \mathrm{Re}(E_x H_y^* - E_y H_x^*)dxdy$$

$$\left(\frac{\omega\mu m^2\pi^2\beta a}{8bh^4}|H_0|^2 + \frac{\omega\mu n^2\pi^2\beta b}{8ah^4}|H_0|^2\right).\exp(-2\alpha z)$$

$$= (\omega\mu\pi^2/8h^4)(m^2a/b + n^2b/a)|H_0|^2.\exp(-2\alpha z)$$

12. Assume that a lossless transmission line is terminated with a load Z_L and excited by a sinusoidal voltage source with phasor $V_0\exp(j0)$, i.e. the input voltage as a function of time is given by $V_0\cos(\omega t)$ with ω being the frequency. Calculate the ratios $\frac{V_{max}}{V_{min}}$ and $\frac{I_{max}}{I_{min}}$ for the line where V_{max} and V_{min} are respectively the maximum and minimum amplitudes of the voltage on the line and I_{max} and I_{min} are respectively the maximum and minimum amplitudes of the current on the line. Assume that the line has characteristic impedance Z_0. Also calculate the input impedance at the point of voltage maximum.

Ans. The voltage phasor at z (distance from the source end) is given by

$$V(z) = V_+\exp(-j\beta z) + V_-\exp(j\beta z)$$

with β real since the line is assumed to be lossless. Likewise, the current is

$$I(z) = \frac{V_+}{Z_0}\exp(-j\beta z) - \frac{V_-}{Z_0}\exp(j\beta z)$$

The terminal conditions are

$$V(0) = V_+ + V_- = V_0,\ V(L) = I(L)\, Z_L$$

The latter condition gives

$$V_+\exp(-j\beta L) + V_-\exp(j\beta L) = \frac{Z_L}{Z_0}(V_+\exp(-j\beta L) - V_-\exp(j\beta L))$$

and hence solving the two equations, we get

$$V_+ = \frac{V_0}{1+\Gamma.\exp(-j2\beta L)},\quad V_- = \frac{V_0\Gamma.\exp(-j2\beta L)}{1+\Gamma.\exp(-j2\beta L)}$$

where

$$\Gamma = \frac{Z_L - Z_0}{Z_L + Z_0}$$

This gives

$$V(z) = V_+(\exp(-j\beta z) + \Gamma.\exp(j\beta(z-2L))) = V_+\exp(-j\beta z)$$

$$(1 + \Gamma.\exp(j2\beta(z-L))$$

The voltage attains its maximum amplitude at the points z where $2\beta(L - z) = Arg(\Gamma) + 2n\pi$ with n being any integer and the corresponding maximum amplitude is given by

$$V_{max} = |V_+|(1+|\Gamma|)$$

Likewise, the voltage attains its minimum amplitude at points z where $2\beta(L - z) = Arg(\Gamma) + (2n + 1)\pi$, where n is an integer and the corresponding minimum amplitude is given by

$$V_{min} = |V_+|(1-|\Gamma|)$$

This gives

$$s = V_{max}/V_{min} = \frac{1+|\Gamma|}{1-|\Gamma|}$$

For the current, wc have the expression

$$I(z)=\frac{V_+}{Z_0}\exp(-j\beta z)(1-\Gamma.\exp(j2\beta(z-L)))$$

This means that the current amplitude attains its maximum when $2\beta(L-z)=Arg(\Gamma)+(2n+1)\pi$ and its minimum when $2\beta(L-z)=Arg(\Gamma)+2n\pi$. Equivalently, points of voltage maxima correspond to points of current minima and vice versa. The ratio of current maximum to current minimum is once again s. At a point z_0 of voltage maximum, we have the expression for the voltage and current phasors

$$V(z_0)=V_+\exp(-j\beta z_0)(1+|\Gamma|)$$

$$I(z_0)=\frac{V_+}{Z_0}\exp(-j\beta z_0)(1+|\Gamma|)$$

This gives for the input impedance at this point

$$Z_{in,\max}=\frac{V(z_0)}{I(z_0)}=Z_0\frac{1+|\Gamma|}{1-|\Gamma|}=sZ_0$$

13. Derive the expression for the power input to a lossless transmission line at a distance z from the source end when the $z=0$ end is excited by a voltage source $V_0\exp(j0)$ and the line is terminated with an impedance Z_L.

Ans. At a distance z from the source, we have the following expressions for the voltage and current phasors:

$$V(z)=V_+\exp(-j\beta z)(1+\Gamma.\exp(j2\beta(z-L))),$$

$$I(z)=\frac{V_+}{Z_0}\exp(-j\beta z)(1-\Gamma.\exp(j2\beta(z-L)))$$

The average power input to the line at z is

$$P(z)=\frac{1}{2}\text{Re}(V(z)I(z)^*)=\frac{|V_+|^2}{2Z_0}\text{Re}((1+\Gamma.\exp(-j2\beta d)).(1-\bar{\Gamma}.\exp(j2\beta d)))$$

$$=\frac{|V_+|^2}{2Z_0}(1-|\Gamma|^2)$$

since $\Gamma.\exp(-j2\beta d)-\bar{\Gamma}.\exp(j2\beta d)$ is a pure imaginary number. Here, $d=L-z$ is the distance of the given point from the load. Recall from the previous problem that

$$V_+ = \frac{V_0}{1+\Gamma.\exp(-j2\beta L)}$$

so that the power input at z can be expressed as

$$P(z) = \frac{|V_0|^2}{2Z_0}\frac{1-|\Gamma|^2}{|1+\Gamma.\exp(-j2\beta L)|^2}$$

Note that this expression is independent of z and corresponds to the fact that for a lossless transmission line, the power input at every point is the same since no power is dissipated in the line.

14. Derive the expression for the power input to a lossy transmission line at the point z.

Ans. The above formulae get modified by changing $j\beta$ to $\gamma = \alpha + j\beta$ where

$$\gamma = \sqrt{(R+j\omega L)(G+j\omega C)}$$

For the lossless case, we took $\beta = \omega\sqrt{LC}$. Correspondingly, for the lossless case, $Z_0 = \sqrt{L/C}$ and for the lossy case, $Z_0 = \sqrt{(R+j\omega L)(G+j\omega C)}$. The expressions for the voltage and current phasor are now

$$V(z) = V_+ \exp(-\gamma z)(1+\Gamma.\exp(-2\gamma d))$$

$$I(z) = \frac{V_+}{Z_0}(\exp(-\gamma z)(1-\Gamma.\exp(-2\gamma d))$$

with $d = L - z$ and

$$V_+ = \frac{V_0}{1+\Gamma.\exp(-2\gamma L)}$$

We define $I_+ = \frac{V_+}{Z_0}$. Then,

$$V(z)I(z)^* = |I_+|^2 Z_0\exp(-2\alpha z)(1+\Gamma.\exp(-2\gamma d))(1-\bar{\Gamma}.\exp(-2\bar{\gamma}d))$$

$$= |I_+|^2 Z_0\exp(-2\alpha z)(1-|\Gamma|^2\exp(-2\alpha d)+\Gamma.\exp(-2\gamma d)-\bar{\Gamma}.\exp(-2\bar{\gamma}d))$$

Hence the power input at z is given by

$$P(z) = \frac{1}{2}\text{Re}(V(z)I(z)^*) = |I_+|^2\exp(-2\alpha z).\text{Re}(Z_0.(1-|\Gamma|^2 exp(-2\alpha d)$$

$$+2jIm(\Gamma.\exp(-2\gamma d))) = |I_+|^2\exp(-2\alpha z)\text{Re}(Z_0)(1-|\Gamma|^2\exp(-2\alpha d)) - 2|I_+|^2$$
$$\exp(-2\alpha z)Im(Z_0)Im(\Gamma.\exp(-2\gamma d))$$

$$= |I_+|^2 \operatorname{Re}(Z_0)\exp(-2\alpha L) - 2|I_+|^2 \operatorname{Im}(Z_0)\exp(-2\alpha L)\operatorname{Im}(\Gamma\exp(-j2\beta d\,|))$$

15. Compute the input impedance at z, ratio of maximum to minimum voltage amplitudes and the points where the voltage maxima and minima are attained when the line is lossless for (a) A shorted transmission line (b) An open transmission line (c) A matched line.

Ans. For a shorted line, the condition is $V(L) = 0$. This gives

$$V_+\exp(-\gamma L) + V_-\exp(\gamma L) = 0$$

or

$$\frac{V_-}{V_+} = -\exp(-2\gamma L)$$

The voltage and current at z are then respectively given by

$$V(z) = V_+(\exp(-\gamma z) - \exp(\gamma(z-2L))) = V_+\exp(-\gamma z)(1-\exp(-2\gamma d))$$

$$I(z) = \frac{V_+}{Z_0}\exp(-\gamma z)(1+\exp(-2\gamma d))$$

where $d = L - z$ so that

$$Z(z) = Z_0\frac{1-\exp(-2\gamma d)}{1+\exp(-2\gamma d)} = Z_0\tanh(\gamma d)$$

For the lossless line, $\gamma = j\beta$ and we get

$$Z(z) = Z_0\frac{1-\exp(-2j\beta d)}{1+\exp(-2j\beta d)} = jZ_0\tan(\beta d)$$

The input impedance at each point on the line is purely reactive. For the lossless case, the voltage maximum has an amplitude

$$V_{max} = |V_+|(1 + 1) = 2|V_+|$$

and this is attained when $2\beta d = ((2n + 1)\pi$ or equivalently, when $d = (n + 1/2)\pi/\beta = (n + 1/2)\lambda/2$. The voltage minimum is zero and is attained when $2\beta d = 2n\pi$ or when $d = n\pi/\beta = n\lambda/2$. The ratio V_{max}/V_{min} is infinity. For the open line, we have $I(L) = 0$ and this gives the condition

$$V_+\exp(-\gamma L) = V_-\exp(\gamma L)$$

or

$$\frac{V_-}{V_+} = \exp(-2\gamma L)$$

Thus, the voltage phasor at z is given by

$$V(z) = V_+\exp(-\gamma z) + \exp(\gamma(z-2L)) = V_+\exp(-\gamma z)(1+\exp(-2\gamma d))$$

and the current at z by

$$I(z) = \frac{V_+}{Z_0}\exp(-\gamma z) - \exp(\gamma(z-2L)) = \frac{V_+}{Z_0}\exp(-\gamma z)(1-\exp(-2\gamma d))$$

This gives for the input impedance at z

$$Z(z) = \frac{V(z)}{I(z)} = Z_0\frac{1+\exp(-2\gamma d)}{1-\exp(-2\gamma d)} = Z_0\coth(\gamma d)$$

For the lossless case, this reduces to

$$Z(z) = Z_0\frac{1+\exp(-2j\beta d)}{1-\exp(-2j\beta d)} = -jZ_0\cot(\beta d)$$

For the lossless case, we have

$$V(z) = V_+\exp(-j\beta z)(1 + \exp(-j2\beta d))$$

and its magnitude attains its maximum value of

$$V_{max} = 2|V_+|$$

when $2\beta d = 2n\pi$, or when $d = n\pi/\beta = n\lambda/2$. The voltage attains its minimum magnitude value of zero when $2\beta d = (2n + 1)\pi$, i.e. when $d = (n + 1/2)\pi/\beta = (n + 1/2)\lambda/2$. The ratio $\frac{V_{max}}{V_{min}} = \infty$ once again. It is instructive to derive these values of the maximum to minimum voltage ratio starting from the expressions for $\Gamma = \frac{Z_L - Z_0}{Z_L + Z_0}$. For the shorted line, $Z_L = 0$ giving $\Gamma = -1$. Thus,

$$\frac{V_{max}}{V_{min}} = \frac{1+|\Gamma|}{1-|\Gamma|} = \infty$$

For the open line, we have $Z_L = \infty$ giving $\Gamma = 1$ and once again

$$\frac{V_{max}}{V_{min}} = \frac{1+|\Gamma|}{1-|\Gamma|} = \infty$$

Note that the input impedance for the open-ended line is $Z_{in,oc}(z) = Z_0\coth(\gamma d)$ and that for the shorted line is $Z_{in,sc}(z) = Z_0\tanh(\gamma d)$. The product of these two impedances is

$$Z_{in,oc}(z).Z_{in,sc}(z) = Z_0^2$$

The load of a matched line is the same as the characteristic impedance of the line, i.e. $Z_L = Z_0$. This gives us the condition $V(L) = I(L)\,Z_0$ or

$$V_+ \exp(-\gamma L) + V_- \exp(\gamma L) = V_+ \exp(-\gamma L) - V_- \exp(\gamma L)$$

so that $V_- = 0$. Thus,

$$V(z) = V_+ \exp(-\gamma z), I(z) = \frac{V_+}{Z_0}\exp(\gamma z)$$

Only for the matched line, we do not have any reflected wave. When $Z_L \neq Z_0$, there is a reflected wave. The voltage maximum and minimum amplitude for the lossless line are clearly the same equal to $|V_+|$ and hence $\frac{V_{max}}{V_{min}} = 1$. The input impedance is given by

$$Z_{in}(z) = \frac{V(z)}{I(z)} = Z_0$$

at each point on the line, we see only the characteristic impedance looking into the line.

16. A transmission line operates at the frequency ω_0 and has a propagation constant γ, characteristic impedance Z_0 and the line is connected to a voltage source having voltage phasor V_g and internal impedance Z_g. Take the load impedance to be Z_L. Find (a) the input impedance, (b) sending end current, (c) current at z.

Ans. Let $V(0)$, $I(0)$ be the voltage and current phasor at the input of the transmission line. Then, by Kirchhoff's voltage equation,

$$V_g - I(0)Z_g = V(0)$$

with

$$V(0) = V_+ + V_-,\ I(0) = \frac{1}{Z_0}(V_+ - V_-)$$

Plugging in these values, we get

$$V_g - \frac{Z_g}{Z_0}(V_+ - V_-) = V_+ + V_-$$

so that

$$V_+(Z_g + Z_0) - V_-(Z_g - Z_0) = Z_0 V_g$$

The load equation $V(L) = I(L)Z_L$ translates to

$$V_+ \exp(-\gamma L) + V_- \exp(\gamma L) = \frac{Z_L}{Z_0}(V_+ \exp(-\gamma L) - V_- \exp(\gamma L))$$

or

$$(Z_L - Z_0)\exp(-\gamma L)V_+ - (Z_L + Z_0)\exp(\gamma L)V_- = 0$$

These two equations can be cast in matrix form as

$$\begin{pmatrix} Z_g + Z_0 & Z_0 - Z_g \\ (Z_L - Z_0)\exp(-\gamma L) & -(Z_L + Z_0)\exp(\gamma L) \end{pmatrix}\begin{pmatrix} V_+ \\ V_- \end{pmatrix} = \begin{pmatrix} Z_0 V_g \\ 0 \end{pmatrix}$$

The solution to these two equations is

$$V_+ = \frac{Z_0(Z_L + Z_0)V_g\exp(\gamma L)}{(Z_L + Z_0)(Z_g + Z_0)\exp(\gamma L) - (Z_L - Z_0)(Z_g - Z_0)\exp(-\gamma L)}$$

$$V_- = \frac{Z_0(Z_L - Z_0)V_g\exp(-\gamma L)}{(Z_L + Z_0)(Z_g + Z_0)\exp(\gamma L) - (Z_L - Z_0)(Z_g - Z_0)\exp(-\gamma L)} =$$

$$= \Gamma.\exp(-2\gamma L)V_+$$

with

$$\Gamma = \frac{Z_L - Z_0}{Z_L + Z_0}$$

As derived earlier,

$$\frac{V(0)}{I(0)} = Z_0\frac{V_+ + V_-}{V_+ - V_-} = Z_0\frac{1 + \Gamma.\exp(-2\gamma L)}{1 - \Gamma.\exp(-2\gamma L)}$$

$$= Z_0\frac{(Z_L + Z_0) + (Z_L - Z_0)\exp(-2\gamma L)}{(Z_L + Z_0) - (Z_L - Z_0)\exp(-2\gamma L)}$$

$$Z_0\frac{Z_L\cosh(\gamma L) + Z_0\sinh(\gamma L)}{Z_L\sin(\gamma L) + Z_0\cosh(\gamma L)}$$

The sending end current is given by

$$I(0) = \frac{V_+}{Z_0}(1 - \Gamma.\exp(-2\gamma L))$$

$$= \frac{V_g.\exp(\gamma L)((Z_L + Z_0)\exp(\gamma L) - (Z_L - Z_0)\exp(-\gamma L))}{(Z_L + Z_0)(Z_g + Z_0)\exp(\gamma L) - (Z_L - Z_0)(Z_g - Z_0)\exp(-\gamma L)}$$

$$= \frac{V_g(Z_L\sinh(\gamma L) + Z_0\cosh(\gamma L))}{(Z_L Z_g + Z_0^2)\sinh(\gamma L) + Z_0(Z_L + Z_g)\cosh(\gamma L)}$$

The current at a distance z from the source is given by

$$I(z) = \frac{V_+}{Z_0}(\exp(-\gamma z) - \Gamma.\exp(\gamma(z-2L)))$$

$$= \frac{V_g(Z_L + Z_0)\exp(\gamma L)}{2(Z_L Z_g + Z_0^2)\sinh(\gamma L) + 2Z_0(Z_L + Z_g)\cosh(\gamma L)}(\exp(-\gamma z) - \Gamma.\exp(\gamma(z-2L))$$

$$= \frac{V_g((Z_L + Z_0)\exp(\gamma(L-z)) - (Z_L - Z_0)\exp(\gamma(z-L))}{2(Z_L Z_g + Z_0^2)\sinh(\gamma L) + 2Z_0(Z_L + Z_g)\cosh(\gamma L)}$$

$$= \frac{V_g(Z_L\sinh(\gamma(L-z)) + Z_0\cosh(\gamma(L-z))}{(Z_L Z_g + Z_0^2)\sinh(\gamma L) + Z_0(Z_L + Z_g)\cosh(\gamma L)}$$

17. Two circular conducting plates of radius R are located parallel to each other at a distance d from each other forming a cylindrical capacitor. Wires connected to the centre of the plates carry a current at frequency ω. Using the Maxwell equations, calculate the capacitance between the two plates.

Ans. The electric field phasor has a $\hat{z}$ component $E_z(\rho)$ and the magnetic field a $\hat{\phi}$ component $H_\phi(\rho)$ from the symmetry of the problem. The region between the plates is free space and hence these fields satisfy the wave or the Helmholtz equations, in particular,

$$\frac{1}{\rho}\frac{d}{d\rho}\rho\frac{dE_z(\rho)}{d\rho} + \omega^2\mu\varepsilon E_z(\rho, z) = 0$$

We can derive this equation starting directly from the Maxwell equations

$$\nabla \times \mathbf{E} = -j\mu\omega\mathbf{H}$$

$$\nabla \times \mathbf{H} = j\omega\varepsilon\mathbf{E}$$

gives

$$\hat{\rho}\frac{d}{d\rho} \times E_z\hat{z} = -j\mu\omega H_\phi\hat{\phi}$$

or

$$E_z'(\rho) = j\mu\omega H_\phi(\rho)$$

for the first and for the second

$$\left(\hat{\rho}\frac{\partial}{\partial\rho} + \hat{\phi}\frac{1}{\rho}\frac{\partial}{\partial\phi}\right) \times H_\phi\hat{\phi} = j\omega\varepsilon E_z\hat{z}$$

Using the identity

$$\frac{\partial \hat{\phi}}{\partial \phi} = -\hat{\rho}$$

this reduces to

$$H'_\phi(\rho) + \frac{1}{\rho} H_\phi(\rho) = j\omega\varepsilon E_z(\rho)$$

Substituting for H_ϕ from the first into the second equation gives

$$E''_z(\rho) + \frac{1}{\rho} E'_z(\rho) + \omega^2 \mu\varepsilon E_z(\rho) = 0$$

which is the required Helmholtz equation. The solution to this equation having no singularity at $\rho = 0$ is

$$E_z(\rho) = AJ_0(\omega\sqrt{\mu\varepsilon}\,\rho)$$

where $J_0(x)$ is the Bessel function of zeroth order. The potential difference between the centre of the plates is

$$V = E_z(0)d = Ad$$

where the normalization of the Bessel function is such that $J_0(0) = 1$. The surface charge density on one plate is given by

$$\sigma(\rho) = \varepsilon . E_z(\rho) = A\varepsilon J_0\left(\omega\sqrt{\mu\varepsilon}\,\rho\right)$$

and hence the total charge on one of the plates is

$$Q = \int_0^R \sigma(\rho) 2\pi\rho d\rho = A\varepsilon \int_0^R J_0(\omega\sqrt{\mu\varepsilon}\,\rho) 2\pi\rho d\rho$$

The capacitance between the plates is

$$C = Q/V = \frac{\varepsilon}{d} \int_0^R J_0(\omega\sqrt{\mu\varepsilon}\,\rho) 2\pi\rho d\rho$$

We leave it as an exercise to calculate an expansion for the capacitance in powers of the frequency ω.

Exercises

[1] Derive the wave equation for the electric and magnetic fields propagating in free space.

[2] Show that if **n** is a unit vector, then $\psi(t, \mathbf{r}) = f(t - \mathbf{n}.\mathbf{r}/c)$ satisfies the three-dimensional wave equation.

[3] Prove that $f(t, \mathbf{r}) = \frac{1}{r} f(t - r/c)$ satisfies the three dimensional wave equation having spherical symmetry.

[4] For each unit vector $\mathbf{n}$, let $f(\mathbf{n}, t)$ be a pulse. Show that

$$\psi(t,\mathbf{r}) = \int f(\mathbf{n}, t - \mathbf{n}.\mathbf{r}/c) d\Omega(\mathbf{n})$$

the integral being over all solid angles satisfies the three-dimensional wave equation. What physical situation does this correspond to. In particular, show that if $\mathbf{n}_1,...,\mathbf{n}_r$ are k unit vectors, then

$$\phi(t,\mathbf{r}) = \sum_{k=1}^{r} A_k f_k (t - \mathbf{n}_k.\mathbf{r}/c)$$

satisfies the three-dimensional wave equation.

[5] Consider the three-dimensional wave equation with harmonic dependence. Show that the equation for the amplitude reduces to the Helmholtz equation

$$\nabla^2 \psi + k^2 \psi = 0$$

Write down the general solution for this using separation of variables technique in all the three co-ordinates, rectangular, cylindrical and spherical polar.

[6] Let $\rho(t,r)$ and $J_k(t,r), k = 1, 2, 3$ be the charge and current density components in space. Assume these to be random fields. Assume further that their correlations are given by

$$< \rho(t,\mathbf{r}).\rho(t',\mathbf{r}') > = R_{\rho\rho}(t-t',\mathbf{r}-\mathbf{r}')$$
$$< \rho(t,\mathbf{r}).J_k(t',\mathbf{r}') > = R_k(t-t',\mathbf{r}-\mathbf{r}'), k = 1,2,3$$
$$< J_k(t,\mathbf{r}).J_m(t',\mathbf{r}') > = R_{km}(t-t',\mathbf{r}-\mathbf{r}'), k, m = 1,2,3$$

Write down the expressions for the auto and cross correlations

$$< V(t,\mathbf{r})V(t',\mathbf{r}') >, < V(t,\mathbf{r})A_k(t',\mathbf{r}') >, < A_k(t,\mathbf{r})A_m(t',\mathbf{r}') >$$

in the scalar potential and vector potential fields.

[7] The charge and current densities corresponding to the motion of a point charge Q moving along the trajectory $\mathbf{R}(t)$ are given respectively by $\rho(t, \mathbf{r}) = Q\delta(\mathbf{r} - \mathbf{R}(t))$ and $\mathbf{J}(t, \mathbf{r}) = Q\mathbf{v}(t)\delta(\mathbf{r} - \mathbf{R}(t))$, where $\mathbf{v}(t) = \mathbf{R}'(t)$ is the particle velocity. Show that

$$V(t,\mathbf{r}) = \frac{1}{4\pi\varepsilon}\int \frac{\delta(t'-t+|\mathbf{r}-\mathbf{r}'|/c)\rho(t',\mathbf{r}')}{|\mathbf{r}-\mathbf{r}'|}dV'dt'$$

$$= \frac{Q}{4\pi\varepsilon}\int \frac{\delta(t'-t+|\mathbf{r}-\mathbf{R}(t')|/c)}{|\mathbf{r}-\mathbf{R}(t')|}dt'$$

Evaluate this integral and show that it equals

$$V(t,\mathbf{r}) = \frac{Q}{4\pi\varepsilon(1-\mathbf{v}(t').\mathbf{n}(t'))|\mathbf{r}-\mathbf{R}(t')|c}$$

where t' is the solution to the equation

$$t' = t - |\mathbf{r} - \mathbf{R}(t')|/c$$

and

$$\mathbf{n}(t') = (\mathbf{r} - \mathbf{R}(t'))/|\mathbf{r} - \mathbf{R}(t')|$$

Give the physical interpretation of t' and $\mathbf{n}(t')$. Similarly, derive an expression for $\mathbf{A}(t, \mathbf{r})$ and by differentiation of the potentials, obtain expressions for the electric and magnetic fields produced by a moving charge.

[8] Prove that the ratios $\frac{E_x}{H_y}$ and $\frac{E_y}{H_x}$ for an air filled rectangular waveguide in the (n, m)th mode corresponding to the TE and TM modes are constant. Evalutate these ratios. These are called the intrinsic impedances of the wave guide. Given that the mode that propagates is TM, calculate the intrinsic impedance when

$$E_z = A.\sin(n\pi x/a).\sin(m\pi y/b).\cos(\omega t - \beta z)$$

Also derive an expression for the average power flow for this wave.

[9] In an air filled rectangular waveguide, a TE mode propagates with

$$E_y = A.\sin(n\pi x/a).\cos(m\pi y/b).\sin(\omega t - \beta z)$$

Determine the cutoff frequency, the intrinsic impedance and E_x, H_x, H_y for this wave.

[10] Consider a transmission line of length L having characteristic impedance Z_0 and propagation constant γ as a two port network with the voltages and currents at the two ends being the variables. Show that if (V_1, I_1) and (V_2, I_2) are the input voltage and current and output voltage and current with the currents being directed into the transmission line, then

$$V_1 = V_+ + V_-, \qquad I_1 = \frac{1}{Z_0}(V_+ - V_-)$$

$$V_2 = V_+\exp(-\gamma L) + V_-\exp(\gamma L), \qquad I_2 = -\frac{1}{Z_0}(V_+\exp(-\gamma L) - V_-\exp(\gamma L))$$

By eliminating V_+, V_- from these equations, obtain the Z parameters of this two port network.

[11] There are three transmission lines having lengths L_1, L_2, L_3 respectively with characteristic impedances Z_{01}, Z_{02}, Z_{03} and propagation constants γ_1, γ_2, γ_3 respectively. The output end of the first transmission line is connected to the input end of the second line and also to the input end of the third line. The ends of the second and the third line are terminated respectively with impedances Z_{L2} and Z_{L3}. Draw a diagram showing the arrangement and calculate the impedances looking into the inputs of the first line.

Hint: If (V_1, I_1) is the voltage current pair at the input of line 1, (V_2, I_2) the voltage and current at the output of line 1, (V_2, I_3) the voltage and current at the input of line 2, (V_2, I_4) the voltage and current at the input of line 3, (V_3, I_5) the voltage and current at the output of line 2 and finally, (V_4, I_6) the voltage and current at the output of line 3, then write down the governing equations, noting that $V_3 = I_5 Z_{L2}$, $V_4 = I_6 Z_{L3}$, $I_2 = I_3 + I_4$. Deduce from these equations the ratio V_1/I_1. This procedure can be used for evaluating the impedance looking into the input of line 1.

[12]. A lossless transmission line has characteristic impedance Z_0. When a load Z_L is applied at one end, the ratio $s = \frac{V_{max}}{V_{min}}$ is measured. Also, the adjacent voltage minima are found to occur at distances d_1, d_2 from the load end. Find the operating wavelength and the load impedance.

Hint: Start with the basic equations

$$V(z) = V_+\exp(-j\beta z) + V_-\exp(j\beta z), \qquad I(z) = \frac{1}{Z_0}(V_+\exp(-j\beta z) - V_-\exp(j\beta z))$$

where z is the distance measured from the source end. The equation $V(L) = I(L)\,Z_L$ gives

$$\frac{V_+\exp(-j2\beta L) + V_-}{V_+\exp(-j2\beta L) - V_-} = \frac{Z_L}{Z_0}$$

so that

$$\frac{V_-}{V_+} = \frac{Z_L - Z_0}{Z_L + Z_0}\exp(-j2\beta L) = \Gamma.\exp(-j2\beta L)$$

Thus,

$$V_{max} = |V_+|(1 + |\Gamma|),\ V_{min} = |V_+|(1 - |\Gamma|)$$

Also since

$$V(z) = V_+(\exp(-j\beta z) + \Gamma\exp(j\beta(z - 2L)))$$

it follows that if $Arg(\Gamma) = \phi$, then minima will occur when

$$\beta(2L - 2z) = (2n + 1)\,\pi + \phi$$

for n any integer. Thus, the distance between successive minima is

$$d_2 - d_1 = \pi/\beta$$

Thus, the operating wavelength can be determined. Also

$$s = \frac{1+|\Gamma|}{1-|\Gamma|}$$

ϕ can be determined from knowledge of the first voltage minimum d_1:

$$2\beta d_1 = \pi + \phi$$

Knowing $|\Gamma|$ and ϕ we can determine the load impedance Z_L. We leave the details to be worked out by the student.

[13] Derive the expression

$$Z_{in} = Z_0\frac{jZ_0\tan(\beta d)+Z_L}{jZ_L\tan(\beta d)+Z_0}$$

for the input impedance to a loss transmission line of length d terminated by a load impedance Z_L and characteristic impedance Z_0. What is Z_{in} when $d = \lambda/4$ where $\lambda = 2\pi/\beta$ is the wavelength of the propagation? This is called a quarter wave transformer. It transforms Z_L to Z_{in}. Explain how by connecting a quarter wave transformer of appropriate characteristic impedance to a load, we can obtain matching to a given transmission line having characteristic impedance Z_0?

[14] Consider an inifinitesimal current element $Icos(\omega t)dl\hat{z}$ located at the origin. Determine an expression for the electric field and magnetic field at large distances. Your answers for the fields should decay down as $1/r$. Higher order decay terms should be neglected. Calculate the Poynting vector at large distances and the average power W flowing out of a sphere of radius R where R is large. Hence obtain an expression for the radiation resistance R_r of the antenna defined by $I^2dl^2R_r/2 = W$.

[15] Assume that a straight wire is placed along the z axis stretching from $z = -L/2$ to $z = L/2$. The wire carries a current $I(z)\cos(\omega t)$. Show that the vector potential phasor can be expressed as

$$\mathbf{A}(\mathbf{r}) = \frac{\mu I}{4\pi}\int_{-L/2}^{L/2} \exp(-ik|\mathbf{r} - z'\hat{z}|)I(z')dz'/|\mathbf{r} - z'\hat{z}|$$

Show that for $r >> L$, this evaluates approximately to

$$\mathbf{A}(\mathbf{r}) \approx \frac{\mu I \exp(-ikr)}{4\pi r}\int_{-L/2}^{L/2} \exp(ikrz'\cos\theta)I(z')dz'$$

Evaluate this for

$$I(z) = I_0(1 - 2|z|/L)$$

and calculate the radiation resistance. This is defined by the condition $W = I_0^2 R_r/2$.

[16] Using the result of problem [15], calculate the far field expressions for a thin current of length L extending along the z axis from $z = -L/2$ to $z = L/2$ when the current is $I(z,t) = I_0\cos(\beta z)\cos(\omega t)$, $\beta = \omega/c$. What do these expressions reduce to when $L = n\lambda$ and when $L = (n + 1/2)\,\lambda$ with $\lambda = 2\pi/\beta$ being the wavelength.

[17] Prove that if a conductor is bent along a curve Γ, then its resistance can be calculated using the formula

$$R = \frac{\int_\Gamma \mathbf{E}.d\mathbf{r}}{\int_S \sigma\mathbf{E}.d\mathbf{S}}$$

where S is a cross section of the conductor and the electric field $\mathbf{E}$ within the body of the conductor is obtained from $\mathbf{E} = -\nabla V$, with V being the solution to Laplace's equation $\nabla^2 V = 0$ within the conductor body with the boundary conditions $V = V_1$ and V_2 at the ends of the conductor. Prove also that the capacitance between two conductors can be expressed as

$$C = \frac{\varepsilon\int_S \mathbf{E}.d\mathbf{S}}{\int \mathbf{E}.d\mathbf{r}}$$

where the surface integral in the numerator is taken over the surface of any one conductor and the line integral in the denominator extends from one surface to the other. $\mathbf{E}$ is determined by solving Laplace's equation $\nabla^2 V = 0$ in the region between the conducting surfaces with the boundary conditions $V = V_0$ and $V = -V_0$ on the two surfaces.

[18] A disc of thickness t and radius b has a central hole of radius a. Taking the conductivity of the disc as σ, find the resistance between (a) the hole and the rim of the disc, (b) the two flat sides of the disc.

Hint: Go a radial distance r, then $r + dr$ and note that the resistance of this piece between the surfaces $\rho = r$ and $\rho = d + dr$ is given by

$$\frac{dr}{\sigma S(r)} = \frac{dr}{\sigma 2\pi rt}$$

The integral of this from $r = a$ to $r = b$ gives the resistance of the conductor between the surfaces $\rho = a$ and $\rho = b$, i.e. between the hole and the rim of the disc. Taking the integral corresponds to adding up the infinitesimal resistances which are in series.

For the second part, the resistance between the top and bottom surfaces of the rings between $\rho = r$ and $\rho = r + dr$ is $dS(r) = 2\pi r dr$. These resistances are in parallel (why ?). Thus, the total resistance between the top and bottom surfaces is

$$R = \left(\int_a^b \frac{\sigma dS(r)}{t} \right)^{-1}$$

[19] A coaxial cable contains an insulating material of conductivity σ_1 in its upper half and another material of conductivity σ_2 in its lower half. Radius of the central wire is a and that of the sheath is b. Find the leakage resistance of the cable per unit length.

Hint: Note that the upper half and lower half form resistances in parallel since the inner surface is one connected piece and so is the outer surface.

The resistance of the upper half is $R_1 = \int_a^b \frac{dr}{\sigma_1 \pi r}$ and that of the lower half is $R_2 = \int_a^b \frac{dr}{\sigma_2 \pi r}$. Use this to calculate the total resistance R by the formula $1/R = 1/R_1 + 1/R_2$.

[20] Consider two concentric spherical surfaces of radii a and b respectively. The region between these two surfaces is filled with a medium of conductivity $\sigma(r)$ where r is the radial distance from the common centre of the spherical surfaces. Show that the resistance between the two surfaces is given by

$$R = \int_a^b \frac{dr}{4\pi r^2 \sigma(r)}$$

[21] The region $a_i \leq r \leq a_{i+1}$ is filled with a material of dielectric constant ε_i and conductivity σ_i, where $i = 1, 2, \ldots, n - 1$. Find the capacitance and resistance between the surfaces $r = a_1$ and $r = a_n$.

Hint: The electric displacement $D(r)$ is directed radially. Moreover it is continuous and from Gauss' law $\nabla . \mathbf{D} = 0$, it satisfies the equation

$$\frac{1}{r^2} \frac{d}{dr} r^2 D(r) = 0$$

Thus, $D(r) = A/r^2$, where A is a constant. The electric field in $a_i \le r \le a_{i+1}$ is then given by

$$E(r) = \frac{A}{\varepsilon_i r^2}$$

The potential difference between the surfaces $r = a_1$ and $r = a_n$ is

$$V = \sum_{i=1}^{n-1} \int_{a_i}^{a_{i+1}} E(r)dr$$

The charge on the surface $r = a_1$ is

$$Q = \varepsilon_1 E(a) 4\pi a^2 = 4\pi A$$

Then, $C = Q/V$ is the capacitance between the surfaces $r = a_1$ and $r = a_n$. For the second part, the resistance between the surfaces $r = a_i$ and $r = a_{i+1}$ is

$$R_i = \int_{a_i}^{a_{i+1}} \frac{dr}{4\pi r^2 \sigma_i}$$

These resistances are in series. So the total resistance between the surfaces $r = a_1$ and $r = a_n$ is $\sum_{i=1}^{n-1} R_i$.

[22] The electromagnetic field in free space is given by $\mathbf{E} = E(x, t)\,\hat{y}$, $\mathbf{B} = B(x,t)\hat{z}$. Show that the Maxwell equations

$$\nabla \times \mathbf{E} = -\frac{\partial \mathbf{B}}{\partial t}, \nabla \times \mathbf{B} = \varepsilon\mu \frac{\partial \mathbf{E}}{\partial t}$$

reduce to

$$\frac{\partial E}{\partial x} = -\frac{\partial B}{\partial t}, \qquad \frac{\partial B}{\partial x} = -\varepsilon\mu \frac{\partial E}{\partial t}$$

Deduce that E and B satisfy the one-dimensional wave equations

$$\frac{\partial^2 E}{\partial x^2} - \varepsilon\mu \frac{\partial^2 E}{\partial t^2} = 0$$

$$\frac{\partial^2 B}{\partial x^2} - \varepsilon\mu \frac{\partial^2 B}{\partial t^2} = 0$$

[23] (a) What condition must the function $f(t,x)$ satisfy for $\mathbf{B} = f(t, x)\,\hat{x}$ to be a magnetic field consistent with the Maxwell equations. (b) Can the field $\mathbf{D} = \psi(y) f(t - y/c)$ be a free space electric displacement field? State your reasons. (c) Can the field $\mathbf{E} = \frac{\sin\theta}{r} f(t - r/c)\hat{\theta}$ be a free space electric field? State your reasons.

Hint for the last two parts: Verify whether the wave equation is satisfied or not. For the last part, in the computation of $\nabla^2\mathbf{E}$, use the identities

$$\frac{\partial\hat{\theta}}{\partial r} = 0, \quad \frac{\partial\hat{\theta}}{\partial\theta} = -\hat{r}, \quad \frac{\partial\hat{\theta}}{\partial\phi} = \cos\theta\hat{\phi}$$

[24] The xz plane has a surface current flowing given by $J_s(t)\,\hat{z}$. Assume electric and magnetic fields $E_{z+}(t, y)\hat{z}$, $H_{x+}(t, y)\hat{x}$ in the region $y > 0$ and $E_{z-}(t, y)\hat{z}$, $H_{x-}(t, y)\hat{x}$ in the region $y < 0$ and write down the Maxwell equations for these fields. Deduce that these fields satisfy the wave equation in the respective regions and obtain the general solution. Then use the boundary condition for the tangential component of the magnetic field and the electric field to complete the solutions to the fields in both the regions.

Hint: The tangential component of the electric field is continuous at the boundary while the discontinuity in the tangential component of the magnetic field at the boundary equals the surface current flowing along the perpendicular direction.

[25] We have seen that the scalar and vector potentials are respectively given by the equations

$$V(t,\mathbf{r}) = \frac{1}{4\pi\varepsilon}\int \frac{\rho(t-|\mathbf{r}-\mathbf{R}|/c,\mathbf{R})}{|\mathbf{r}-\mathbf{R}|}\,dV(\mathbf{R})$$

$$\mathbf{A}(t,\mathbf{r}) = \frac{\mu}{4\pi}\int \frac{\mathbf{J}(t-|\mathbf{r}-\mathbf{R}|/c,\mathbf{R})}{|\mathbf{r}-\mathbf{R}|}\,dV(\mathbf{R})$$

By evaluating the partial derivatives of these potentials, derive integral expressions for the electric and magnetic fields produced by a current and charge density. Whereever prossible, make simplifications by using the charge conservation equation

$$\frac{\partial\rho}{\partial t} + \nabla.\mathbf{J} = 0$$

[26] A charge Q moves along the $\hat{z}$ axis according to the law $z = \xi(t)$. Show that the vector potential is given by

$$\mathbf{A}(t,\mathbf{r}) = \frac{\mu p(t')\hat{z}}{4\pi|r-\xi(t')\hat{z}|}$$

where t' is the solution to the equation

$$t - |\mathbf{r} - \xi(t')\hat{z}|/c = t'$$

Show that up to $O(1/r)$, the vector potential can be expressed as

$$\mathbf{A}(t,\mathbf{r}) \approx \frac{\mu p(t - r/c)\hat{z}}{r}$$

Use this expression to determine the electric field and the magnetic field up to $O(1/r)$ terms.

Hint: Calculate the scalar potential using the Lorentz gauge condition $\nabla.\mathbf{A} = -\frac{1}{c^2}\frac{\partial \phi}{\partial t}$.

[27] Consider a waveguide having arbitrary cross section. Let the direction along which propagation takes place be the $\hat{z}$ direction. Decompose the electromagnetic field into a transverse and a longitudinal component. $E_z\hat{z}$, $H_z\hat{z}$ are the longitudinal components while $\mathbf{E}_\perp$ and $\mathbf{H}_\perp$ are the transverse components, i.e.

$$\mathbf{E}_\perp = E_x\hat{x} + E_y\hat{y}$$
$$\mathbf{H}_\perp = H_x\hat{x} + H_y\hat{y}$$

Show assuming sinusoidal dependence that the second and third Maxwell equations can be expressed as

$$\left(\nabla_\perp + \hat{z}\frac{\partial}{\partial z}\right)\times(\mathbf{E}_\perp + E_z\hat{z}) = -i\omega\mu(\mathbf{H}_\perp + H_z\hat{z})$$
$$\left(\nabla_\perp + \hat{z}\frac{\partial}{\partial z}\right)\times(\mathbf{H}_\perp + H_z\hat{z}) = (\sigma + i\omega\varepsilon)(\mathbf{E}_\perp + E_z\hat{z})$$

where $\nabla_\perp$ denotes the transverse component of the gradient operator, i.e.

$$\nabla_\perp = \hat{x}\frac{\partial}{\partial x} + \hat{y}\frac{\partial}{\partial y}$$

Show assuming $\exp(-\gamma z)$ dependence on z that these equations decouple to give

$$\nabla_\perp \times \mathbf{E}_\perp = -i\omega\mu H_z\hat{z}$$
$$-\gamma\hat{z}\times\mathbf{E}_\perp + \nabla_\perp E_z \times \hat{z} = -i\omega\mu\mathbf{H}_\perp$$
$$\nabla_\perp \times \mathbf{H}_\perp = (\sigma + i\omega\varepsilon)E_z\hat{z}$$
$$-\gamma\hat{z}\times\mathbf{H}_\perp + \nabla_\perp H_z \times \hat{z} = (\sigma + i\omega\varepsilon)\mathbf{E}_\perp$$

Explain how $\mathbf{E}_\perp$ and $\mathbf{H}_\perp$ can be expressed in terms of E_z and H_z using these equations. Also derive the equation

$$(\nabla_\perp^2 + \gamma^2 + \omega^2 \varepsilon\mu) E_z = 0$$
$$(\nabla_\perp^2 + \gamma^2 + \omega^2 \varepsilon\mu) H_z = 0$$

Explain how the TM modes can be obtained as the solution of an eigenvalue problem by applying the boundary condition $E_z = 0$ on the boundary of the waveguide. Explain how the TE modes can be obtained by applying the boundary condition $\mathbf{H}_\perp.\mathbf{n} = 0$ on the boundary where $\mathbf{n}$ is the unit normal to the boundary.

[28] A transmission line having characteristic impedance R_0 (real) is terminated by a load R_L (real). If a pulse $v(t)$ is sent out by a generator connected to the line, write down the expressions for $I(z, t)$. Assume that the generator has impedance R_g (real). Assume the length of the line is L and that the speed of propagation of the signals along the line is u.

Hint: Let $V(\omega)$ be the Fourier transform of $v(t)$ and $I(z,\omega)$ that of $I(z,t)$. Let $V(z,t)$ be the voltage along the line and $V(z, \omega)$ its Fourier transform. Then,

$$V(z,\omega) = V_+ \exp(-j\beta z) + V_- \exp(j\beta z)$$

where $\beta = \omega.u$. Here, V_+, V_- are functions of frequency. Also

$$I(z,\omega) = \frac{1}{R_0}(V_+ \exp(-j\beta z) - V_- \exp(j\beta z))$$

$$V(\omega) - I(0,\omega) R_g = V(\omega) - \frac{R_g}{R_0}(V_+ - V_-) = V(0,\omega) = V_+ + V_-$$

The next equation is the load equation

$$V(L,\omega) = V_+ \exp(-j\beta L) + V_- \exp(j\beta L) = R_L I(L,\omega) = \frac{R_L}{R_0}(V_+ \exp(-j\beta L) - V_- \exp(j\beta L))$$

These two equations can be cast in matrix form as

$$\begin{pmatrix} 1 + \dfrac{R_g}{R_0} & 1 - \dfrac{R_g}{R_0} \\ \exp(-j\beta L)\left(1 - \dfrac{R_L}{R_0}\right) & \exp(-j\beta L)\left(1 + \dfrac{R_L}{R_0}\right) \end{pmatrix} \begin{pmatrix} V_+ \\ V_- \end{pmatrix} = \begin{pmatrix} V \\ 0 \end{pmatrix}$$

The solution to these equations is

$$V_+(\omega) = \frac{V(\omega)\exp(-j\omega L/u)(1+R_L/R_0)}{(1+R_L/R_0)(1+R_g/R_0)\exp(j\omega L/u)-(1-R_L/R_0)(1-R_g/R_0)\exp(-j\omega L/u)}$$

$$V_-(\omega) = -\frac{V(\omega)\exp(-j\omega L/u)(1-R_L/R_0)}{(1+R_L/R_0)(1+R_g/R_0)\exp(j\omega L/u)-(1-R_L/R_0)(1-R_g/R_0)\exp(-j\omega L/u)}$$

To express this in a convenient form, we set

$$a = (1+R_g/R_0)^{-1}$$

$$b = \frac{(1-R_L/R_0)(1-R_g/R_0)}{(1+R_L/R_0)(1+R_g/R_0)}$$

$$c = \frac{(1-R_L/R_0)}{(1+R_L/R_0)(1+R_g/R_0)}$$

Then,

$$V_+(\omega) = \frac{a.\exp(-j2\omega L/u)V(\omega)}{1-b.\exp(-j2\omega L/u)}$$

$$V_-(\omega) = -\frac{C.\exp(-j2\omega L/u)V(\omega)}{1-b.\exp(-j2\omega L/u)}$$

$$I(z,\omega) = \frac{1}{R_0}(V_+(\omega)\exp(-j\omega z/u) - V_-(\omega)\exp(j\omega z/u))$$

$$= \frac{V(\omega)}{R_0(1-b.\exp(-j2\omega L/u)}(a.\exp(-j\omega(z+2L)/u) - b.\exp(j\omega(z-2L)/u))$$

$$= \frac{V(\omega)}{R_0}\left(a\sum_{n=0}^{\infty} b^n \exp(-j\omega(z+2(n+1)L)/u) + \right.$$

$$\left. + c\sum_{n=0}^{\infty} b^n \exp(j\omega(z-2(n+1)L)/u)\right)$$

The inverse Fourier transform of this phasor is

$$I(z,t) = \frac{a}{R_0}\sum_{n=0}^{\infty} b^n v(t-(z+(2n+1)L)/u) + \frac{c}{R_0}\sum_{n=0}^{\infty} b^n v(t+(z-2(n+1)L)/u)$$

using the fact that the inverse Fourier transform of $V(\omega)\exp(j\omega t)$ is $v(t+\tau)$.

6

CHAPTER

FUNDAMENTALS OF QUANTUM MECHANICS

1. The wave function
2. Stern-Gerlach experiments on spin states in a magnetic field
3. The principle of superposition
4. Observables and events in quantum mechanics
5. Spin and rotation operators
6. Linear momentum operator and translations of space
7. Angular momentum: Computation of their eigenvalues and eigenstates
8. Angular momentum in spherical polar co-ordinates
9. Hamiltonian, Schrodinger equation, wave and matrix mechanics
10. The stationary Schrodinger equation. Solutions for
    (a) Particle in a box
    (b) Particle in a three-dimensional box
    (c) Tunnelling problems
    (d) Harmonic oscillator
11. Variational method for determining ground state energy eigenvalue
12. Simulation of quantum systems
13. Electron in a magnetic field with spin neglected

INTRODUCTION

Thus far, we have studied various aspects of the electromagnetic field, how these are produced by charges and currents and they propagate as waves in space and through waveguides. We have also studied reflection and refraction of electromagnetic waves at a boundary separating two media and the problem of wave propagation in a transmission line. The question arises why at all has this chapter on quantum mechanics been included in this book. There are three reasons for this. The first is that electromagnetic theory and quantum theory form the two great theories of the last two centuries that have succeeded Newtonian mechanics. The great leaps in man's understanding of nature have been (1) Newtonian mechanics, (2) Maxwell's electromagnetic theory, (3) Quantum theory, (4) Einstein's special and general theories of relativity. So in order to understand nature, the next natural step after understanding the classical theory of electromagnetism is the quantum theory. The second reason is the strong interplay between electromagnetism and quantum theory which has led to the subject of quantum electrodynamics. The classical Maxwell theory of electromagnetism assumes that light is a wave comprising an electric and magnetic field oscillating perpendicular to each other. In the quantum theory of electrodynamics, one writes down the expression for the energy of the radiation in a box expressing it as a quadratic form in the Fourier components of the vector potential. Thus, one is able to draw a correspondence between the electromagnetic field and an ensemble of harmonic oscillators. The usual quantization rules can then be applied to the radiation field. Such a treatment involves regarding the radiation field as an ensemble of bosons called photons and is the particle treatment of electromagnetic field theory. Although this chapter does not give any description of how the electromagnetic field is quantized, it does provide a fairly concise introduction of quantum mechanics which is essential to the understanding of quantum electrodynamics. This chapter can thus be regarded as a bridge between the Maxwell theory of electromagnetism and quantum electrodynamics. Some problems related to quantum theory and electrodynamics are discussed in this chapter like the splitting of the energy levels in a magnetic field, the spinning electrom in a magnetic field and nuclear magnetic resonance. The problem of the hydrogen atom involves considering the quantum motion of an electron in the coulomb electrostatic field of the nucleus. This has not been treated in this chapter but the reader armed with the tools of this chapter can easily understand Schrodinger's solution to this problem from any standard textbook on quantum mechanics.

1. The wave function

If the configuration of a quantum system is a manifold M having a measure μ (for example, $\mathbf{R}^{3N}$ with Lebesgue measure specifies the configuration of a system of N particles in position space), then a wave function is a function $\psi(\mathbf{x})$ defined on this manifold. Its interpretation according to Max Born is that $|\psi(\mathbf{x})|^2 d\mu(\mathbf{x})$ determines the probability of the configuration falling in the vicinity of $\mathbf{x}$ having measure $d\mu(\mathbf{x})$. For example the wave function for two electrons when we are only interested in their positions is given by $\psi(\mathbf{r}_1, \mathbf{r}_2)$. Then, we say that $\psi(\mathbf{r}_1, \mathbf{r}_2)$ is the probability amplitude for the two particles being respectively at the positions $\mathbf{r}_1$ and $\mathbf{r}_2$ and the magnitude square of this function gives us the joint probability density for the two particles. Suppose that the configuration of a quantum system can assume only one of a finite number N of distinct values. This happens for example while dealing with the spin of the electron which assumes only two values or for a system of k electrons when we are interested in only their spin. The possible values of the configuration are 2^k in number. The state space here is $\mathbb{C}^2 \otimes \mathbb{C}^2 \otimes ... \otimes \mathbb{C}^k$, the k-fold tensor product of the spin state space of each electron. We can conceive of a situation in which we are dealing with k independent quantum systems. The state space of the ith particle is the Hilbert space $\mathbb{H}_i$ and hence the state space of the entire system of k particles is $\mathbb{H}_1 \otimes ... \otimes \mathbb{H}_k$. One of the basic postulates of the quantum theory is that if ψ and ϕ are two wave functions in the Hilbert space $\mathbb{H}$, then $| < \phi | \psi > |^2$ gives us the probability of finding the system in the state ϕ when actually the system is in the state ψ.

2. The Stern-Gerlach experiment

consists in preparing the electrons coming from a device in the state having spin along a given direction $\mathbf{n}$ as 1/2. This state is the $|S_\mathbf{n} = 1/2 >$ state. Here, our units are such that $h/2\pi$ equals 1. The filtered beam is then passed through another filter that produces electrons in the state $S_\mathbf{m} = 1/2$, that is, the spin along the $\mathbf{m}$ direction is 1/2. We now ask for the relative proportion of electrons that exit out of the second filter in comparison with the electrons coming out of the first filter. Now, the spin operators in the quantum theory are described by the operators $\frac{1}{2}(\sigma_x, \sigma_y, \sigma_z)$. Thus, the spin along the direction $\mathbf{n} = (n_x, n_y, n_z)$ is defined by the operator $\frac{1}{2}\sigma_n$, where

$$\sigma_n = n_x\sigma_x + n_y\sigma_y + n_z\sigma_z$$

The spin matrices are 2 × 2 matrices satisfying the commutation rules

$$[\sigma_x, \sigma_y] = 2i\sigma_z,\ [\sigma_y, \sigma_z] = 2i\sigma_x,\ [\sigma_z, \sigma_x] = 2i\sigma_y$$

One choice for these matrices is

$$\sigma_x = \begin{pmatrix} 0 & 1 \\ 1 & 0 \end{pmatrix},\ \sigma_y = \begin{pmatrix} 0 & -i \\ i & 0 \end{pmatrix}$$

$$\sigma_z = \begin{pmatrix} 1 & 0 \\ 0 & -1 \end{pmatrix}$$

We calculate

$$\sigma_n = \begin{pmatrix} n_z & n_x - in_y \\ n_x + in_y & -n_z \end{pmatrix}$$

The eigenvalues of σ_n are ± 1. In terms of polar co-ordinates, we let $n_x = \cos\phi.\sin\theta$, $n_y = \sin\phi.\sin\theta$, $n_z = \cos\theta$. Then, the normalized eigenvector corresponding to the eigenvalue 1 is

$$|u> = \begin{pmatrix} \cos(\theta/2).\exp(-i\phi) \\ \sin(\theta/2) \end{pmatrix}$$

and the normalized eigenvector corresponding to the eigenvalue -1 is

$$|v> = \begin{pmatrix} -\sin\theta/2 \\ \cos\theta/2\ \exp(i\phi) \end{pmatrix}$$

If the particle is in the state $|u>$, then it has got a spin of 1/2 along the direction **n**. We also let $\mathbf{m} = (\cos\alpha.\sin\beta, \sin\alpha.\sin\beta, \cos\beta)$. When the system is in the state

$$|w> = \begin{pmatrix} \cos(\beta/2).\exp(-i\alpha) \\ \sin(\beta/2) \end{pmatrix}$$

then it has a spin of 1/2 along the direction **m**. When the actual spin of the particle is 1/2 along the direction **n**, then the probability that a measurement will show it to have a spin 1/2 along the direction **m** is given by $|<w/u>|^2$. We leave the evaluation of this probability to the student. This probability is also the relative intensity of the electron beam coming out of the second filter with respect to that coming out of the first filter. In particular, taking $\theta = 0$ gives $|u>$ as the upspin state for the z component of spin. Taking $\theta = \pi/2$, $\phi = \pi/2$ gives $|u>$ as the upspin state for the y component of spin. Taking $\theta = \pi/2$, $\phi = 0$ gives $|u>$ as the upspin state for the x component of the spin. These states are respectively represented by $|S_z = 1/2>$, $|S_y = 1/2>$ and $|S_x = 1/2>$. It is easily verified that

$$|<S_x=1/2|S_z=1/2>|^2 = |<S_y=1/2|S_z=1/2>|^2 = |<S_x=1/2|S_y=1/2>|^2 = 1/2$$

This means that if the first filter produces electrons in the $S_z = 1/2$ state, then the proportion of electrons that pass through a filter that polarizes in the $S_x = 1/2$ direction is 1/2. It should be noted that if a beam of electrons comes out of the gun with the spin of each electron being either $S_z = 1/2$ or $S_z = -1/2$, then the two kinds of spins can be separated by applying a magnetic field $B_z(z)$ in the z direction having a nonzero gradient $\frac{dB_z}{dz}$

along the z direction. Suppose the spin is 1/2. Then, we know that the spin magnetic moment of the electron is $-eh/4\pi m$ along the $\hat{z}$ axis and hence the energy of the electron in the magnetic field is $U(z) = ehB_z(z)/4\pi m$. This means that the electron will experience a force $-\frac{dU(z)}{dz} = -\frac{eh}{4\pi m}\frac{dB_z(z)}{dz}$ in the $\hat{z}$ direction and assuming that $\frac{dB_z}{dz}$ is negative, it follows that the spin 1/2 particles will get deflected upward and form a spot on the screen above the level of the incoming electron beam. On the other hand the electrons having spin $S_z = -1/2$ will have a magnetic moment $eh/4\pi m$ and hence will experience a force $\frac{eh}{4\pi m}\frac{dB_z}{dz}$ which is downward. The $S_z = -1/2$ particles will thus get deflected downwards and form a spot below the level of the incoming beam on the screen. In this way, we can separate out the $S_z = 1/2$ and $S_z = -1/2$ particles.

3. The principle of superposition

In its simplest form states that if $|\psi_1>, ..., |\psi_n>$ are vectors in a Hilbert space $\mathcal{H}$ that specify n different possible states of a quantum system (This means that the normalization condition $<\psi_i|\psi_i>= 1$ holds), then for any complex constants $c_1,...,c_m |\psi> = \Sigma_{i=1}^n c_i|\psi_i>$ is also a possible state, provided that the normalization condition

$$<\psi|\psi>= \sum_{i,j=1}^{n} \bar{c}_i c_j <\psi_i|\psi_j>= 1$$

holds. Suppose that we are dealing with position space wave functions i.e. $\psi_i(\mathbf{r})$, $i = 1, 2, ..., n$ are normalized wave functions: $\int |\psi_i(\mathbf{r})|^2 d^3\mathbf{r} = 1$. Then, the above result states that $\psi(\mathbf{r}) = \Sigma_{i=1}^n c_i\psi_i(\mathbf{r})$ is also a position space wave function provided that

$$\sum_{i,j=1}^{n} \bar{c}_i c_j \int \bar{\psi}_i(\mathbf{r})\psi_j(\mathbf{r})d^3\mathbf{r} = 1$$

In particular suppose that the states $|\psi_i>$, $i = 1, 2,..., n$ are mutually orthogonal, i.e. $<\psi_j|\psi_i>= \delta_{ij}$. Then, $|\psi>= \Sigma_{i=1}^n c_i|\psi_i>$ is a state provided that $\Sigma_{i=1}^n |c_i|^2 = 1$. This state is such that if the system is in this state, then the probability of a measurement showing to be in the state $|\psi_i>$ is

$$|<\psi_i|\psi>|^2 = |c_i|^2$$

4. Observables and events in quantum mechanics

In classical mechanics, an observable is a physical quantity that can be measured. Examples are positions of the particles, momenta of the particles, angular momentum and energy. A system in classical mechanics is described by specifying at each time instant, the generalized co-ordinates and momenta of the particles. If it has n degrees of freedom, then the generalized co-ordinates may be represented by $(q_1, ..., q_n)$ which are n functions of time. The generalized momenta are represented by $(p_1, ..., p_n)$, which are once again n functions of time. The state of the system at time t is then the $2n$ tuple, $(q_1(t), ..., q_n(t), p_1(t), ..., p_n(t))$. For example, the motion of a system of N particles moving freely in space with some interaction is described by specifying the x, y, z co-ordinates and momenta of each particle. Thus, the state vector is $(x_i, y_i, z_i, p_{x_i}, p_{y_i}, p_{z_i}, i = 1,2, ..., N)$. This means that there are $3N$ generalized co-ordinates and $3N$ generalized momenta for the given system of particles. In the quantum theory, we have seen that the state of system is specified by a wave function which implies that the co-ordinates and momenta are not determined to full accuracy. We shall later see that there is an uncertainty between position and momentum, that is if we are able to measure the position with complete accuracy, then there is no accuracy in the measurement of the momentum and vice versa. We've noted that the state of a quantum system is a vector ψ in a Hilbert space. This vector is usually denoted by $|\psi>$ and read as *ket*ψ. We also have Bra vectors $<\psi|$ which are formally the complex conjugate adjoints of the ket vectors. Given a ket vector $|\psi>$ and a Bra vector $<\phi$, we can multiply the former by the latter from the left to get the complex scalar $<\phi|\psi>$. For the position representation, for example, in one dimension, $|x>$, $x \in \mathbb{R}$ form a complete orthogonal set of ket vectors. These are not however normalizable. They satisfy the orthogonality relations $<x'|x> = \delta(x - x')$, where $\delta(x)$ is the Dirac delta function. Given a ket vector $|\psi>$, in the position representation, it is represented formally by the function $\psi(x) =< x|\psi>$. Thus the bracket of the two ket vectors $|\psi>$ and $|\phi>$ is given by

$$<\phi|\psi>= \int_{-\infty}^{\infty} <\phi|x> dx <x|\psi>= \int_{-\infty}^{\infty} \bar{\phi}(x)\psi(x)dx$$

We can also consider the momentum representation of the same state vector $|\psi>$. This is given by the function $\hat{\psi}(p) = <p|\psi>$ and is related to the position representation $\psi(x) =< x|\psi>$ by the transformation formula

$$\hat{\psi}(p) =< p|\psi>= \int_{-\infty}^{\infty} <p|x> dx <x|\psi>= \int_{-\infty}^{\infty} <p|x> dx\psi(x)$$

To determine this transformation rule, we must therefore determine the function $<p|x>$. Note that $<\phi|\psi>$ for normalizable kets $|\psi>$, $|\phi>$

equals the probability amplitude of the system showing up in the state $|\phi>$ after a measurement given that it is originally in the state $|\psi>$. Thus, according to Max Born, $|<\phi|\psi>|^2$ is the probability of the system showing up to be in the state $|\phi>$ after the measurement given that prior to the measurement, the system is in the state $|\psi>$. To arrive at the amplitude $<x|p>$, we need to go a little deep into the notion of an observable. An observable in the quantum theory is formally a Hermitian operator on the Hilbert space of quantum states. Thus for a finite dimensional quantum system, the Hilbert space of states can be taken as $\mathbb{C}^n$ with n being a finite integer. The space of observables then consists of all $n \times n$ complex Hermitian matrices. Given a Hermitian operator A on a separable Hilbert space H which is our space of quantum states, let $\{|e_1>,..., |e_n>,...\}$ be a complete orthonormal basis of its eigenvectors. By orthonormal we mean that $<e_i|e_j>=\delta_{ij}$ for all i, j. Let λ_i be the eigenvalue of A associated with the eigenvector $|e_i>$, that is $A|e_i>=\lambda_i|e_i>$. Then, if $|\psi>$ is the state of our quantum system, the probability that a measurement of A will show it to be in the state $|e_i>$ and hence assume a value of λ_i is $|<e_i|\psi>|^2$. When this value shows up, the state of the system collapses to the state $|e_i>$. The average value of A in the state $|\psi>$ is therefore given by

$$<A>_\psi=\sum_i \lambda_i |<e_i|\psi>|^2=<\psi|\sum \lambda_i |e_i><e_i|\psi>=<\psi|A|\psi>$$

where, the spectral representation $A = \Sigma_i \lambda_i|e_i><e_i|$ has been used. This result carries over to infinite dimensions. In $L^2(\mathbb{R})$, for example, we may adopt the position representation. Then a ket $|\psi>$ is represented by the wave function $\psi(x)=<x|\psi>$. An observable A is represented by a kernel $A(x, x')=<x|A|x'>$ which satisfies the Hermiticity condition: $A(x', x) = \bar{A}(x', x)$. This means that A can be viewed as an infinite dimensional Hermitian matrix. Then,

$$<\psi|A|\psi>=\int <\psi|x>dx<x|A|x'>dx'<x'|\psi>=\int \bar{\psi}(x)A(x,x')\psi(x')dxdx'$$

Or regarding A as an operator on $L^2(\mathbb{R})$,

$$<\psi|A|\psi>=\int_{-\infty}^{\infty} \bar{\psi}(x)A\psi(x)dx$$

In the separable situation, if $\{|e_i>\}$ is a complete orthonormal basis for $\mathbb{H}$, then

$$<\phi|\psi>=\sum_i <\phi|e_i><e_i|\psi>$$

In the nonseparable situation, we have a system $\{|\lambda>: \lambda \in \mathbb{R}\}$ such that $<\lambda|\lambda'> = \delta(\lambda - \lambda')$ and the representation of the inner product is

$$<\phi|\psi> = \int <\phi|\lambda> d\lambda <\lambda|\psi> = \int_{\mathbb{R}} \bar{\phi}(\lambda)\psi(\lambda)d\lambda$$

5. Spin and rotation

The state of a spin 1/2 particle like an electron can be expressed as $c_1|1> + c_2|0>$, where $|1>$ is the state corresponding to $S_z = 1/2$ and $|0>$ is the state corresponding to $S_z = -1/2$. We may then say that the particle has $S_z = 1/2$ with probability $|c_1|^2$ and $S_z = -1/2$ with probability $|c_2|^2$. The spin operator $S_z = \sigma_z/2$ acts on the two states according to the rule $S_z|1> = (1/2)|1>$, $S_z|0> = -(1/2)|0>$. Rotation of the spin states about the z axis by an angle ϕ is described by the matrix $U_z(\phi) = \exp(iS_z\phi)$. We say that S_z is the generator of the spinor representation of rotations about the $\hat{z}$ axis and U_z is the spinor representation of rotations about the $\hat{z}$ axis. In general, the spinor representation of rotations about any axis with direction cosines $\mathbf{n} = (n_x, n_y, n_z)$ by an angle ϕ is given by

$$U_{\mathbf{n}}(\phi) = \exp(i(n_x\sigma_x + n_y\sigma_y + n_z\sigma_z)\phi/2) = \exp(i\mathbf{n}.\sigma\phi/2)$$

To see the connection between the spinor representation and rotations of R^3, we consider a point $\mathbf{r} = (x, y, z)$ in space and the matrix

$$\Phi(\mathbf{r}) = \mathbf{r}.\sigma = x\sigma_x + y\sigma_y + z\sigma_z = \begin{pmatrix} z & x - iy \\ x + iy & -z \end{pmatrix}$$

Note that $\Phi(\mathbf{r})$ is a Hermitian 2×2 matrix of zero trace and

$$U_{\mathbf{n}}(\delta\phi).(\mathbf{r},\sigma).U_{\mathbf{n}}(-\delta\phi) = (1 + i(\mathbf{n},\sigma)\delta\phi/2)).(\mathbf{r},\sigma).(1 - i(\mathbf{n},\sigma)\delta\phi/2)$$
$$= (\mathbf{r},\sigma) - i[(\mathbf{r},\sigma), (\mathbf{n},\sigma)]\delta\phi/2$$

Now,

$$[(\mathbf{r}, \sigma),(\mathbf{n}, \sigma)] = (xn_y - yn_x)[\sigma_x, \sigma_y] + (yn_z - zn_y)[\sigma_y,\sigma_z] + (zn_x - xn_z)$$
$$[\sigma_z,\sigma_x] = 2i(xn_y - yn_x)\sigma_z + 2i(yn_z - zn_y)\sigma_x + 2i(zn_x - xn_z)\sigma_y = 2i(\mathbf{r} \times \mathbf{n}, \sigma)$$

Thus,

$$U_{\mathbf{n}}(\delta\phi).(\mathbf{r},\sigma)\ U_{\mathbf{n}}(-\phi) = (\mathbf{r}, \sigma) + \delta\phi(\mathbf{r} \times \mathbf{n}, \sigma)$$

This fact can be paraphrased as follows. Let

$$\Phi(\mathbf{r}) = (\mathbf{r}, \sigma)$$

Then,

$$U_{\mathbf{n}}(\delta\phi)\Phi(\mathbf{r})U_{\mathbf{n}}(-\delta\phi) = \Phi(\mathbf{r}')$$

where

$$\mathbf{r}' = \mathbf{r} - \mathbf{n} \times \mathbf{r}\delta\phi = R_{\mathbf{n}}(-\delta\phi)$$

is obtained from **r** by applying an infinitesimal rotation about the axis **n** by the angle $\delta\phi$. Successive applications of the infinitesimal operator $U_n(\delta\phi)$ and $U_n(-\delta\phi)$ respectively to the left and right of $\Phi(\mathbf{r})$ gives the identity

$$U_n(\phi)\Phi(\mathbf{r})U_n(-\phi) = \Phi(R_n(\phi)\mathbf{r})$$

This means the following: The map $\mathbf{r} \to \Phi(\mathbf{r})$ transforms $\mathbb{R}^3$ onto the space $\mathbb{H}$ of 2×2 Hermitian matrices of trace zero. Rotations act on $\mathbb{R}^3$ and corresponding to each rotation $R_n(\phi)$, we have two 2×2 unitary matrices $U_n(\phi)$ and $U_n(-\phi)$ which act on H according to the rule $X \to UXU^*$ and these represent the rotation. This is also called the spinor representation of the rotation group.

6. Linear Momentum and translations of space

Let $\psi(x)$ be a wave function of a particle executing motion in one dimension relative to the position representation. From intuitive considerations, if under infinitesimal translations of the system, the wave function changes by a constant phase factor, the state remains unaltered and hence linear momentum must be conserved. Now under an infinitesimal translation δx, the wave function changes to

$$\psi(x - \delta x) = \psi(x) - \psi'(x)\delta x$$

This must equal $\exp(i\delta\alpha)\psi(x) = \psi(x) + i\delta\alpha\psi(x)$ for momentum to be conserved where $\delta\alpha$ is a constant. Thus,

$$\psi'(x) = i\frac{\delta\alpha}{\delta x}\psi(x)$$

and this means that $\frac{\delta\alpha}{\delta x}$ is an eigenvalue of the operator $-i\frac{d}{dx}$ with $\psi(x)$ as the associated eigenfunction. Furthermore $-i\frac{d}{dx}$ is a Hermitian operator as follows from

$$<\phi|-i\frac{d}{dx}|\psi> = \int \bar{\phi}(x)(-i)\psi'(x)dx = i\int \bar{\phi}'(x)\psi(x)dx$$
$$= \int (-i\phi(x))^*\psi(x)dx = <-i\frac{d}{dx}\phi|\psi>$$

Thus, it is natual to regard a real constant times $-i\frac{d}{dx}$ as the momentum operator along the x direction, with the value the same constant times $\frac{\delta\alpha}{\delta x}$

as the value of the momentum in the state ψ. ψ here is a state in which the particle has a definite momentum. The constant is chosen as $\frac{h}{2\pi}$ to get agreement with experiment. The vector operator of momentum in space is thus $\mathbf{p} = -\frac{ih}{2\pi}\nabla$ and its three Cartesian components are $-\frac{ih}{2\pi}$ $(d/dx, d/dy, d/dz)$. Let $\psi(\mathbf{r})$ be the position representation of a wave function $\mathbb{R}^3$. An infinitesimal translation of this wave function by the vector $(\varepsilon_1, \varepsilon_2, \varepsilon_3)$ in space produces the wave function

$$\psi'(\mathbf{r}) = \psi(x-\varepsilon_1, y-\varepsilon_2, z-\varepsilon_3) = \psi(x,y,z) - \varepsilon_1\frac{\partial\psi(\mathbf{r})}{\partial x} - \varepsilon_2\frac{\partial\psi(\mathbf{r})}{\partial y} - \varepsilon_3\frac{\partial\psi(\mathbf{r})}{\partial z}$$

$$= \psi(\mathbf{r}) + \frac{2\pi i}{h}(\varepsilon, \mathbf{p})\psi(\mathbf{r})$$

This is the same as saying that $2\pi i\mathbf{p}/h$ is the generator of translations in space. A finite translation by the vector $\mathbf{a}$ sends the wave function $\psi(\mathbf{r})$ to

$$\psi(\mathbf{r} - \mathbf{a}) = \exp(-(\mathbf{a}, \nabla))\psi(\mathbf{r}) = \exp(-2\pi i(\mathbf{a}, \mathbf{p})/h)\psi(\mathbf{r})$$

We've already noted in the section on observables that if X is an observable and $|\lambda>$ an eigenket of X corresponding to the eigenvalue λ, then when the quantum system is in the state $|\lambda>$, the observable X has the definite value λ. Thus, a state of definite momentum is a ket specified as $|\mathbf{p}'>$, where $\mathbf{p}' = (p'_x, p'_y, p'_z)$ is a tuple of three real numbers. In this state, the momentum of the particle has definite values for its three components $\mathbf{p}' = (p'_x, p'_y, p'_z)$. Likewise, the ket $|\mathbf{r}'>$, where $\mathbf{r}' = (x', y', z')$ is a state in which the position operator $\mathbf{r}$ has definite values for its three components namely (x', y', z'). The position operator corresponds to multiplication of wave functions in position space by the vector $\mathbf{r}$. Now consider, the evaluation of the bracket $<\mathbf{r}'|\mathbf{p}'>$. This can be evaluated by using the eigen-equation $\mathbf{p}|\mathbf{p}'> = \mathbf{p}'|\mathbf{p}'>$. This gives

$$<\mathbf{r}'|\mathbf{p}|\mathbf{p}'> = \mathbf{p}'<\mathbf{r}'|\mathbf{p}'>$$

Now,

$$<\mathbf{r}'|\mathbf{p}|\mathbf{p}'> = <\mathbf{r}'|-i\frac{h}{2\pi}\nabla|\mathbf{p}'> = -\frac{ih}{2\pi}\nabla'<\mathbf{r}'|\mathbf{p}'>$$

and hence we get the eigen-equation

$$-\frac{ih}{2\pi}\nabla'<\mathbf{r}'|\mathbf{p}'> = \mathbf{p}'<\mathbf{r}'|\mathbf{p}'>$$

so that

$$< \mathbf{r}'|\mathbf{p}' > = C.\exp(i2\pi(\mathbf{p}',\ \mathbf{r}')/h)$$

This when written in full notation, is

$$< x',\ y',\ z'|p'_x,\ p'_y,\ p'_z > = C.\exp(i2\pi(p'_x x' + p'_y y' + p'_z z')/h)$$

The value of the constant C is determined from the normalization condition

$$< x',y',z' \mid x'',y'',z'' > = \int < x',y',z' \mid p'_x, p'_y, p'_z > dp'_x dp'_y dp'_z < p'_x, p'_y, p'_z$$
$$\mid x'',y'',z'' > = \delta(x'-x'')\delta(y'-y'')\delta(z'-z'')$$

This gives

$$\mid C \mid^2 \int exp(i2\pi(\mathbf{p}',\mathbf{r}'-\mathbf{r}'')/h)d\mathbf{p}' = \delta(\mathbf{r}'-\mathbf{r}'')$$

However, the left side evaluates to $|C|^2h^3\delta(\mathbf{r}'-\mathbf{r}'')$. Thus, $|C| = 1/h^{3/2}$. We may multiply this by an arbitrary phase factor to obtain the value of C. Note that $< \mathbf{r}'|\mathbf{p}' >$ is the position space representation of a state having definite momentum $\mathbf{p}'$. It is thus as displayed above and eigen function of the momentum operator corresponding to a definite momentum $\mathbf{p}'$.

7. Angular momentum

Just as the linear momentum operator of the quantum theory is related to the infinitesimal generator of translations in space, likewise, the angular momentum operator is intimately connected to the generators of rotation matrices. In what follows, we first deduce the commutation relation that exists between the infinitesimal generators of rotations in space and then obtain expressions for the angular momentum operator in the quantum theory. These operators appear as the representations of the generators of the rotation in the space $L^2(\mathbf{R})$ of wave functions.

1. Rotation matrices: Rotation about the $\hat{x}$ axis has the following matrix in the standard ordered basis:

$$R_x(\phi) = \begin{pmatrix} 1 & 0 & 0 \\ 0 & \cos\phi & -\sin\phi \\ 0 & \sin\phi & \cos\phi \end{pmatrix}$$

This is the same as saying that under a rotation about the x axis by an angle ϕ, a point $(X,\ Y,\ Z)$ gets transformed to $(X',\ Y',\ Z')$ where

$$X' = X,\ Y' = Y.\cos\phi - Z.\sin\phi,\ Z' = Y.\sin\phi + Z.\cos\phi$$

Rotation about the $\hat{y}$ axis has the following matrix in the standard ordered basis:

$$R_y(\phi) = \begin{pmatrix} \cos\phi & 0 & \sin\phi \\ 0 & 1 & 0 \\ -\sin\phi & 0 & \cos\phi \end{pmatrix}$$

which is equivalent to the transformation equations

$$X' = X.\cos\phi + Z.\sin\phi,\ Y' = Y,\ Z' = -X.\sin\phi + Z.\cos\phi$$

Finally, rotation about the $\hat{z}$ axis has the matrix

$$R_z(\phi) = \begin{pmatrix} \cos\phi & -\sin\phi & 0 \\ \sin\phi & \cos\phi & 0 \\ 0 & 0 & 1 \end{pmatrix}$$

which is equivalent to the transformation equations

$$X' = X.\cos\phi - Y.\sin\phi,\ Y' = X.\sin\phi + Y.\cos\phi,\ Z' = Z$$

Infinitesimal rotations about these three axes by angle $\delta\phi$ are, up to $O(\delta\phi^2)$ given by using the expansions: $\sin\delta\phi = \delta\phi + O(\delta\phi^3)$, $\cos\delta\phi = 1 - \delta\phi^2/2$. Thus, with neglect of $O(\delta\phi^3)$ terms, these rotations are given by

$$R_x(\delta\phi) = \begin{pmatrix} 1 & 0 & 0 \\ 0 & 1-\delta\phi^2/2 & -\delta\phi \\ 0 & \delta\phi & 1-\delta\phi^2/2 \end{pmatrix}$$

$$R_y(\delta\phi) = \begin{pmatrix} 1-\delta\phi^2/2 & 0 & \delta\phi \\ 0 & 1 & 0 \\ -\delta\phi & 0 & 1-\delta\phi^2/2 \end{pmatrix}$$

$$R_z(\delta\phi) = \begin{pmatrix} 1-\delta\phi^2/2 & -\delta\phi & 0 \\ \delta\phi & 1-\delta\phi^2/2 & 0 \\ 0 & 0 & 1 \end{pmatrix}$$

Elementary algebra shows that

$$R_x(\delta\phi)R_y(\delta\phi) = \begin{pmatrix} 1-\delta\phi^2/2 & 0 & \delta\phi \\ \delta\phi^2 & 1-\delta\phi^2/2 & -\delta\phi \\ -\delta\phi & \delta\phi & 1-\delta\phi^2 \end{pmatrix}$$

$$R_y(\delta\phi)R_x(\delta\phi) = \begin{pmatrix} 1-\delta\phi^2/2 & \delta\phi^2 & \delta\phi \\ 0 & 1-\delta\phi^2/2 & -\delta\phi \\ -\delta\phi & \delta\phi & 1-\delta\phi^2 \end{pmatrix}$$

$$R_y(\delta\phi)R_z(\delta\phi) = \begin{pmatrix} 1-\delta\phi^2 & -\delta\phi & \delta\phi \\ \delta\phi & 1-\delta\phi^2/2 & 0 \\ -\delta\phi & \delta\phi^2 & 1-\delta\phi^2/2 \end{pmatrix}$$

$$R_z(\delta\phi)R_y(\delta\phi) = \begin{pmatrix} 1-\delta\phi^2/2 & -\delta\phi & 0 \\ \delta\phi & 1-\delta\phi^2/2 & 0 \\ 0 & 0 & 1 \end{pmatrix}$$

$$R_z(\delta\phi).R_x(\delta\phi) = \begin{pmatrix} 1-\delta\phi^2/2 & -\delta\phi & \delta\phi^2 \\ \delta\phi & 1-\delta\phi^2 & -\delta\phi \\ 0 & \delta\phi & 1-\delta\phi^2/2 \end{pmatrix}$$

$$R_x(\delta\phi).R_z(\delta\phi) = \begin{pmatrix} 1-\delta\phi^2/2 & -\delta\phi & 0 \\ \delta\phi & 1-\delta\phi^2 & -\delta\phi \\ \delta\phi^2 & \delta\phi & 1-\delta\phi^2/2 \end{pmatrix}$$

These lead us to the following commutation relations for the infinitesimal generators:

$$[R_x(\delta\phi), R_y(\delta\phi)] = \begin{pmatrix} 0 & -\delta\phi^2 & 0 \\ \delta\phi^2 & 0 & 0 \\ 0 & 0 & 0 \end{pmatrix}$$

$$[R_y(\delta\phi), R_z(\delta\phi)] = \begin{pmatrix} 0 & 0 & 0 \\ 0 & 0 & -\delta\phi^2 \\ 0 & \delta\phi^2 & 0 \end{pmatrix}$$

$$[R_z(\delta\phi), R_x(\delta\phi)] = \begin{pmatrix} 0 & 0 & \delta\phi^2 \\ 0 & 0 & 0 \\ -\delta\phi^2 & 0 & 0 \end{pmatrix}$$

The commutation relations are valid up to $O(\delta\phi^2)$.

The usual way to state these commutation relations is by introducing the generators of the rotations about the three axes. We have from the above identities,

$$R_x(\delta\phi) = I + i\delta\phi.L_x + O(\delta\phi^2)$$
$$R_y(\delta\phi) = I + i\delta\phi.L_y + O(\delta\phi^2)$$
$$R_z(\delta\phi) = I + i\delta\phi.L_z + O(\delta\phi^2)$$

where

$$iL_x = \begin{pmatrix} 0 & 0 & 0 \\ 0 & 0 & -1 \\ 0 & 1 & 0 \end{pmatrix}$$

$$iL_y = \begin{pmatrix} 0 & 0 & 1 \\ 0 & 0 & 0 \\ -1 & 0 & 0 \end{pmatrix}$$

$$iL_z = \begin{pmatrix} 0 & -1 & 0 \\ 1 & 0 & 0 \\ 0 & 0 & 0 \end{pmatrix}$$

From these, it follows that $[L_x, L_y] = -iL_z$, $[L_y, L_z] = -iL_x$, $[L_z, L_x] = -iL_y$

Total magnitude square of the angular momentum is defined by $L^2 = L_x^2 + L_y^2 + L_z^2$. Any set of three linear operators L_x, L_y, L_z on a vector space satisfying these commutation relations qualify to be angular momentum operators. One such set is obtained using $L^2(\mathbb{R})$ as our representation space. For any $\psi(\mathbf{r}) \in L^2(\mathbb{R})$, we have under an infinitesimal rotation about the x axis its transformation into $\psi'(\mathbf{r})$, where

$$\delta\psi(\mathbf{r}) = \psi'(\mathbf{r}) - \psi(\mathbf{r}) = \psi(R_x(-\delta\alpha)\mathbf{r}) - \psi(\mathbf{r}) = \psi(x, y + z.\delta\alpha, z - y.\delta\alpha) - \psi(\mathbf{r})$$

$$= (z.\frac{\partial\psi}{\partial y} - y.\frac{\partial\psi}{\partial z})\delta\alpha$$

$$= -i(yp_z - zp_y)\psi\delta\alpha = -iL_x\psi\delta\alpha$$

where

$$p_y = -i\frac{\partial}{\partial y}, p_z = -i\frac{\partial}{\partial z}$$

are the operators of linear momentum along the y and z directions. The linear momentum vector operator is $-i\nabla$ and it satisfies the following commutation relations with the position operator:

$$[x,p_x] = i,[y,p_y] = 1, [z,p_z] = i$$

with all the other commutators for the variables $\{x, y, z, p_x, p_y, p_z\}$ being zero. This can be expressed alternately as

$$[x_a, p_b] = i\delta_{a,b}, [x_a, x_b] = 0, [p_a, p_b] = 0$$

The operator of linear momentum that we just introduced is in fact the representation of the linear momentum in the space $L^2(\mathbb{R})$. The angular momentum operators in the space $L^2(\mathbb{R})$ are

$$L_x = yp_z - zp_y, L_y = zp_x - xp_z, L_z = xp_y - yp_x$$

or in a vectorial form,

$$\mathbf{L} = \mathbf{r} \times \mathbf{p} = -i\mathbf{r} \times \nabla$$

where we choose our units such that $h/2\pi = 1$. Their commutation relations can be obtained as

$$[L_x,L_y] = [yp_z - zp_y, zp_x - xp_z] = y[p_z, z]\, p_x + x[z, p_z]p_y = -iyp_x + ixp_y = iL_z$$

$$[L_y,L_z] = [zp_x - xp_z, xp_y - yp_x] = z[p_x, x]\, p_y + y[x,p_x]p_z = -izp_y + iyp_z = iL_x$$

Finally,

$$[L_z,L_x] = [xp_y - yp_x, yp_z - zp_y] = x[p_y, y]\, p_z + z[y, p_y]\, p_x = izp_x - ixp_z = iL_y$$

These can be expressed in the more handy notation by introducing ε_{abc} which is 1 when (abc) = (123) or any cyclic permutation of this and -1 when it is either (213), (321) or (132) and zero when any two indices of a, b, c are equal. In terms of this tensor, we can write

$$[L_a,L_b] = i\varepsilon_{abc}L_c$$

Under an infinitesimal rotation about the y axis by angle α, the change in the wave function is

$$\delta\psi(\mathbf{r}) = \psi(R_y(-\delta\alpha)\mathbf{r}) - \psi(\mathbf{r}) = \psi(x - z\delta\alpha, y, z + x\delta\alpha) - \psi(\mathbf{r})$$

$$= -\delta\alpha.z\frac{\partial\psi}{\partial x} + \delta\alpha.x\frac{\partial\psi}{\partial z}$$

$$= -i\delta\alpha(zp_x - xp_z) = -i\delta\alpha.L_y\psi$$

Finally an infinitesimal rotation about the z axis by α results in the wave function changing by

$$\delta\psi(\mathbf{r}) = \psi(R_z(-\delta\alpha)\mathbf{r}) - \psi(\mathbf{r}) = \psi(x + y\delta\alpha, y - x\delta\alpha, z) - \psi(\mathbf{r})$$

$$= \delta\alpha.\left(y\frac{\partial\psi}{\partial x} - x\frac{\partial\psi}{\partial y}\right)$$

$$= i\delta\alpha.(yp_x - xp_y)\psi = -i\delta\alpha L_z\psi$$

The operator L^2 is seen to commute with L_z:

$$[L^2, L_z] = [L_x^2, L_z] + [L_y^2, L_z]$$

$$=[L_x, L_z]L_x + L_x[L_x, L_z] + [L_y, L_z]L_y + L_y[L_y, L_z] = -iL_yL_x - iL_xL_y + iL_xL_y + iL_yL_x = 0$$

Thus, these two operators can be diagonalized simultaneously. Let $|j, m>$ denote an eigenket of these two corresponding to the eigenvalue $j(j + 1)$ for L^2 and the eigenvalue m for L_z, i.e. $L^2|j, m> = j(j + 1)|j, m>$, $L_z|j, m> = m|j, m>$. We want to show that j takes half integer values ≥ 0 and that for each j, m takes values $\{-j, -j + 1, ..., j\}$. Define the ladder operators $L_+ = L_x + iL_y$, $L_- = L_x - iL_y$.

Then,

$$L_+L_- = L_x^2 + L_y^2 - i[L_x, L_y] = L_x^2 + L_y^2 + L_z = L^2 - L_z^2 + L_z$$

$$L_-L_+ = L_x^2 + L_y^2 + i[L_x, L_y] = L_x^2 + L_y^2 - L_z = L^2 - L_z^2 - L_z$$

$$[L_+, L_z] = [L_x, L_z] + i[L_y, L_z] = -iL_y - L_x = -L_+, [L_-, L_z]$$

$$= [L_x, L_z] - i[L_y, L_z] = -iL_y + L_x = L_-$$

Thus,

$$L_zL_+|j, m> = (L_+L_z + L_+)|j, m> = (m + 1)L_+|j, m>$$

$$L_zL_-|j, m> = (L_-L_z - L_-)|j, m> = (m - 1)L_-|j, m>$$

Thus, L_+ raises the value of m by one and L_- lowers it by one for any eigenket of L_z. For any given j, there is a maximum and a minimum value of m. This follows since $L_-^* = L_+$ implies

$$(j(j + 1) - m(m + 1)) = <j, m|L^2 - L_z^2 - L_z|j, m> = <j, m|L_-L_+|j, m>$$

$$= \| L_+|j, m>\|^2 \geq 0$$

implying $m \leq j$ and

$$(j(j+1)-m(m-1)) = < j,m \mid L^2 - L_z^2 + L_z \mid j,m > = < j,m \mid L_+ L_- \mid j,m >$$
$$= \| L_- \mid j,m > \|^2 \geq 0$$

implying $m \geq -j$. The ket $|j, j>$ is an eigenket of L^2 with eigenvalue $j(j+1)$ and an eigenket of L_z with eigenvalue j. L_+ operating on this ket gives zero. This can be verified from the fact that the norm square of $L_+| j, j >$ is $j, j|L_-L_+| j, j > = < j, j|L^2 - L_z^2 - L_z| j, j > = < j, j| j(j+1) - j^2 - j| j, j > = 0$. Thus, $| j, j >$ satisfies the eigenequation $L_+| j, j > = 0$. After determining this ket in as the solution of this differential equation, we determine successively the kets $| j, m >$, $m = j-1, ..., -j$ by applying $L-$ with an appropriate normalizing factor. For example,

$$L_-| j, m > = c| j, m-1>$$

Take the norm square on both sides to get

$$<j, m |L_+ L_-| j, m > = |c|^2$$

Now,

$$L_+L_- = (L_x + iL_y)(L_x - iL_y) = L_x^2 + L_y^2 - i[L_x, L_y] = L^2 - L_z^2 + L_z$$

Thus,

$$|c|^2 = j(j+1) - m(m-1)$$

and this gives

$$|j, m-1> = (j(j+1) - m(m-1))^{-1/2} L_- |j, m >$$

up to an arbitrary phase factor. Equivalently, we could calculate the wave function $|j, -j>$ using $L_-|j, -j> = 0$ and then use the raising operator L_+ to calculate $| j, m >$, $m = -j+1, ..., j$. The normalization constant can be determined using the following scheme:

$$L_+|j, m > = c|j, m+1 >$$

Taking norm on both sides gives

$$|c|^2 = < j, m|L_-L_+|j, m >$$

and since $L_-L_+ = (L_x - iL_y)(L_x + iL_y) = L_x^2 + L_y^2 + i[L_x, L_y] = L^2 - L_z^2 - L_z$, it follows that

$$|c|^2 = j(j+1) - m(m+1)$$

and thus

$$|j, m+1 > = (j(j+1) - m(m+1))^{-1/2} L_+|j, m >$$

up to an arbitrary phase factor. To see that $L_-|j, -j> = 0$, we need only note that its norm square is given by $< j, -j|L_+L_-|j, -j >=< j, -j|L^2 - L_z^2 + L_z|j, -j >=< j, -j|j(j+1) - j^2 - j|j, -j > = 0$. Thus, L_- when operating $2j$ times on the ket $|j, j>$ produces the ket $|j, -j>$ and this means that $2j$ must be a nonnegative integer. Equivalently, j must be a half integer or an integer. j is called the spin of the particle. Particles with half integral spins are called Fermions and particles with integral spins are called Bosons.

8. Angular momentum in spherical polar co-ordinates

We consider a wave function $\psi(r,\theta,\phi)$ expressed in terms of polar co-ordinates. We then compute the angular momentum operator in terms of these co-ordinates. First we need to compute $L_x = yp_x - zp_y$ in terms of polar angles. This means that $\frac{\partial}{\partial x}$ and $\frac{\partial}{\partial y}$ must be expressed in polar angles.

$$\frac{\partial\psi}{\partial x} = \frac{\partial r}{\partial x}\frac{\partial\psi}{\partial r} + \frac{\partial\theta}{\partial x}\frac{\partial\psi}{\partial\theta} + \frac{\partial\phi}{\partial x}\frac{\partial\psi}{\partial\phi}$$

Now,

$$\frac{\partial r}{\partial x} = x/r = \cos\phi.\sin\theta$$

$$\frac{\partial\theta}{\partial x} = \frac{\partial}{\partial x}\cos^{-1}(z/r) = \frac{1}{r}\cos\phi.\cos\theta$$

$$\frac{\partial\phi}{\partial x} = \frac{\partial}{\partial x}\tan^{-1}(y/x) = -\frac{y}{x^2+y^2} = -\frac{\sin\phi}{r.\sin\theta}$$

This gives

$$\frac{\partial\psi}{\partial x} = \cos\phi.\sin\theta\frac{\partial\psi}{\partial r} + \frac{1}{r}\cos\phi.\cos\theta\frac{\partial\psi}{\partial\theta} - \frac{\sin\phi}{r.\sin\theta}\frac{\partial\psi}{\partial\phi}$$

$$\frac{\partial\psi}{\partial y} = \frac{\partial r}{\partial y}\frac{\partial\psi}{\partial r} + \frac{\partial\theta}{\partial y}\frac{\partial\psi}{\partial\theta} + \frac{\partial\phi}{\partial y}\frac{\partial\psi}{\partial\phi}$$

Now,

$$\frac{\partial r}{\partial y} = y/r = \sin\phi.\sin\theta$$

$$\frac{\partial \theta}{\partial y} = \frac{1}{r}\sin\phi.\cos\theta$$

$$\frac{\partial \phi}{\partial y} = \frac{\cos\phi}{r.\sin\theta}$$

Thus,

$$\frac{\partial \psi}{\partial y} = \sin\phi.\sin\theta\frac{\partial \psi}{\partial r} + \frac{1}{r}\sin\phi.\cos\theta\frac{\partial \psi}{\partial \theta} + \frac{\cos\phi}{r.\sin\theta}\frac{\partial \psi}{\partial \phi}$$

Thus,

$$L_z\psi = (xp_y - yp_x)\psi = -\frac{ih}{2\pi}\left(x\frac{\partial \psi}{\partial y} - y\frac{\partial \psi}{\partial x}\right)$$

The quantity within brackets is

$$x\frac{\partial \psi}{\partial y} - y\frac{\partial \psi}{\partial x} =$$

$$= r\cos\phi.\sin\theta.\left(\sin\phi\sin\theta\frac{\partial \psi}{\partial r} + \frac{1}{r}\sin\phi.\cos\theta\frac{\partial \psi}{\partial \theta} + \frac{\cos\phi}{r.\sin\theta}\frac{\partial \psi}{\partial \phi}\right)$$

$$- r\sin\phi.\sin\theta.\left(\cos\phi\sin\theta\frac{\partial \psi}{\partial r} + \frac{1}{r}\cos\phi.\cos\theta\frac{\partial \psi}{\partial \theta} - \frac{\sin\phi}{r.\sin\theta}\frac{\partial \psi}{\partial \phi}\right)$$

$$= \frac{\partial \psi}{\partial \phi}$$

This means that the operator of angular momentum about the $\hat{z}$ axis in spherical polar co-ordinates is given by

$$L_z = -\frac{ih}{2\pi}\frac{\partial}{\partial \phi}$$

We next compute the operator of angular momenta about the $\hat{x}$ and $\hat{y}$ axes. To do this, we need the expression for $\frac{\partial}{\partial z}$ in spherical polars.

$$\frac{\partial\psi}{\partial z} = \frac{\partial r}{\partial z}\frac{\partial\psi}{\partial r} + \frac{\partial\theta}{\partial z}\frac{\partial\psi}{\partial\theta} + \frac{\partial\phi}{\partial z}\frac{\partial\psi}{\partial\phi}$$

$$\frac{\partial r}{\partial z} = z/r = \cos\theta$$

$$\frac{\partial\theta}{\partial z} = \frac{\partial}{\partial z}\cos^{-1}(z/r) = -\frac{\sin\theta}{r}$$

$$\frac{\partial\phi}{\partial z} = \frac{\partial}{\partial z}\tan^{-1}(y/x) = 0$$

Thus,

$$\frac{\partial\psi}{\partial z} = \cos\theta\frac{\partial\psi}{\partial r} - \frac{\sin\theta}{r}\frac{\partial\psi}{\partial\theta}$$

Now,

$$L_x\psi = (yp_z - zp_y)\psi = -\frac{ih}{2\pi}\left(y\frac{\partial\psi}{\partial z} - z\frac{\partial\psi}{\partial y}\right)$$

The quantity within brackets is

$$y\frac{\partial\psi}{\partial z} - z\frac{\partial\psi}{\partial y} =$$

$$r.\sin\phi,\sin\theta\left(\cos\theta\frac{\partial\psi}{\partial r} - \frac{\sin\theta}{r}\frac{\partial\psi}{\partial\theta}\right)$$

$$-r.\cos\theta.\left(\sin\phi.\sin\theta\frac{\partial\psi}{\partial r} + \frac{1}{r}\sin\phi.\sin\theta\frac{\partial\psi}{\partial\theta} + \frac{\cos\phi}{r.\sin\theta}\frac{\partial\psi}{\partial\phi}\right)$$

$$= -\sin\phi.\frac{\partial\psi}{\partial\theta} - \cot\theta.\cos\phi\frac{\partial\psi}{\partial\phi}$$

Thus, the operator of angular momentum about the x axis in spherical polars is

$$L_x = \frac{ih}{2\pi}\left(\sin\phi\frac{\partial}{\partial\theta} + \cot\theta.\cos\phi\frac{\partial}{\partial\phi}\right)$$

Finally,

$$L_y\psi = (zp_x - xp_z)\psi = -\frac{ih}{2\pi}\left(z\frac{\partial\psi}{\partial x} - x\frac{\partial\psi}{\partial z}\right)$$

The quantity within brackets is

$$z\frac{\partial\psi}{\partial x}-x\frac{\partial\psi}{\partial z}$$

$$= r\cos\theta\left(\cos\phi.\sin\theta\frac{\partial\psi}{\partial r}+\frac{1}{r}\cos\phi.\cos\theta\frac{\partial\psi}{\partial\theta}-\frac{\sin\phi}{r.\sin\theta}\frac{\partial\psi}{\partial\phi}\right)$$

$$-r.\cos\phi.\sin\theta.\left(\cos\theta\frac{\partial\psi}{\partial r}-\frac{\sin\theta}{r}\frac{\partial\psi}{\partial\theta}\right)$$

$$= \cos\phi\frac{\partial\psi}{\partial\theta}-\cot\theta.\sin\phi\frac{\partial\psi}{\partial\phi}$$

Thus,

$$L_y = \frac{ih}{2\pi}\left(-\cos\phi\frac{\partial}{\partial\theta}+\cot\theta.\sin\phi\frac{\partial}{\partial\phi}\right)$$

We leave it as an exercise to show that up to a constant multiplicative factor, the squared angular momentum operator $L^2 = L_x^2 + L_y^2 + L_z^2$ is given by the angular part of the Laplacian operator, i.e.

$$L^2 = -\frac{h^2}{4\pi^2}\left(\frac{1}{\sin\theta}\frac{\partial}{\partial\theta}\sin\theta\frac{\partial}{\partial\theta}+\frac{1}{\sin^2\theta}\frac{\partial^2}{\partial\phi^2}\right)$$

9. Hamiltonian, Schrodinger Equation, Wave and matrix mechanics

The energy of a system of n particles moving in a potential field $V(\mathbf{r}_1, ..., \mathbf{r}_n)$ according to classical mechanics is given by

$$H(\mathbf{r}_1,...,\mathbf{r}_n,\mathbf{p}_1,...,\mathbf{p}_n) = \sum_{i=1}^{n}\frac{|\mathbf{p}_i|^2}{2M_i}+V(\mathbf{r}_1,...,\mathbf{r}_n)$$

Here, $\mathbf{p}_i = M_i\,\frac{dr_i}{dt}$ is the momentum of the ith particle. Our formulation of quantum mechanics implies that every observable is represented by an operator. Thus, to obtain the energy operator, we replace $\mathbf{p}_i$ by $-\frac{ih}{2\pi}\nabla_i$, where ∇_i denotes the gradient with respect to the position vector $\mathbf{r}_i$ of the ith particle. This gives us the energy operator as

$$H = -\frac{h^2}{4\pi^2}\sum_{i=1}^{n}\frac{\nabla_i^2}{2M_i} + V(\mathbf{r}_1, \ldots, \mathbf{r}_n)$$

Just as position and momentum are dual to each other, energy and time are also dual to each other. This means that the energy operator must be of the form a constant times $\frac{\partial}{\partial t}$. This can also be seen from the wave viewpoint of De Broglie and Max Planck. A plane wave of frequency ω radians per second and wave number k has a spatio-temporal variation of amplitude given by $\psi = \exp(-i(\omega t - kx))$ assuming propagation to occur in the x direction. Planck's theory tells us that $E = h\omega/2\pi$ should be the energy of the particle associated with this wave and De Broglie's theory tells us that the momentum of the particle associated with this wave must be $p = hk/2\pi$. Now, $\frac{\partial \psi}{\partial t} = -i\omega\psi$ and $\frac{\partial \psi}{\partial x} = ik\psi$. This is the same as saying that

$$\frac{ih}{2\pi}\frac{\partial \psi}{\partial t} = E\psi, \; -\frac{ih}{2\pi}\frac{\partial \psi}{\partial x} = p\psi$$

These two formulas suggest that it is natural to regard $\frac{ih}{2\pi}\frac{\partial}{\partial t}$ as the energy operator and $-\frac{ih}{2\pi}\frac{\partial}{\partial x}$ as the momentum operator. This leads us to the Schrodinger equation for the state of our quantum particle:

$$\frac{ih}{2\pi}\frac{\partial \psi}{\partial t} = H\psi$$

or written in full,

$$\frac{ih}{2\pi}\frac{\partial \psi}{\partial t} = -\frac{h^2}{4\pi^2}\sum_{i=1}^{n}\frac{\nabla_i^2\psi}{2M_i} + V(\mathbf{r}_1, \ldots, \mathbf{r}_n)\psi$$

where $\psi(t, \mathbf{r}_1, \ldots, \mathbf{r}_n)$ is a function of time and the position co-ordinates of the different particles. For a single free particle, this equation reduces to

$$\frac{ih}{2\pi}\frac{\partial \psi(t,\mathbf{r})}{\partial t} = -\frac{h^2}{8\pi^2 M}\nabla^2\psi(t,\mathbf{r})$$

the solution to which is a superposition of the functions

$$\psi_{\mathbf{p}}(t,\mathbf{r}) = \exp(-i(\omega t - \mathbf{k}.\mathbf{r}))$$

with $h\omega/2\pi = \mathbf{p}^2/2M$ and $h\mathbf{k}/2\pi = \mathbf{p}$. Now, the solution to the general Schrodinger equation is usually accomplished by first determining the solutions to the stationary Schrodinger equation given by

$$H\psi(\mathbf{r}_1,\ldots,\mathbf{r}_n|E) = E\psi(\mathbf{r}_1,\ldots,\mathbf{r}_n|E)$$

where E is a constant parameter called the energy eigenvalue associated with the function $\psi(\mathbf{r}_1,\ldots,\mathbf{r}_n|E)$. This equation will generally not admit square integrable solutions for all values of the real parameter E but only for a discrete set. This discrete set $\{E_n\}$ is called the allowed set of energy levels of the quantum system. By square integrable we mean that $\int|\psi|^2 d^3\mathbf{r}_1\ldots d^3\mathbf{r}_n = 1$. When such a normalization is possible for a given value of E, we say that the corresponding wave function $\psi(\mathbf{r}_1, \ldots, \mathbf{r}_n|E)$ is a stationary state of the particle corresponding to the energy level E. Then, $|\psi(\mathbf{r}_1, \ldots, \mathbf{r}_n|E)|^2 d^3\mathbf{r}_1\ldots d^3\mathbf{r}_n$ gives us the joint probability of finding the ith particle in the differential volume element $d^3\mathbf{r}_i$. When such a normalization is not possible, ψ may still correspond to a quantum state but not of a single particle, rather of a stream of particles having density $|\psi|^2$ in the $3n$ dimensional space. Suppose that the allowed spectrum is discrete say $\{E_k\}$ and the corresponding stationary states $\{\psi_k\}$ are all normalized. Then, we can express the general solution to our time dependent Schrodinger equation as

$$\psi(t,\mathbf{r}_1,\ldots,\mathbf{r}_n) = \sum_{k=1}^{\infty} c_k \exp(-i2\pi E_k t/h)\psi_k(\mathbf{r}_1,\ldots,\mathbf{r}_n)$$

For we get on differentiating this expression with respect to time

$$\frac{ih}{2\pi}\frac{\partial\psi}{\partial t} = \sum c_k \exp(-i2\pi E_k t/h)E_k\psi_k = \sum c_k \exp(-i2\pi E_k t/h)H\psi_k = H\psi$$

We can show that if $E_k \neq E_m$, then ψ_k and ψ_m are orthogonal in the sense that

$$\int \bar{\psi}_k(\mathbf{r}_1, \ldots, \mathbf{r}_n)\psi_m(\mathbf{r}_1, \ldots, \mathbf{r}_n)d^3\mathbf{r}_1 \ldots d^3\mathbf{r}_n = 0$$

Indeed, integration by parts shows that for any two functions f, g of the variables $\mathbf{r}_1, \ldots, \mathbf{r}_n$, one has

$$\int \bar{f}(\mathbf{r}_1, \ldots, \mathbf{r}_n)\nabla_i^2 g(\mathbf{r}_1, \ldots, \mathbf{r}_n)d^3\mathbf{r}_1\ldots d^3\mathbf{r}_n = \int (\nabla_i^2 \bar{f}(\mathbf{r}_1, \ldots,\mathbf{r}_n))g(\mathbf{r}_1, \ldots, \mathbf{r}_n)d^3\mathbf{r}_1\ldots d^3\mathbf{r}_n$$

assuming appropriate vanishing boundary conditions at ∞. It follows that

$$\int \bar{f}(H_g)dV = \int (\overline{Hf})g dV$$

or in terms of the inner product $< f|g >= \int \bar{f} g dV$, $< f|Hg >=< Hf|g >$. When an operator X on $L^2(\mathbb{R}^{3m})$ satisfies the condition $< f|Xg >=< Xf|g >$ we write $X^* = X$ and say that X is self-adjoint or Hermitian. The quantity $< f|Xg >$ is then denoted by $< f|X|g >$. It is verified that for X Hermitian,

$$< f|X|g >^* =< Xg|f >=< g|X|f >$$

We are now in a position to prove that eigenfunctions corresponding to distinct eigenvalues of a Hermitian operator are orthogonal. Let X be Hermitian and let $X|\psi_1 >=\lambda_1|\psi_1 >$, $X|\psi_2 >= \lambda_2|\psi_2 >$ with λ_1, λ_2 distinct. Then,

$$<\psi_1|X|\psi_1 >=< \psi_1|\lambda_1|\psi_1 >=\lambda_1 < \psi_1|\psi_1 >$$

on the one hand. On the other hand,

$$<\psi_1|X|\psi_1 >^* =< \psi_1|X|\psi_1 >=\bar{\lambda}_1 < \psi_1|\psi_1 >$$

Thus, $\lambda_1 = \bar{\lambda}_1$, which means that λ_1 is real. In other words, we have proved that all the eigenvalues of a Hermitian operator are real. To prove orthogonality of $|\psi_1 >$ and $|\psi_2 >$, note that

$$<\psi_2|X|\psi_1 >=< \psi_2|\lambda_1|\psi_1 >=\lambda_1 < \psi_2|\psi_1 >$$

on the one hand and on the other hand,

$$<\psi_2|X|\psi_1 >=< \psi_1|X|\psi_2 >^* = \bar{\lambda}_2 < \psi_1|\psi_2 >^* = \bar{\lambda}_2 < \psi_2|\psi_1 >$$

Comparing the above two relations and recalling that λ_2 is real, we get

$$(\lambda_2 - \lambda_1) < \psi_2|\psi_1 > = 0$$

which implies that $< \psi_2|\psi_1 > = 0$. This means that given a Hermitian operator X, we determine its eigenspectrum $\{\lambda_n\}$ and the corresponding eigenfunctions or eigenvectors or eigenstates $|\psi_n >$, which are assumed to be normalized. Then, $< \psi_n|\psi_m > = \delta_{n,m}$ and any state vector $|\psi >$ can be expanded as

$$|\psi >= \sum_n |\psi_n >< \psi_n|\psi >$$

Then,

$$X|\psi >= \sum_n X|\psi_n >< \psi_n|\psi >= \sum_n \lambda_n|\psi_n >< \psi_n|\psi >$$

which amounts to saying that X has the spectral resolution

$$X = \sum_n \lambda_n |\psi_n><\psi_n|$$

Suppose however that X also has a continuous spectrum. Assume that its discrete spectrum is $\{\lambda_n\}$ with eigenstates $|\psi_n>$ and that its continuous spectrum is $\{\mu : \mu \in I\}$ where I is a subset of $\mathbb{R}$ with eigenstates $|\eta_\mu>$. The $|\eta_\mu>$'s are orthogonal but not orthonormal. In fact, they will in general not even be normalizable. One imposes the normalization condition $<\eta_\mu'|\eta_\mu> = \delta(\mu - \mu')$ for the eigenstates corresponding to the continuous spectrum. Thus, the complete set of orthogonality relations can be expressed as

$$<\psi_n|\psi_m> = \delta_{n,m}, \qquad <\psi_n|\eta_\mu> = 0, \qquad <\eta_\mu|\eta_{\mu'}> = \delta(\mu - \mu')$$

Then, an arbitrary ket $|\psi>$ admits the expansion

$$|\psi> = \sum_n |\psi_n><\psi_n|\psi> + \int_I |\eta_\mu> d\mu <\eta_\mu|\psi>$$

Taking the inner product on the right side with $|\psi_m>$ gives in view of the orthogonality relations $<\psi_m|\psi>$ as required. Taking the inner product with $|\eta_{\mu'}>$ gives again in view of the orthogonality relations

$$\int <\eta_{\mu'}|\eta_\mu> d\mu <\eta_\mu|\psi> = \int \delta(\mu' - \mu) d\mu <\eta_\mu|\psi> = <\eta_{\mu'}|\psi>$$

as required. Thus the consistency of the expansion is verified. We operate on both sides of this relation by X to get

$$X|\psi> = \sum_n \lambda_n |\psi_n><\psi_n|\psi> + \int_I \mu|\eta_\mu> d\mu <\eta_\mu|\psi>$$

which is the same as

$$X = \sum_n |\psi_n>\lambda_n<\psi_n| + \int_I \mu|\eta_\mu> d\mu <\eta_\mu|$$

This is the required spectral resolution of the operator X. To write this down in the language of functional analysis, we define the discrete spectral projections $E_n = |\psi_n><\psi_n|$ and the continuous spectral measure $dE(\mu) = |\eta_\mu> d\mu <\eta_\mu|$ Then,

$$I = \sum_n |\psi_n><\psi_n| + \int_I |\eta_\mu> d\mu <\eta_\mu|$$

$$X = \sum_n \lambda_n E_n + \int_I \mu.dE(\mu)$$

Suppose X is an operator on a Hilbert space and $\{|e_n>\}$, $\{|f_\mu>: \mu \in I\}$ is a complete orthonormal basis for the Hilbert space in the sense that every ket $|\psi>$ admits the expansion

$$|\psi>=\sum_n |e_n><e_n|\psi>+\int_I |f_\mu>d\mu<f_\mu|\psi>$$

then we may expand the ket $X|\psi>$ and take the inner product of this expansion with $<\phi|$ to get

$$<\phi|X|\psi>=\sum_{n,m} <\phi|e_n><e_n|X|e_m><e_m|\psi>+\int_{I\times I} <\phi|f_\mu>$$

$$>d\mu<f_\mu|X|f_{\mu'}>d\mu'<f_{\mu'}|\psi>+\sum_n \int_I <\phi|e_n><e_n|X|f_\mu>$$

$$>d\mu<f_\mu|\psi>+\sum_n \int_I <\phi|f_\mu>d\mu<f_\mu|X|e_n><e_n|\psi>$$

The matrix elements of X relative to this basis are $<e_n|X|e_m>$ corresponding to the nth row and the mth column of the discrete spectrum, $<e_n X|f_\mu>$ corresponding to the nth row of the discrete spectrum and the μth column of the continuous spectrum, $<f_\mu|X|e_n>$ corresponding to the μth row of the continuous spectrum and the nth column of the discrete spectrum and $<f_\mu|X|f_{\mu'}>$ corresponding to the μth row and the μ'th column of the continuous spectrum. These matrix elements play a fundamental role in the theory of transitions in quantum theory. Now given the Hamiltonian, which is assumed to be time independent, we have seen that H is a Hermitian operator. In general, it will have both a discrete and a continuous spectrum. Let $\{E_n\}$ be its discrete spectrum and $\{E_\lambda : \lambda \in I\}$ its continuous spectrum of energy eigenvalues. Let $|\psi_n>$ and $|\psi_\lambda>$ be the corresponding eigenstates with the appropriate normalizations. We can expand an arbitrary ket $|\psi>$ using these eigenkets as

$$|\psi>=\sum c_n|\psi_n>+\int_I c(\lambda)|\psi_\lambda>d\lambda$$

Choose another ket $|\phi>$ and effect the same expansion

$$|\phi>=\sum_n d_n|\psi_n>+\int_I d(\lambda)|\psi_\lambda>d\lambda$$

Then,

$$<\phi|\psi>=\sum_n \bar{d}_n c_n+\int_I \bar{d}(\lambda)c(\lambda)d\lambda$$

This is implied by the normalization

$$<\psi_n | \psi_m >= \delta_{n,m}, \qquad <\psi_\mu | \psi_\lambda >= \delta(\mu - \lambda), \qquad <\phi_n | \psi_\lambda >= 0$$

In particular,

$$<\psi | \psi >= \sum_n |c_n|^2 + \int_I |c(\lambda)|^2 \, d\lambda$$

and the condition for this to be normalized is that this quantity must equal unity. Thus, the general solution to the time dependent Schrodinger equation

$$\frac{ih}{2\pi} \frac{\partial |\psi_t >}{\partial t} = H |\psi_t >$$

can be expressed as

$$|\psi_t >= \sum c_n \exp(-2\pi i E_n t / h) |\psi_n > + \int_I c(\lambda) \exp(-2\pi i E_\lambda t / h) |\psi_\lambda > d\lambda$$

where

$$\sum_n |c_n|^2 + \int_I |c(\lambda)|^2 \, d\lambda = 1$$

$$c_n \exp(-2\pi i E_n t / h) = <\psi_n | \psi_t >$$

is the probability amplitude for the system at time t being in the state $|\psi_n >$. Its magnitude squared $|c_n|^2$ gives the probability that at time t, the state of the system is $|\psi_n >$. Likewise, $c(\lambda)\exp(-2\pi i E_\lambda t/h) = <\psi_\lambda|\psi>$ is the probability amplitude that at time t, the state of the system has energy in the continuous range $[E_\lambda, E_{\lambda+d\lambda})$. $|c(\lambda)|^2 d\lambda = |<\psi_\lambda|\psi>|^2 d\lambda$ is the probability that the energy of the system is in the above continuous range. Note that this is independent of time but dependent only upon the initial state of the system. Given an observable X, we can compute its average value in the state $|\psi_t >$. It is given by

$$<X>_t = <\psi_t | X | \psi_t >=$$

$$\sum_{n,m} \bar{c}_n c_m exp(-2\pi i(E_m - E_n)t / h) <\psi_n | X | \psi_m >$$

$$+ \sum_n \bar{c}_n \int_I c(\lambda) \exp(+2\pi i(E_n - E_\lambda)t / h) <\psi_n | X | \psi_\lambda > d\lambda$$

$$+ \sum_n \bar{c}_n \int_I \overline{c(\lambda)} \exp(-2\pi i(E_n - E_\lambda)t / h) <\psi_\lambda | X | \psi_n > d\lambda$$

$$+ \int_{I \times I} \overline{c(\lambda)}\, c(\lambda') \exp(-2\pi i(E_\lambda - E_{\lambda'})t / h) <\psi_\lambda | X | \psi_\lambda > d\lambda d\lambda'$$

According to Heisenberg's matrix mechanics, an observable can also be regarded as evolving in time. The Schrodinger view treats the observables as constant in time and the states as evolving with time while Heisenberg's matrix mechanics treats the state as being constant in time and the observables as evolving with time. Both the views should however yield the same value for the average value of an observable in a state as a function of time for these are the actually measured quantities. According to Schrodinger,

$$\frac{d}{dt}<X>_t=\frac{d}{dt}<\psi_t|X|\psi_t>=\left(\frac{d}{dt}<\psi_t|\right)X|\psi_t>+<\psi_t|X\frac{d}{dt}|\psi_t>$$

$$=\frac{2\pi i}{h}<\psi_t|HX|\psi_t>-\frac{2\pi i}{h}<\psi_t|XH|\psi_t>=\frac{2\pi i}{h}<\psi_t|HX-XH|\psi_t>$$

At $t = 0$, we have the rate of change of $< X >_t$ given by $\frac{2\pi i}{h}$ $<\psi_0|HX - XH|\psi_0>$. If however, the state is regarded as being $|\psi_0>$ at all times and the observable time dependent, the rate of change of $< X >_t$ at $t = 0$ would be given by $<\psi_0|\frac{dX}{dt}|\psi_0>$. Hence for both these expressions to agree, we must have

$$\frac{dX}{dt}=\frac{2\pi i}{h}(HX-XH)$$

This is the Heisenberg's equation for matrix mechanics. Another way to see this is to write down the solution to Schrodinger's equation

$$\frac{ih}{2\pi}\frac{d}{dt}|\psi_t>=H|\psi_t>$$

as

$$\psi_t> = U_t|\psi_0>,\ U_t = \exp(-2\pi iHt/h)$$

$\{U_t\}$ is a one parameter group of unitary operators on the Hilbert space of quantum states. Unitarity follows since H is Hermitian. The group porperty is $U_0 = I$, $U_tU_s = U_{t+s}$, that is evolution in time s followed by evolution in time t is the same as evolution in time $s + t$. The unitarity property is essential for otherwise an initially normalized state would not be normalized after some time has elapsed: $<\psi_t|\psi_t>=<U_t\psi_0|U_t\psi_0>=<\psi_0|\psi_0>$. We can also see that if $|\psi_0>$ and $|\phi_0>$ are two normalized states then initially the probability of making a transition from one to the other is $|<\psi_0|\phi_0>|^2$. After time t, the states evolve respectively to $|\phi_t> = U_t|\phi_0>$ and $|\psi_t>= U_t|\psi_0>$. The transition probability remains invariant:

$$<\psi_t|\phi_t>=<U_t\psi_0|U_t\psi_0>=<\psi_0|\phi_0>$$

in view of the unitarity of the evolution. Now

$$< X >_t =< U_t \psi_0 | X | U_t \psi_0 > = < \psi_0 | U_t^* X U_t | \psi_0 >$$

This is according to the Schrodinger picture. In the Heisenberg picture, the state remains a constant $|\psi_0>$ while the observable changes to X_t. Thus, $< X >_t =< \psi_0 | X_t | \psi_0 >$. If these two expressions are to coincide, we must have

$$X_t = U_t^* X U_t = \exp(2\pi i H t / h) \, X \, \exp(-2\pi i H t / h)$$

Here, $X = X_0$ is the Heisenberg observable at time zero. Taking the temporal derivative of this equation, we get

$$\frac{dX_t}{dt} = \frac{2\pi i}{h}(HX_t - X_t H)$$

which is the desired equation of motion. We end this section with a proof of the uncertainty principle. Let A, B be two operators and $|\psi>$ a state of a quantum system. The average values of A, B in this state are respectively given by

$$< A >=< \psi | A | \psi >, \quad < B >=< \psi | B | \psi >$$

The variances of A, B in this state are respectively given by

$$\sigma_A^2 =< (A- < A >)^2 >=< \psi | (A - < A >)^2 | \psi >, \sigma_B^2 =< (B- < B >)^2$$
$$>=< \psi | (B - < B >)^2 | \psi >$$

Since the operators $A - < A >$ and $B - < B$ are Hermitian, these can be expressed as

$$< (A - < A >)^2 >= \| (A - < A >)\psi \|^2, < (B- < B >)^2 >= \| (B- < B >)\psi \|^2$$

Therefore, by the Cauchy Schwarz inequality,

$$\sigma_A^2 \sigma_B^2 \geq | < (A- < A >)\psi | (B- < B >)\psi > |^2 \geq | Im < (A- < A >)\psi | (B- < B >)\psi > |^2$$

For two observables C, D

$$Im < C\psi | D\psi >= \frac{1}{2i}(< C\psi | D\psi > - < D\psi | C\psi >) = \frac{1}{2i}(< \psi | CD - DC |$$
$$\psi >= \frac{1}{2i} < \psi | [C, D]\psi >$$

Thus,

$$\sigma_A^2 \sigma_B^2 \geq \frac{1}{4} | < \psi | [A- < A >, B- < B >] | \psi > |^2 = \frac{1}{4} | < \psi | [A, B] | \psi > |^2$$

Taking $A = x$, $B = p_x$ gives on recalling that $[x, p_x] = ih/2\pi$, the equation

$$\sigma_x^2 \sigma_{p_x}^2 \geq \frac{h^2}{16\pi^2}$$

Thus the product of the position and momentum uncertainties (dispersion) in any state always exceeds a constant $\frac{h^2}{16\pi^2}$. This means that if one tries to reduce the position uncertainty by taking measurements in a different state, the momentum uncertainty will correspondingly increase and vice versa, such that the product of the two uncertainties always exceeds the lower limit.

10. Solution to the Stationary Schrodinger Equation for a variety of cases

(a) Particle in a one-dimensional box. Here, the particle moves in a one-dimensional box of length L. The Hamiltonian is simply given by that for a free particle: $H = p^2/2m$. In the position representation, $p = -\frac{ih}{2\pi}\frac{d}{dx}$ so that the stationary Schrodinger equation is

$$-\frac{h^2}{8\pi^2 m}\frac{d^2}{dx^2}\psi(x) = E\psi(x),\ 0 \leq x \leq L$$

We define $k^2 = 8\pi^2 mE/h^2$. The solution can then be expressed as

$$\psi(x) = A.\cos(kx) + B.\sin(kx),\ 0 \leq x \leq L$$

The wave function outside the box must be zero since the particle is constrained to move inside the box. Thus from continuity requirement of the wave function, we must have $\psi(0) = \psi(L) = 0$. This gives $A = 0$, $k = n\pi/L$ where n is a positive integer. Normalizing the wave function, we get

$$\psi_n(x) = \sqrt{2/L}\sin(n\pi x/L),\ E_n = n^2h^2/8mL^2,\ n = 1,2,...$$

for the allowed states and the corresponding allowed energy levels. The general solution to the time dependent Schrodinger equation

$$\frac{ih}{2\pi}\frac{\partial\psi(t,x)}{\partial t} = -\frac{h^2}{8\pi^2 m}\frac{\partial^2\psi(t,x)}{\partial x^2}$$

with the boundary conditions $\psi(t, 0) = \psi(t, L) = 0$ is the superposition

$$\psi(t,x) = \sum_{n=1}^{\infty} c_n \exp(-i2\pi E_n t/h)\psi_n(x) = \sum_{n=1}^{\infty} c_n \exp(-2\pi i n^2 h^2 t/8mL^2)\psi_n(x)$$

We can calculate the average value of the position and momentum in the stationary states. The average value of the position in the nth stationary state is given by

$$<x>_n = \int_{-\infty}^{\infty} x\,|\psi_n(x)|^2\,dx = \frac{2}{L}\int_0^L x.\sin^2(n\pi x/L)dx = L/2$$

The average value of the momentum is given by

$$<p>_n = \int_0^L \psi_n(x)(-\frac{ih}{2\pi})\psi_n'(x)dx = -\frac{ih}{2\pi}\frac{2}{L}(n\pi/L)\int_0^L sin(n\pi x/L)$$
$$\cos(n\pi x/L)dx = 0$$

(b) Particle in a three-dimensional box. Here, the Hamiltonian for the particle is given by

$$H = \frac{p_x^2 + p_y^2 + p_z^2}{2m}$$

and its quantum operator is obtained by replacing p_x by $-\frac{ih}{2\pi}\frac{\partial}{\partial x}$, p_y by $-\frac{ih}{2\pi}\frac{\partial}{\partial y}$ and finally p_z by $-\frac{ih}{2\pi}\frac{\partial}{\partial z}$. This gives our stationary Schrodinger equation as

$$-\frac{h^2}{8\pi^2 m}\left(\frac{\partial^2\psi}{\partial x^2} + \frac{\partial^2\psi}{\partial y^2} + \frac{\partial^2\psi}{\partial z^2}\right) = E\psi$$

Our box is the region $0 \le x \le a$, $0 \le y \le b$, $0 \le z \le c$. Effecting the separation of variables

$$\psi(x, y, z) = X(x)\ Y(y)\ Z(z)$$

we get

$$\frac{X''(x)}{X(x)} + \frac{Y''(y)}{Y(y)} + \frac{Z''(z)}{Z(z)} = -k^2$$

where $k^2 = 8\pi^2 mE/h^2$. This means that $\frac{X''(x)}{X(x)}, \frac{Y''(y)}{Y(y)}$ and $\frac{Z''(z)}{Z(z)}$ must separately equal constants that sum up to $-k^2$. The boundary condition that $\psi(x, y, z) = 0$ when $x = 0, a$, $y = 0, b$, $z = 0, c$ amounts to requiring that $X(0) = X(a) = Y(0) = Y(b) = Z(c) = 0$ giving for the allowed stationary state wave functions

$$\psi(x,y,z) = \sqrt{8/abc}\sin(n\pi x/a)\sin(m\pi y/b)\sin(l\pi z/c), 0 \le x, \le a,$$
$$0 \le y \le b, 0 \le z \le c$$

where n, m, l assume values 1, 2,.... The value of k^2 is seen to be given by

$$k^2 = \pi^2(n^2/a^2 + m^2/b^2 + l^2/c^2)$$

and hence the allowed energy levels are

$$E_{n,m,l} = \frac{h^2}{8m^2}(n^2/a^2 + m^2/b^2 + l^2/c^2)$$

with n, m, l assuming positive integer values.

(c) Tunnelling problems: Consider a quantum mechanical particle or rather a stream of such particles moving in a potential $V(x) = 0$, $x < 0$, $V(x) = V_0$, $0 \le x \le a$, $V(x) = 0$, $x > a$. The result of motion in this potential field will be a distribution of particles over the entire space. The motion is assumed to take place in one dimension. The stationary Schrodinger equation to be solved for the energy E is

$$-\frac{h^2}{8\pi^2 m}\frac{d^2\psi(x)}{dx^2} = E\psi(x),\ x < 0,\ x > a$$

$$-\frac{h^2}{8\pi^2 m}\frac{d^2\psi(x)}{dx^2} + V_0\psi(x) = E\psi(x),\ 0 \le x \le a$$

We do not make any assumption at the start regarding the relative values of V_0 and E. Define the following constants having dimensions of *length*$^{-1}$: $k = \sqrt{8\pi^2 mE/h^2}$, $\alpha = \sqrt{8\pi^2 m(V_0 - E)/h^2}$. Then, the above differential equations can be cast in the form

$$\frac{d^2\psi}{dx^2} = -k^2\psi,\ x < 0, x > a$$

$$\frac{d^2\psi}{dx^2} = \alpha^2\psi,\ 0 \le x \le a$$

The solutions in the three regions can be expressed as

$$\psi(x) = A_1\exp(ikx) + A_2\exp(-ikx), x < 0$$

$$\psi(x) = B_1\exp(ikx) + B_2\exp(-ikx), x > a$$

$$\psi(x) = C_1\exp(\alpha x) + C_2\exp(-\alpha x), 0 \le x \le a$$

If we assume that a stream of particles arrives from the left i.e. from $x = -\infty$, tunnels through the potential barrier and produces a wave on the right, then we cannot have a reflected wave on the right. This means that particles on the side $x > a$ of the barrier have only a positive value of the momentum. This means that $B_2 = 0$. Note that the wave $\exp(ikx)$ has a momentum of value $hk/2\pi$ as follows by applying the momentum operator $p = -\frac{ih}{2\pi}\frac{d}{dx}$ to it and noting the eigenvalue:

$$p.\exp(ikx) = -\frac{ih}{2\pi}\frac{d}{dx}\exp(ikx) = \frac{hk}{2\pi}\exp(ikx)$$

Likewise, the component $\exp(-ikx)$ corresponds to momentum $-\frac{hk}{2\pi}$, i.e. a negative value of the momentum. The former corresponds to a wave travelling to the right and the latter to a wave travelling to the left. We assume that on the right of the barrier, we do not have any wave travelling to the left as all particles are incident on the barrier from $x = -\infty$. The probability density or particle density is proportional to $|\psi|^2$ and this quantity, from physical considerations, should be continuous everywhere. However, the derivative of ψ, $\frac{d\psi}{dx}$ should also be continuous everywhere. This can be seen from the following mathematical argument. Schrodinger's equation reads

$$-\frac{h^2}{8\pi^2 m}\psi''(x) + V(x)\psi(x) = E\psi(x), \; -\infty < x < \infty$$

Integrating this equation from $x = x_0-$ to $x = x_0+$ and using the fact that boundedness of V and ψ implies $\int_{x_0-}^{x_0+} V\psi dx = 0$ and $\int_{x_0-}^{x_0+} \psi dx = 0$, this gives

$$\psi'(x_0-) = \psi'(x_0+)$$

which is the desired relation. Applying continuity of ψ and ψ' at $x = 0$ and $x = a$, we therefore get

$$A_1 + A_2 = C_1 + C_2, \; ik(A_1 - A_2) = \alpha(C_1 - C_2),$$
$$B_1\exp(ika) = C_1\exp(\alpha a) + C_2\exp(-\alpha a)$$
$$ikB_1\exp(ika) = \alpha(C_1\exp(\alpha a) - C_2\exp(-\alpha a))$$

We cast these equations in matrix form:

$$\begin{pmatrix} 1 & -1 & -1 & 0 \\ ik & \alpha & -\alpha & 0 \\ 0 & \exp(\alpha a) & \exp(-\alpha a) & -\exp(ika) \\ 0 & \alpha\exp(\alpha a) & -\alpha\exp(-\alpha a) & -ik\exp(-ka) \end{pmatrix} \begin{pmatrix} A_2 \\ C_1 \\ C_2 \\ B_1 \end{pmatrix} = \begin{pmatrix} -A_1 \\ ikA_1 \\ 0 \\ 0 \end{pmatrix}$$

We leave it as an exercise to calculate the ratios $\frac{A_2}{A_1}, \frac{C_1}{A_1}, \frac{C_2}{A_1}, \frac{B_1}{A_1}$ using these equations. It should be noted that even when $V_0 > E$, there is a finite quantum mechanical amplitude for the particle to be in the region $0 < x < a$ and $x > a$. Classically, the particle cannot cross over the potential barrier, but in quantum mechanics, it can.

(d) Harmonic Oscillator: The potential in which the particle moves is $V(x) = \frac{1}{2}kx^2$. The position space representation of the quantum mechanical Hamiltonian is

$$H = -\frac{h^2}{8\pi^2 m}\frac{d^2}{dx^2} + \frac{1}{2}kx^2$$

The stationary Schrodinger equation is thus given by

$$E\psi(x) = -\frac{h^2}{8\pi^2 m}\psi''(x) + \frac{1}{2}kx^2\psi(x)$$

We change the variable x to u where $u = ax$. a is a constant that will be chosen appropriately. We set $\psi(x) = \psi(u/a) = \phi(u)$.

$$\frac{d^2}{dx^2} = a^2\frac{d^2}{du^2}$$

and hence the stationary Schrodinger equation becomes

$$E\phi(u) = -\frac{h^2 a^2}{8\pi^2 m}\phi''(u) + \frac{1}{2a^2}ku^2\phi(u)$$

a is now chosen so that

$$\frac{k}{2a^2} = \frac{h^2 a^2}{8\pi^2 m}$$

or

$$a^4 = \frac{4\pi^2 mk}{h^2}$$

or

$$a = \left(\frac{4\pi^2 mk}{h^2}\right)^{1/4}$$

We also define

$$\varepsilon = \frac{2a^2 E}{k} = \frac{4\pi E}{h}\sqrt{m/k}$$

Then, the equation assumes a simplified form

$$\phi''(u) + (\varepsilon - u^2)\phi(u) = 0$$

As $u \to \pm\infty$, the solution to this equation has the asymptotic dependence $\exp(-u^2/2)$, indeed

$$\frac{d}{du}\exp(-u^2/2) = -u\exp(-u^2/2)$$

$$\frac{d^2}{du^2}\exp(-u^2/2) = -\exp(-u^2/2) + u^2\exp(-u^2/2) \approx u^2\exp(-u^2/2)$$

Thus, we look for a solution of the form

$$\phi(u) = f(u)\exp(-u^2/2)$$

where f does not tend to ∞ as fast as $\exp(u^2/2)$ as $u \to \pm\infty$. This amounts to saying that $f(u)\exp(-u^2/2)$ should tend to zero as $u \to \pm\infty$. We substitute this into the equation:

$$\phi'(u) = (f'(u) - uf(u))\exp(-u^2/2)$$

$$\phi''(u) = (f''(u) - 2uf'(u) + (u^2 - 1)f(u))\exp(-u^2/2)$$

Thus,

$$f''(u) - 2uf'(u) + \lambda f(u) = 0$$

where $\lambda = \varepsilon - 1$. We look for a power series solution to this differential equation

$$f(u) = \sum_{n=0}^{\infty} c_n u^n$$

Substituting this into the equation, we get

$$\sum_{n=2}^{\infty} n(n-1)c_n u^{n-2} - 2\sum_{n=1}^{\infty} nc_n u^n + \lambda \sum_{n=0}^{\infty} c_n u^n = 0$$

Equating coefficients of equal powers of u on both sides, we get

$$(n + 2)(n + 1)c_{n+2} - 2nc_n + \lambda c_n = 0,\ n \geq 0$$

or

$$c_{n+2} = \frac{(2n-\lambda)c_n}{(n+2)(n+1)}, n \geq 0$$

Suppose that this series does not terminate, i.e. no coefficient is zero. Then, for large n, we have

$$c_{n+2} \approx \frac{c_n}{n+2}$$

which implies that c_{2r} is, for large r, proportional to $(2r(2r - 2) \ldots 2)^{-1} = (2^{-r}(r!)^{-1})$. So the series $\Sigma\, c_{2r}u^{2r}$ behaves as $\exp(u^2/2)$ which means that $f(u)\exp(u^2/2)$ will not decay down to zero as $u \to \pm\infty$. The only option is therefore that some c_n vanish. Suppose we choose λ so that $c_n = 0$, i.e. $\lambda = 2(n - 1)$. Then, $c_{n+2} = c_{n+4} = \ldots = c_{n+2m} = \ldots = 0$. Hence, if n is odd, we may take $c_0 = 0$ and the series for $f(u)$ becomes an odd polynomial of degree n, while if n is even, we may take $c_1 = 0$ and the series for $f(u)$ becomes an even polynomial of degree n. n can take values 1, 2, 3,... and the possible values of λ for $f(u)$ to terminate into a polynomial are $\lambda = 2(n - 1)$, $n = 1, 2, \ldots$ This gives for the possible values of ε:

$$\varepsilon = \lambda + 1 = 2n - 1,\ n = 1, 2, \ldots$$

and the possible values of E are

$$E = \sqrt{k/m}\,\frac{h}{4\pi}(2n-1),\ n = 1,2, \ldots$$

or equivalently,

$$E = \sqrt{k/m}\,\frac{h}{2\pi}(n+1/2),\ n = 0, 1, 2, \ldots$$

This is the energy spectrum of the harmonic oscillator. The corresponding possible choices of the scaled wave function $\phi(u)$ are given by $\phi(u) = f_n(u)\exp(-u^2/2)$, $n = 0, 1, 2, \ldots$, where $f_n(u)$ satisfies the differential equation

$$f_n''(u) - 2uf_n'(u) + 2nf_n(u) = 0,\ n = 0,1,2,\ldots$$

This is called the Hermite differential equation and the solutions are the Hermite polynomials. We shall show that but for a normalization factor,

$$f_n(u) = H_n(u) = \exp(u^2)\frac{d^n}{du^n}\exp(-u^2)$$

First note that

$$H_n'(u) = 2uH_n(u) + \exp(u^2)D^{n+1}\exp(-u^2) = 2uH_n(u) - 2\exp(u^2)D^n u\exp$$
$$(-u^2) = 2uH_n(u) - 2\exp(u^2)\left(uD^n\exp(-u^2) + nD^{n-1}\exp(-u^2)\right) = -2nH_{n-1}(u).$$

Hence

$$H_n''(u) = -2nH_{n-1}'(u) = 4n(n-1)H_{n-1}(u)$$

So, we are to prove that

$$4n(n-1)\ H_{n-2}(u) + 4nuH_{n-1}(u) + 2nH_n(u) = 0$$

which amounts to proving that

$$4n(n-1)D^{n-2}\exp(-u^2) + 4nuD^{n-1}\exp(-u^2) + 2nD^n\exp(-u^2) = 0$$

Multiply both sides by $t^n/n!$ and sum from $n = 2$ to ∞ to get the following to establish

$$4t^2\exp(tD)\exp(-u^2) + 4ut(exp(tD) - 1)\ exp\ (-u^2) + 2tD\ (exp(tD) - 1)exp(-u^2) = 0$$

since $exp(tD)g(x) = g(x+t)$, this amounts to establishing that

$$4t^2\exp(-(u+t)^2) + 4ut\exp(-(u+t)^2) - \exp(-u^2) - 4t(u+t)\ \exp(-(u+t)^2) - u\exp(-u^2) = 0$$

which is indeed an identity. It remains to compute the normalization factor for the Hermite polynomials. We consider the generating function

$$P(t,u) = \sum_{n=0}^{\infty} t^n H_n(u)/n! = \sum_{n=0}^{\infty} t^n\exp(u^2)D^n\exp(-u^2)/n! = \exp(u^2)$$
$$\exp(tD)\exp(-u^2) = \exp(u^2 - (u+t)^2) = \exp(-t^2 - 2ut)$$

Thus,

$$\int_{-\infty}^{\infty} P(t,u)P(s,u)\exp(-u^2)du/\sqrt{\pi} = \exp(-t^2 - s^2)\int_{-\infty}^{\infty}\exp(-2u(t+s))$$
$$\exp(-u^2)du/\sqrt{\pi} = \exp(-t^2 - s^2 + (t+s)^2) = \exp(2ts)$$

It follows from this expression that

$$\int_{-\infty}^{\infty} H_n(u)H_m(u)\exp(-u^2)du = 2^n\sqrt{\pi}\, n!\,\delta_{n,m}$$

We are now in a position to write down the wave functions for the state corresponding to the energy eigenvalue $\frac{h}{2\pi}(n+1/2)\sqrt{k/m}$. It is given by

$$\psi_n(x) = A_n H_n(u)\exp(-u^2/2)$$

where A_n is a normalizing constant determined from the condition

$$\int_{-\infty}^{\infty} |\psi_n(x)|^2\, dx = 1$$

which amounts to

$$1 = |A_n|^2 \int_{-\infty}^{\infty} H_n(u)^2\exp(-u^2)du/a = (|A_n|^2/a)2^n\sqrt{\pi}\, n!$$

or

$$A_n = 2^{-n/2}(n!)^{-1/2}\pi^{-1/4}\sqrt{a}$$

The wave functions are clearly orthogonal as follows from the orthogonality of the Hermite polynomials with respect to the Gaussian measure $\exp(-u^2)du$:

$$\int_{-\infty}^{\infty} \psi_n(x)\psi_m(x)dx = A_n A_m \int_{-\infty}^{\infty} H_n(u)H_m(u)\exp(-u^2)du = 0, n \neq m$$

There is another approach to the solution of the harmonic oscillator problem. We start with the form of the Hamiltonian $H = \frac{p^2}{2m} + \frac{kx^2}{2}$. For convenience set $\omega = \sqrt{k/m}$, so that

$$H = \frac{p^2}{2m} + m\omega^2x^2/2 = \frac{1}{2m}(p^2 + (m\omega x)^2)$$

Define the operators

$$a = p - im\omega x,\ a^* = p + im\omega x$$

Note that a^* is the formal adjoint of a since x and p are self-adjoint. a is called the annihilation operator and a^* the creation operator. The commutation relation between x and p is the usual one: $[x, p] = xp - px = ih/2\pi$. This gives us for the commutation relation between a and a^*

$$[a, a^*] = aa^* - a^*a = [p - im\omega x, p + im\omega x] = im\omega[p,x] - im\omega[x,p]$$

$$= mh\omega/2\pi + m\omega h/2\pi = mh\omega/\pi$$

Also,

$$aa^* = (p - im\omega x)(p + im\omega x) = p^2 + m^2\omega^2x^2 + im\omega[p, x]$$
$$= p^2 + m^2\omega^2x^2 + mh\omega/2\pi$$

Thus,

$$\frac{1}{2m}aa^* = \frac{p^2}{2m} + \frac{m\omega^2x^2}{2} + h\omega/4\pi = H + h\omega/4\pi$$
$$a^*a = aa^* - [a, a^*] = aa^* - mh\omega/\pi$$

giving in view of the previous identity

$$\frac{1}{2m}a^*a = H - h\omega/4\pi$$

Let $|E>$ denote an eigenstate of the energy operator H corresponding to the energy eigenvalue E, i.e. $H|E>= E|E>$. Then, $< E|a^*a|E >$ equals the norm square of the state $a|E>$ and is therefore, a non-negative number. It follows that

$$0 \le < E|H - h\omega/4\pi|E > = E - h\omega/4\pi$$

since $< E | E > = 1$. It follows that $E \ge h\omega/4\pi$. Hence, no energy eigenvalue of H can be smaller than $h\omega/4\pi$. Again, consider the equation

$$Ha|E>=\left(\frac{1}{2m}(aa^* - h\omega/4\pi)a|E>= \frac{1}{2m}aa^*a|E> - \frac{h\omega}{4\pi}a|E>=\right.$$
$$\left.= a(H - h\omega/4\pi)|E> - \frac{h\omega}{4\pi}a|E>\right) = aE|E> \frac{h\omega}{2\pi}a|E>=\left(E - \frac{h\omega}{2\pi}\right)a|E>$$

This equation shows that $a|E>$ is an eigenstate of H corresponding to the eigenvalue $E - \frac{h\omega}{2\pi}$. Thus, a can be regarded as an energy lowering operator. We shall now show that likewise, a^* is an energy raising operator.

$$Ha^*|E>=\left(\frac{1}{2m}a^*a + \frac{h\omega}{4\pi}\right)a^*|E>= \frac{1}{2m}a^*aa^*|E> + \frac{h\omega}{4\pi}a^*|E>$$
$$= a^*\left(H + \frac{h\omega}{4\pi}\right)|E> + \frac{h\omega}{4\pi}a^*|E>= a^*\left(E + \frac{h\omega}{4\pi}\right)|E> + \frac{h\omega}{4\pi}a^*|E>$$
$$= \left(E + \frac{h\omega}{2\pi}\right)a^*|E>$$

which shows that $a^*|E>$ is an eigenstate of H corresponding to the energy eigenvalue $E + \frac{h\omega}{2\pi}$. Suppose that $|0>$ is a ket vector satisfying $a|0> = 0$.

Then, $a^*a|0> == 0$ and hence $H|0> = \frac{h\omega}{4\pi}|0>$. Thus $|0>$ is the minimum

possible energy eigenstate. Its energy is given by $\frac{h\omega}{4\pi}$. By the energy raising principle, $a^{*n}|0>$ is an energy eigenstate corresponding to the energy level

$$E_n = \frac{h\omega}{4\pi} + \frac{nh\omega}{2\pi} = \left(n + \frac{1}{2}\right)\frac{h\omega}{2\pi}, \; n = 0, 1, 2, ...$$

In this way we get all the eigenstates of energy. Note that if $E \geq \frac{h\omega}{4\pi}$ is not of the form E_n for some $n = 0, 1, 2, ...,$ then by successive application of a, we can arrive at an energy eigenket having an energy lower than $\frac{h\omega}{4\pi}$, which would not be permissible. Specifically, suppose $(n - 1/2) h\omega/2\pi < E < (n + 1/2)h\omega/2\pi$ for some $n \geq 1$ and suppose $|E>$ is an energy eigenket. Then, $a|E>$ is an energy eigenket having energy E' where $(n - 3/2)h\omega/2\pi < E' < (n - 1/2)h\omega/2\pi$. Moreover, this cannot be the zero eigenket because otherwise $a^*a|E> = 0$ which would imply that $H|E> = \frac{h\omega}{4\pi}|E>$ in contradiction to our hypothesis. In this way, we successively operate by a n times to arrive at the ket $a^n|E>$ having energy α with $\alpha < \frac{h\omega}{4\pi}$, contradicting the minimality of the energy. We note that adopting the position representation, the equation $a|0> = 0$ can be solved to get the ground state energy level: $a|0> = 0$ is the same as $(p - im\omega x)|0> = 0$ or

$$-\frac{ih}{2\pi}\frac{d}{dx} < x \,|\, 0 > - im\omega x < x \,|\, 0 > = 0$$

so that the position space representation of the ground state wave function is given by

$$< x|0 > = \psi_0(x) = C_0.\exp(-\pi m\omega x^2/h)$$

where C_0 is a normalization constant. The wave function for the nth energy state is given by operating on this ground state function by $a^{*n} = (p + im\omega x)^n$ and incorporating an appropriate normalizing factor:

$$\psi_n(x) = C_n\left(\frac{h}{2\pi}\frac{d}{dx} - m\omega x\right)^n \exp(-\pi m\omega x^2/h)$$

We leave it as an exercise to show that this coincides with the wave function obtained using the Hermite polynomial approach discussed earlier.

11. A variational method for determining the ground state wave function of a quantum system

We want to cast the problem of determining the ground state eigenvalue and eigenfunction of the Hamiltonian operator of a quantum system in variational form.

Ans. Let H be the Hamiltonian operator of the system. We seek to minimize $< \psi \mid H \mid \psi >$ subject to the constraint $< \psi \mid \psi > = 1$. In terms of Lagrange multipliers, the function to be minimized is

$$L = \int \bar{\psi}(\mathbf{r}) H \psi(\mathbf{r}) dV - E(\int \bar{\psi}(\mathbf{r}) \psi(\mathbf{r}) dV - 1)$$

We treat $\psi(\mathbf{r})$ and $\bar{\psi}(\mathbf{r})$ as independent functions over which the minimization is to be carried out. Taking the variation with respect to these functions and the Lagrange multiplier E gives

$$\int \delta\bar{\psi}(\mathbf{r}) H \psi(\mathbf{r}) dV + \int \bar{\psi}(\mathbf{r}) H \delta\psi(\mathbf{r}) dV - E \int \delta\bar{\psi}(\mathbf{r}) \psi(\mathbf{r}) dV - E$$

$$\int \bar{\psi}(\mathbf{r}) \delta\psi(\mathbf{r}) dV - \delta E \int \bar{\psi}(\mathbf{r}) \psi(\mathbf{r}) dV + \delta E = 0$$

Equating the coefficient of δE to zero gives the normalization condition $\int \bar{\psi}(\mathbf{r})\, \psi(\mathbf{r})\, dV = 1$. Equating the coefficient of $\delta\bar{\psi}(\mathbf{r})$ to zero gives

$$H\psi(\mathbf{r}) - E\psi(\mathbf{r}) = 0$$

which means that the function that minimizes the functional L is an eigenfunction of H.

12. Simulation of some quantum systems

A quantum system corresponding to a particle is defined by a state vector which is a function ψ in the Hilbert space $L^2(\mathbb{R})$. At time $t = 0$ if one assumes that the state of the particle is specified, then the dynamics of this state is described by a unitary evolution, i.e. a family $\{U_t\}$ of unitary operators acting on the initial state ψ_0. This means that at time t, the state of the particle is given by $\psi_t = U_t\psi_0$, where $\{U_t : t \in \mathbb{R}\}$ is a family of operators acting on $L^2(\mathbb{R})$ satisfying the unitarity constraint $< U_t f, U_t g > = < f, g >$ for all $t \in \mathbb{R}$ and $U_t U_s = U_{t+s}$ for all $t, s \in \mathbb{R}$. Schrodinger discovered that the unitary evolution can be described by a Hermitian operator called the energy of the system H so that $U_t = \exp(-2\pi it\, H/h)$ with h being Planck's constant. We may thus derive from the equation $\frac{dU_t}{dt} = -\frac{2\pi i}{h} HU_t$, the following differential equation that describes the evolution of the wave function of the system in time: $\frac{ih}{2\pi}\frac{d\psi_t}{dt} = H\psi_t$. Schrodinger discovered by studying the wave particle duality set up by Louis De-Broglie that for a particle moving in a potential $V(x)$, the energy operator is the classical one $H = \frac{p^2}{2m} + V(x)$, where now p, the momentum operator is given by $p = -\frac{ih}{2\pi}\frac{d}{dx}$. The Schrodinger wave equation that describes the quantum mechanical motion of the particle in one dimension is therefore given by

$$\frac{ih}{2\pi}\frac{\partial\psi(t,x)}{\partial t} = -\frac{h^2}{8\pi^2 m}\frac{\partial^2\psi(t,x)}{\partial x^2} + V(x)\psi(t,x) \tag{1}$$

The solution can be expressed using the separation of variables as a superposition of waves corresponding to different energies. For a definite energy E, the wave function $\psi(x,E)$ (called the stationary state corresponding to energy E) satisfies the eigen equation

$$E\psi(x,E) = -\frac{h^2}{8\pi^2 m}\frac{\partial^2\psi(x,E)}{\partial x^2} + V(x)\psi(x,E) \tag{2}$$

and the corresponding time dependent wave function that satisfies the time dependent Schrodinger equation is given by

$$\psi(t,r) = \sum_E \psi(x,E)\exp(-2\pi iEt/h) \tag{3}$$

the sum being over the allowed values of E. For bounded motion, the integral $\int|\psi(x,E)|^2dx$ must be unity. This imposes a constraint on the allowed values of E, making it assume discrete values. For unbounded motion corresponding to a current of particles, we do not require to impose this constraint and the allowed values of E can vary over a continuous range. According to Born's interpretation of the wave function $|\psi(t,x)|^2dx$ determines the probability of the particle occupying the region $[x, x + dx]$ in position space at time t. Thus, for each $x \in \mathbb{R}$, $\psi(t,x)$ can be regarded as determining a probability amplitude for the particle to be in the range $[x, x + dx]$. The wave function can be regarded as being composed of a continuous infinity of such probability amplitudes, one corresponding to each position x.

We can also conceive of systems like the ammonia molecule or the spinning electron in a magnetic field wherein, the state of the particle is only finite in number. The ammonia molecule has three hydrogen atoms in a plane and a Nitrogen atom that can occupy two positions, either above the plane or below the plane. It therefore defines a two state system. If we are interested in only the spin of the electron, then it is once again a two state system, in which either the electron can have an upspin or a down spin. In general, an n state quantum system is defined by specifying a set of n complex numbers $[c_1(t),...,c_n(t)]^T$ at each instant of time t. $|c_i(t)|^2$ determines the probability of the state being the ith one at time t and for the sake of consistency, we must have $\Sigma_{i=1}^n |c_i(t)|^2 = 1$. The dynamical equations satisfied by the complex amplitudes $\{c_i(t)\}$ are first order given by

$$\frac{ih}{2\pi}\frac{dc_i(t)}{dt} = \sum_{j=1}^{n} H_{ij}(t)c_j(t) \tag{4}$$

Here, $\mathbf{H} = ((H_{ij}))$ is called the Hamiltonian or energy matrix of the n state quantum system and our equations indicate that in general, these could be time dependent. The earlier equation that we wrote down for a particle moving in one dimensions in the potential $V(x)$ can also be expressed in the same form

$$\frac{ih}{2\pi}\frac{d\psi}{dt} = H\psi \tag{5}$$

where now rather than $\mathbb{C}^n$, our vector space is $L^2(\mathbb{R})$ and H is an operator on $L^2(\mathbb{R})$ defined by

$$H = -\frac{h^2}{8\pi^2 m}\frac{d^2}{dx^2} + V(x) \tag{6}$$

We can express it in the form of a continuous matrix $H(x, x')$ given by

$$H(x,x') = -\frac{h^2}{8\pi^2 m}\delta''(x-x') + V(x)\delta(x-x') \tag{7}$$

Integration by parts reveals that the equation

$$\frac{ih}{2\pi}\frac{d\psi(x)}{dt} = \int_{-\infty}^{\infty} H(x,x')\psi(x')dx' \tag{8}$$

coincides with the Schrodinger equation (1). Consider first the n state system in which the energy is time independent but depends upon an unknown parameter θ, which we seek to estimate. Our estimation procedure is based on measuring the average value of an observable as a function of time. For this, we need first of all, to describe what we mean by an observable in the quantum theory and how its probability distribution and average in a particular quantum state are calculated. An observable in the quantum theory is described by a Hermitian operator X acting on the state space. When the state space is $L^2(\mathbb{R})$ as it is for a particle moving in one dimensions in a potential $V(x)$, the observable X is a Hermitian operator on this Hilbert space. When the state space is $\mathbb{C}^n$ as it is for an n state quantum system, then X is a Hermitian matrix. Consider the latter situation first. Let $\lambda_1, ..., \lambda_n$ denote the eigenvalues of X with $\psi_1, ..., \psi_n$ as the associated set of orthonormal eigenvectors. When the state is $\mathbf{c}(t)$, the probability that a measurement of X will reveal the value λ_j at time t is then $p_j(t) = |< \psi_j | \mathbf{c}(t) >|^2$ and hence the average value of X at time t is given by

$$< X >_t = \sum_j \lambda_j |< \psi_j | \mathbf{c}(t) >|^2 = < \mathbf{c}(t) | X | \mathbf{c}(t) > = \mathbf{c}(t)^* X \mathbf{c}(t)$$

on making use of the spectral resolution

$$X = \sum_j \lambda_j \psi_j \psi_j^*$$

of the Hermitian operator X. The rate of change of the average value of the observable at time t is thus given by

$$\frac{d < X >_t}{dt} = \frac{d\mathbf{c}(t)^*}{dt} X\mathbf{c}(t) + \mathbf{c}(t)^* X \frac{d\mathbf{c}(t)}{dt}$$

$$= \frac{2\pi i}{h}\mathbf{c}(t)^* HX\mathbf{c}(t) - \frac{2\pi i}{h}\mathbf{c}(t)^* XH\mathbf{c}(t)$$

$$\frac{2\pi i}{h}\mathbf{c}(t)^*(HX - XH)\mathbf{c}(t) = \frac{2\pi i}{h}\mathbf{c}(t)^*[H, X]\mathbf{c}(t) = \frac{2\pi i}{h} < \mathbf{c}(t) | [H, X] | \mathbf{c}(t) >$$

This equation is sometimes called the Heisenberg dynamical equation. With the Hamiltonian matrix $\mathbf{H}(\theta)$ dependent upon θ and time independent, we have the solution to the state

$$\mathbf{c}(t, \theta) = exp(-2\pi i H(\theta) t / h)\mathbf{c}(0)$$

and the average value of the observable at time t can be expressed as

$$< X >_t = \sum_{i=1}^{n} \lambda_i |< \psi_i | \mathbf{c}(t) >|^2$$

θ can be estimated from measurement on $< X >_t$ at various times by the least squares method. Suppose that we assume that the initial state $\mathbf{c}(0)$

of the system is unknown. Let $\mathbf{v}_1(\theta), \ldots, \mathbf{v}_n(\theta)$ denote the eigenvectors of $\mathbf{H}(\theta)$ with $a_1(\theta), \ldots, a_n(\theta)$ as the associated eigenvalues. Both the eigenvectors and eigenvalues depend upon the unknown parameter θ. The functional form of this dependence is known but the parameter itself is unknown. Then,

$$\mathbf{c}(t,\theta) = \exp(-2\pi i t H(\theta)/h)\mathbf{c}(0) = \sum_{k=1}^{n} \exp(-2\pi i t a_k(\theta)/h)\mathbf{v}_k(\theta)\mathbf{v}_k(\theta)^* \mathbf{c}(0)$$

using the spectral resolution of $H(\theta)$. Thus,

$$< X >_t = \sum_{i=1}^{n} \lambda_i \,|\sum_{k=1}^{n} \exp(-2\pi i t a_k(\theta)/h)\psi_i^* \mathbf{v}_k(\theta)\mathbf{v}_k(\theta)^* \mathbf{c}(0)|^2$$

or using the bra-ket notation,

$$< X >_t = \sum_{i=1}^{n} \lambda_i \,|\sum_{k=1}^{n} \exp(-2\pi i a_k(\theta)t/h) < \psi_i \,|\, \mathbf{v}_k(\theta) >< \mathbf{v}_k(\theta) \,|\, \mathbf{c}(0) > |^2$$

We note that $< X >_t$ depends upon both the unknown parameter θ as well as the unknown initial state $\mathbf{c}(0)$. We need to estimate both of these parameters from measurement on $\{< X >_t\}$ at different times.

The problem of estimating e/m **for an electron:** Consider the electron in a magnetic field directed along the z axis. Let the magnetic field be B_z. The magnetic moment of the electron due to its spin alone is $\mu = -\frac{eh\sigma_z}{4\pi m}$, where $\sigma_z = \begin{pmatrix} 1 & 0 \\ 0 & -1 \end{pmatrix}$ is the z component of the Pauli spin matrix. The interaction Hamiltonian of the magnetic moment of the electron with the magnetic field is given by

$$H = -\mu B_z = \frac{ehB_z\sigma_z}{4\pi m} = h\omega.\sigma_z/2\pi$$

where $\omega = \frac{eB_z}{2m}$ is the Larmor frequency of precession of the electron in the magnetic field. The eigen states of this Hamiltonian are $\begin{pmatrix} 1 \\ 0 \end{pmatrix}$ *and* $\begin{pmatrix} 0 \\ 1 \end{pmatrix}$ corresponding respectively to the electron having an upspin or a down spin. We set $\theta = h\omega/2\pi$. The Hamiltonian can be expressed as $H = \theta.\sigma_z$. When the electron is in the upspin state, its energy is $+\theta$ and when it is in the downspin state, its energy is $-\theta$. Let $\mathbf{c}(0) = [c_1(0), \mathbf{c}_2(0)]^T$ be initial state of the electron. Then, after time t, its state is

$$\mathbf{c}(t) = \exp(-2\pi i t H / h)\mathbf{c}(0) = \begin{pmatrix} \exp(-i\omega t)c_1(0) \\ \exp(+i\omega t)c_2(0) \end{pmatrix}$$

We may choose to measure an observable X, for example, the spin of the electron in a certain direction

$$X = p\sigma_x + q\sigma_y + r\sigma_z = \begin{pmatrix} r & p-iq \\ p+iq & -r \end{pmatrix}$$

Its average value at time t is given by

$$< X >_t = (\mathbf{c}(t) \mid X \mid \mathbf{c}(t) >= r(\mid c_1(0)\mid^2 - \mid c_2(0)\mid^2) + 2.\mathrm{Re}\{c_1(0)\bar{c}_2(0)$$
$$(p+iq)\exp(-2i\omega t)\}$$

Assuming that both $c_1(0)$ and $c_2(0)$ are non-zero (which means that initially, the electron is in a state where there is nonzero probability for finding it in upspin as well as in downspin state), it follows that the frequency of oscillation of $< X >_t$ is given by 2ω. Thus, by measuring the frequency of oscillation of the average value of an observable, we can determine ω and hence the ratio $\frac{e}{m}$ for the electron.

Simulation of an *n* state quantum system

Apparently, one might be let to the following iterative equation for the simulation of an n state quantum system: $\mathbf{c}[n+1] = (I - \frac{2\pi i}{h}\mathbf{H}\Delta)\mathbf{c}[n]$. This follows by replacing the temporal derivative in the Schrodinger equation by a finite difference with time step Δ. The problem with this discretized equation is that it does not preserve the norm of the state vector. We require that the state vector have unit norm at every instant of time in order that it describes a valid probability distribution. This is satisfied by the dynamical equations (4) in view of the Hermitianity of $\mathbf{H}$. In fact, if $\mathbf{H}(t)$ is any possibly time dependent Hermitian matrix, and $\| \mathbf{c}(0) \| = 1$, then $\| \mathbf{c}(t) \| = 1$ for all t whenever $\mathbf{c}(t)$ satisfies the equation $\frac{ih}{2\pi}\frac{d\mathbf{c}(t)}{dt} = \mathbf{H}(t)\mathbf{c}(t)$. This is easily seen by noting that

$$\frac{d}{dt}\| \mathbf{c}(t) \|^2 = \frac{d\mathbf{c}^*(t)}{dt}\mathbf{c}(t) + \mathbf{c}(t)^* \frac{d\mathbf{c}(t)}{dt}$$
$$\frac{2\pi i}{h}\mathbf{c}(t)^* \mathbf{H}(t)\mathbf{c}(t) - \frac{2\pi i}{h}\mathbf{c}(t)^* H\mathbf{c}(t) = 0$$

i.e., the norm of $\mathbf{c}(t)$ does not vary with time. For the time invariant energy, this may also be seen to follow easily from the unitarity of $U_t = \exp(-2\pi it\mathbf{H}/h)$ whenever $\mathbf{H}$ is a Hermitian matrix. The corresponding difference equation derived above fails to satisfy this unitarity constraint and hence must be replaced by another equation. In fact, the difference equation $\mathbf{c}[n+1] = \mathbf{U}\mathbf{c}[n]$ with $\mathbf{U} = \exp(-2\pi i\Delta.H/h)$ gives a norm preserving solution $\mathbf{c}[n]$ exactly coincides with $\mathbf{c}(n\Delta)$. When $\mathbf{H}$ depends on time, we may replace it with $\mathbf{H}[n] = \mathbf{H}(n\Delta)$ in the above discrete time equation but now $\mathbf{c}[n]$ will only approximately coinçide with $\mathbf{c}(n\Delta)$.

Electron in a time varying magnetic field

We describe this system for an electron in a constant magnetic field along the z axis and a sinusoidally varying weak magnetic field along the $\mathbf{x}$ axis. The Hamiltonian is given by

$$H(t) = \frac{eh\sigma_z B_z}{4\pi m} + \frac{eh\sigma_x B_x(t)}{4\pi m} = \frac{eh}{4\pi m}\begin{pmatrix} B_z & B_x(t) \\ B_x(t) & -B_z \end{pmatrix}$$

We find that our equations for this two state quantum system are

$$i\frac{dc_1(t)}{dt} = \frac{e}{2m}(B_z c_1(t) + B_x(t)c_2(t))$$

$$i\frac{dc_2(t)}{dt} = \frac{e}{2m}(B_x(t)c_1(t) - B_z c_2(t))$$

When $B_x(t)$ is a constant function, this problem reduces to the one tackled earlier once we change our co-ordinate system, i.e. assume that the z direction coincides with the direction of the magnetic field. We therefore assume that $B_x(t)$ is not a constant function. We assume that $B_x(t)$ is much smaller in magnitude in comparison with B_z. Let $c_1(t) = \gamma_1(t)\exp(-i\omega t)$, $c_2(t) = \gamma_2(t)\exp(i\omega t)$. We expect $\gamma_1(t)$ and $\gamma_2(t)$ to be slowly time varying functions. In fact, if $B_x(t) = 0$, then γ_1 and γ_2 are constants corresponding to the fact that the upspin and downspin states are energy eigen-states. Plugging in the values of $c_1(t)$ and $c_2(t)$ into the dynamical equations gives

$$i\frac{d\gamma_1(t)}{dt} = \frac{eB_x(t)}{2m}.\gamma_2(t).\exp(2i\omega t)$$

$$i\frac{d\gamma_2(t)}{dt} = \frac{eB_x(t)}{2m}.\gamma_1(t).\exp(-2i\omega t)$$

Assuming that $c_1(0) = A$, $c_2(0) = B$, meaning that at $t = 0$, the probability of finding the electron in the upspin state is $|A|^2$ and of finding it in the downspin state is $|B|^2$, we have up to first order of smallness.

$$\gamma_1(t) = -iB \int_0^t \omega_1(\tau)\exp(2i\omega\tau)d\tau$$

$$\gamma_2(t) = -iA \int_0^t \omega_1(\tau)\exp(-2i\omega\tau)d\tau$$

where $\omega_1(t) = \dfrac{eB_x(t)}{2m}$. This means that up to first order terms for small values of t, if the electron starts in the upspin state, i.e., $A = 1$, $B = 0$, then the probability of it making a transition to the down spin state in the time interval $[0, t]$ is approximately given by

$$P(t) = |\gamma_2(t)|^2 = |\int_0^t \omega_1(\tau)\exp(-2i\omega\tau)d\tau|^2$$

Suppose that $B_x(t) = B_0\cos(\omega_0 t)$. Then, this transition probability is negligible when $\omega_0 \neq 2\omega$. When $\omega_0 = 2\omega$, we get $P(t) \approx |\frac{eB_0 t}{4m}|^2$. This is called magnetic resonance. When the frequency of the x component of the magnetic field approaches the Bohr frequency associated with the transition of the electron from upspin to downspin state, then the transition probability increases significantly. The value of $\frac{e}{m}$ can be calculated using this resonance. We keep varying the frequency of the x component of the magnetic field until a sudden increase in the number of electrons making a transition from upspin to downspin state occurs. By measuring the relative number P of electrons making such a transition in $[0, t]$ we get $\frac{e}{m}$ from the relation $|\frac{eB_0 t}{4m}|^2 = P$. We may consider alternately the determination of this parameter from measurement of the average value of an observable $X = \begin{pmatrix} a & b \\ \bar{b} & c \end{pmatrix}$ for an electron in the above magnetic field. The approximate amplitudes are given by

$$c_1(t) = \exp(-i\omega t), \qquad c_2(t) = -i.\frac{e}{2m}. \int_0^t B_x(\tau)\exp(i\omega(t-2\tau))d\tau$$

When $B_x = 0$, we find that $< X >_t = a$ which is a constant in time. When $B_x \neq 0$, we have approximately up to linear orders in B_x

$$< X >_t = a - \frac{e}{m}.\text{Re}(ib.\int_0^t B_x(\tau)\exp(i2\omega\tau)d\tau) = a + \frac{e}{m}.Im$$

$$(b.\int_0^t B_x(\tau)d\tau.\exp(i2\omega\tau))$$

When $B_x(t) = B_0.\cos(2\omega t)$, the second term almost vanishes and we have to consider second order corrections to $< X >_t$. This gives

$$< X >_t = a + c\,|\frac{eB_0}{4m}|^2$$

which does not vary with time. Thus by measurement of $< X >_t$ at any time, the ratio $\frac{e}{m}$ may be determined.

Discrelization of the two state equations

$$i\frac{dc_1(t)}{dt} = \frac{e}{2m}(B_z c_1(t) + B_x(t)c_2(t))$$

$$i\frac{dc_2(t)}{dt} = \frac{e}{2m}(B_x(t)c_1(t) - B_z c_2(t))$$

Discretizing this with a time step Δ gives

$$c_1[n+1] = c_1[n] - i\omega\Delta c_1[n] - i\omega_1[n]\Delta c_2[n]$$

$$c_2[n+1] = c_2[n] + i\omega\Delta c_2[n] - i\omega_1[n]\Delta c_1[n]$$

with $\omega = \frac{eB_z}{2m}$ and $\omega_1[n] = \frac{eB_x(n\Delta)}{2m}$. We may introduce the time varying Hermitian matrix

$$\mathbf{H}[n] = \begin{pmatrix} \omega & \omega_1[n] \\ \omega_1[n] & -\omega \end{pmatrix}$$

in conjunction with the evolution operator

$$\mathbf{U}[n] = \mathbf{I} - i\Delta.\mathbf{H}[n]$$

Then, the above evolution equation is given by

$$\mathbf{c}[n + 1] = \mathbf{U}[n]\mathbf{c}[n]$$

Since $\mathbf{U}[n]$ is generally not unitary, this does not correspond to a valid quantum evolution. That is, $\| \mathbf{c}[n] \| = 1$ does not necessarily imply $\| \mathbf{c}[n + 1] \| = 1$. To make the evolution unitary, we replace $\mathbf{U}$ by its approximate version $\mathbf{U}_d = \exp(-i\Delta.\mathbf{H}_d[n])$.

Simulation of the motion of a Particle in a Potential

Assume that the particle is moving in a one-dimensional box extending from $x = 0$ up to $x = L$. The discretized Schrodinger equation is given by

$$\frac{ih}{2\pi}\frac{\psi[n+1,k]-\psi[n,k]}{\tau} = -\frac{h^2}{8\pi^2 m\Delta^2}(\psi[n,k+1]-2\psi[n,k]+ +\psi[n,k-1])+V[k]\psi[n,k]$$

where $\psi[n, k] = \psi(n\tau, k\Delta)$, $V[k] = V(k\Delta)$. Here, time has been discretized with a step size of τ and space with a step size of Δ. Assuming that k varies from 0 to N (i.e. the box length is $L = N\Delta$) and that $\psi[n, k] = 0$ when $k < 0$ and $k > N$, we can express the above evolution equation as

$$\psi_{n+1} = \mathbf{U}\psi_n \tag{9}$$

where $\psi_n = [\psi[n, 0], \psi[n, 1], ..., \psi[n, N]]^T$ and

$$\mathbf{U} = \begin{pmatrix} 1-\frac{ih\tau}{2\pi m\Delta^2}-\frac{2\pi i\tau V[0]}{h} & \frac{ih\tau}{4\pi m\Delta^2} & 0 & \cdots & 0 \\ \frac{ih\tau}{4\pi m\Delta^2} & 1-\frac{ih\tau}{2\pi m\Delta^2}-\frac{2\pi i\tau V[1]}{h} & \frac{ih\tau}{4\pi m\Delta^2} & \cdots & 0 \\ \cdots & \cdots & \cdots & \cdots & \end{pmatrix}$$

We can write $U = \mathbf{I} - \frac{2\pi i\tau\mathbf{H_d}}{h}$, where H_d is a Hermitian matrix defined by

$$\mathbf{H}_d = \begin{pmatrix} 2a+V[0] & -a & 0 & \cdots & 0 \\ -a & 2a+V[1] & -a & \cdots & 0 \\ \cdots & \cdots & \cdots & \cdots & \end{pmatrix}$$

with $a = \frac{h^2}{8\pi^2 m\Delta^2}$. The discrete time evolution equation is not a unitary evolution since U is non-unitary. Thus, even if $|\psi[n, k]|^2\Delta$, $k = 0, 1, ..., N$ is a probability distribution at $n = 0$, for $n > 0$, it no longer remains a valid probability distribution. To make the evolution unitary, we approximate U by $\mathbf{U}_d = \exp(-2\pi i\tau\mathbf{H}_d/h)$. Simulations for a harmonic oscillator potential based on this technique can be made. The energy levels of this n state quantum system may be approximated by the eigenvalues of H_d.

13. Electron in a Magnetic Field with spin neglected

A spinning electron in a magnetic field with spin neglected is described by the usual Schrodinger equation with the momentum operator **p** replaced by **p** + e**A**. Here, **A** is the vector potential corresponding to the field in which the electron executes motion. The Hamiltonian is

$$H = \frac{(\mathbf{p}+e\mathbf{A})^2}{2m} - eV = \frac{\mathbf{p}^2}{2m} + \frac{e^2\mathbf{A}^2}{2m} + \frac{e}{2m}(\mathbf{p}.\mathbf{A}+\mathbf{A}.\mathbf{p}) - eV$$

$$= -\frac{h^2}{8\pi^2 m}\nabla^2 - \frac{ieh}{4\pi m}(\nabla.\mathbf{A}+\mathbf{A}.\nabla) - eV$$

With neglect of terms involving A2. These terms can be neglected provided the field is weak.
where $\nabla.\mathbf{A}$ is defined by its action on a wave function ψ via the rule

$$\nabla.(\mathbf{A}\psi) = (\nabla.\mathbf{A})\psi + \mathbf{A}.\nabla\psi$$

The time dependent Schrodinger equation for the electron is thus

$$\frac{ih}{2\pi}\frac{\partial\psi}{\partial t} = H\psi = -\frac{h^2}{8\pi^2 m}\nabla^2\psi - \frac{ieh}{2\pi m}(\nabla.\mathbf{A})\psi - \frac{ieh}{4\pi m}\mathbf{A}.\nabla\psi - eV\psi$$

Before mentioning how this partial differential equation is discretized, we should supply an argument that shows that the Hamiltonian $\frac{\mathbf{p}+e\mathbf{A}(\mathbf{r}))^2}{2m} - e\,V(\mathbf{r})$ indeed does give the correct equations of motion for classical mechanics, namely, the Lorentz equation

$$m\frac{d\mathbf{v}}{dt} = -e\mathbf{v}\times\mathbf{B}(\mathbf{r})$$

The Hamilton equation of motion for the x, p_x variables are

$$v_x = \frac{dx}{dt} = \frac{\partial H}{\partial p_x} = \frac{p_x + eA_x}{m}$$

$$\frac{dp_x}{dt} = -\frac{\partial H}{\partial x} = -e\frac{\partial A_x}{\partial x}\frac{p_x+eA_x}{m} - e\frac{\partial A_y}{\partial x}\frac{p_y+eA_y}{m} - e\frac{\partial A_z}{\partial x}\frac{p_z+eA_z}{m} + e\frac{\partial V}{\partial x}$$

The first of these equations gives $p_x = mv_x - eA_x$. Substituting this in the second gives

$$m\frac{dv_x}{dt} - e\frac{dA_x}{dt} = -ev_x\frac{\partial A_x}{\partial x} - ev_y\frac{\partial A_y}{\partial x} - ev_z\frac{\partial A_z}{\partial x} + e\frac{\partial V}{\partial x}$$

Now

$$\frac{dA_x}{dt} = v_x \frac{\partial A_x}{\partial x} + v_y \frac{\partial A_x}{\partial y} + v_z \frac{\partial A_x}{\partial z} + \frac{\partial A_x}{\partial t}$$

and plugging this into the above equation we get

$$m\frac{dv_x}{dt} = -e(\mathbf{v}\times\nabla\times\mathbf{A})_x + e\frac{\partial A_x}{\partial t} + e\frac{\partial V}{\partial x}$$
$$= -e(\mathbf{v}\times\mathbf{B})_x - eE_x$$

which is the correct Lorentz equation of motion. Analogous results hold for the y and z components. For a uniform magnetic field **B**, the vector potential may be taken as $\mathbf{A} = \frac{1}{2}\mathbf{B}\times\mathbf{r}$. In terms of components,

$$A_x = \frac{1}{2}(B_y z - B_z y),\ A_y = \frac{1}{2}(B_z x - B_x z),\ A_z = \frac{1}{2}(B_x y - B_y x)$$

For example,

$$\frac{\partial A_z}{\partial y} - \frac{\partial A_y}{\partial z} = \frac{1}{2}B_x + \frac{1}{2}B_x = B_x$$

It is left to verify the identity $\mathbf{B} = \nabla\times\mathbf{A}$ for the other components. Assuming that the magnetic field is weak, we can neglect terms involving B^2 in the Hamiltonian. This gives for the approximate Hamiltonian

$$H\psi = -\frac{h^2}{8\pi^2 m}\nabla^2\psi - \frac{ieh}{4\pi m}(\nabla.(\psi\mathbf{A}) + \mathbf{A}.\nabla\psi) - eV\psi$$

with

$$\nabla.(\psi\mathbf{A}) = \mathbf{A}.\nabla\psi + (\nabla.\mathbf{A})\psi$$
$$2\mathbf{A}.\nabla\psi = \mathbf{B}\times\mathbf{r}.\nabla\psi = \mathbf{B}.\mathbf{r}\times\nabla\psi$$
$$\nabla.\mathbf{A} = \frac{1}{2}\nabla.(\mathbf{B}\times\mathbf{r}) = -\frac{1}{2}\mathbf{B}.\nabla\times\mathbf{r} = 0$$

Thus,

$$H = -\frac{h^2}{8\pi^2 m}\nabla^2 - \mathbf{B}.\frac{ieh}{4\pi m}\mathbf{r}\times\nabla - eV$$

or

$$H = \frac{p^2}{2m} + \mathbf{B}.\frac{e\mathbf{L}}{2m} - eV$$

where $\mathbf{L} = \mathbf{r}\times\mathbf{p}$ is the angular momentum operator. Assume without loss of generality that the magnetic field is oriented along $\hat{z}$, i.e. $\mathbf{B} = B_z \hat{z}$. Then,

$$\mathbf{H} = H_0 + \frac{eL_z B_z}{2m}$$

where

$$H_0 = \frac{p^2}{2m} - eV$$

is the energy of the electron in the absence of the magnetic field. Now, assume that V is radially symmetric, i.e. depends only upon the radial coordinate r. It is clear that L_z commutes with r. This can be seen directly from the fact that L_z is an operator involving only the angular variables θ, ϕ. Alternately,

$$[L_z, r] = [xp_y - yp_x, r] = x[p_y, r] - y[p_x, r] = -\frac{ih}{2\pi} x \frac{\partial r}{\partial y} + \frac{ih}{2\pi} y \frac{\partial r}{\partial x}$$

$$= -\frac{ih}{2\pi}\frac{xy}{r} + \frac{ih}{2\pi}\frac{yx}{r} = 0$$

By symmetry, all components of the angular momentum commute with r. It can also be seen that L_z commutes with p^2. To see this, we simply note that

$$[L_z, p_x^2] = [xp_y - yp_x, p_x^2] = [x, p_x^2]p_y = \frac{ihp_x p_y}{\pi}$$

$$[L_z, p_y^2] = [xp_y - yp_x, p_y^2] = -[y, p_y^2]p_x = -\frac{ihp_y p_x}{\pi}$$

$$[L_z, p_z^2] = [xp_y - yp_z, p_z^2] = 0$$

We have used the following facts about commutators

$$[x, f(p_x)] = \frac{ih}{2\pi} f'(p_x)$$

because in the momentum representation, x can be expressed as $x = \frac{ih}{2\pi}\frac{\partial}{\partial p_x}$. We have also used the relation

$$[AB, C] = [A, C]B + B[A, C]$$

for three operators A, B, C. It follows from the fact that H_0 commutes with L^2 and L_z and L^2 commutes with L_z that the eigenstates of H_0 are labelled by three quantum numbers n, l, m i.e. $| n, l, m >$ where $H_0 | n, l,$

$m> = E_{n,l}|\, n,\, l,\, m>$, $L^2|\, n,\, l,\, m> = l\,(l+1)\dfrac{h^2}{4\pi^2}|\, n,\, l,\, m>$ and $L_z|\, n,$ $l,\, m> = \dfrac{mh}{2\pi}|\, n,\, l,\, m>$. In other words $|\, n,\, l,\, m>$ is a simultaneous eigenket of the three operators H_0, L^2 and L_z. This is the characteristic feature of quantum mechanical motion in any central potential. Note that l assumes values 0, 1, 2, ... and for each l, m assumes values $-l$, $-l+1$, ..., l. Of course, the bound on the values assumed by l is decided by the quantum number n. This bound is specific to the form of the potential $V(r)$. Now,

$$H\,|\,n,l,m> = (E_{n,l} + \frac{eB_z mh}{4\pi m_0})\,|\,n,l,m>$$

It follows that for a given n, l (principal and azimuthal quantum number), the energy states have a $2l + 1$ fold degeneracy. However, introduction of the magnetic field lifts this degeneracy splitting $E_{n,l}$ into $2l + 1$ distinct energy levels. To see the significance of the quantum number n and the fact that the energy $E_{n,l}$ of the unmagnetized system depends on two quantum numbers n, l but not on m we have got to express H_0 in terms of spherical polar co-ordinates:

$$H_0\psi = -\frac{h^2}{8\pi^2 m}\nabla^2\psi + V(r)\psi = -\frac{h^2}{8\pi^2 m}\frac{1}{r}\frac{d^2}{dr^2}(r\psi) + \frac{L^2}{2mr^2}\psi + V(r)\psi$$

where

$$L^2 = -\frac{h^2}{4\pi^2}\left(\frac{1}{\sin\theta}\frac{\partial}{\partial\theta}\sin\theta\frac{\partial}{\partial\theta} + \frac{1}{\sin^2\theta}\frac{\partial^2}{\partial\phi^2}\right)$$

The stationary Schrodinger equation for the unmagnetized system is

$$H_0\psi = E\psi$$

By adopting separation of variables of the wave function into the radial and angular co-ordinates, we get

$$\psi\,(r,\,\theta,\,\phi) = u(r)\,Y_{lm}\,(\theta,\,\phi)$$

where Y_{lm} are the spherical harmonics, i.e. eigenfunctions of L^2 and L_z:

$$L^2 Y_{lm} = l(l+1)\frac{h^2}{4\pi^2}Y_{lm},\; L_z Y_{lm} = \frac{mh}{2\pi}Y_{lm}$$

This shows that u satisfies the radial equation

$$-\frac{h^2}{8\pi^2 mr}(ru(r))'' + \frac{l(l+1)h^2}{8\pi^2 m}u(r) + V(r)u(r) = Eu(r)$$

It is clear that for a given value of l, we can have several values of E and hence the energy eigenvalues can be labelled as $E_{n,l}$, $n = 1, 2, \ldots$ Note that the quantum number m does not appear in the above equation and hence the energy of the unmagnetized system is independent of m.

Solved problems

1. If $\psi(x)$ is the position space wave function of a particle moving in one dimension, then determine the momentum space wave function and hence the probability distribution of the momentum.

Solution: We can write $\psi(x) =< x|\psi >$ as the position space representation of the state $|\psi >$. The momentum space wave function is then

$$\hat{\psi}(p) =< p|\psi >= \int_{-\infty}^{\infty} < p|x > dx < x|\psi >= C\int_{-\infty}^{\infty} \exp(-2\pi ipx/h)\psi(x)dx$$

where C is a normalization constant determined by the condition

$$\int_{-\infty}^{\infty} |\psi(p)^2 dp = 1$$

That is

$$1 = |C|^2 \int \exp(-2\pi ip(x-x')/h)\psi(x)\overline{\psi}(x')dpdxdx' = |C|^2 \int \delta((x-x')/h)$$

$$\psi(x)\overline{\psi}(x')dxdx' = |C|^2\, h\int |\psi(x)|^2\, dx$$

giving $C = h^{-1/2}$. Thus, the momentum space wave function is

$$\hat{\psi}(p) = h^{-1/2}\int_{-\infty}^{\infty} \exp(-2\pi ipx/h)\psi(x)dx$$

and the probability of the momentum assuming values in the range $[p, p + dp]$ when the state of the system is $|\psi >$ is given by

$$h^{-1}\left|\int_{-\infty}^{\infty} \psi(x)\exp(-2\pi ipx/h)dx\right|^2 dp$$

2. Let a wave packet be described at time $t = 0$ by the equation

$$\psi(x,0) = A.\exp(-x^2/2a^2 + ik_0 x)$$

(a) Express ψ, $(x, 0)$ as a superposition of plane waves.

(b) Determine the relation between the width of the wave packet in configuration space and its width in k space.

(c) Find $\psi(x, t)$ using the De-Broglie dispersion equation.

(d) Find $|\psi(x, t)|^2$, the probability density at time t of the position of the particle.

(e) Determine the constant A so that $|\psi(x, t)|^2$ defines a probability density.

Ans. (a)

$$\psi(x,0) = \int_{-\infty}^{\infty} \hat{\psi}(k,0)\exp(ikx)dk$$

where by the inverse Fourier transform equation,

$$\hat{\psi}(k,0) = \frac{1}{2\pi}\int_{-\infty}^{\infty} \psi(x,0)\exp(-ikx)dx = \frac{1}{2\pi}\int_{-\infty}^{\infty} A.\exp(-x^2/2a^2)$$

$$\exp(i(k_0 - k)x)dx = \frac{Aa}{\sqrt{2\pi}}\exp(-a^2(k-k_0)^2/2)$$

The momentum space wave function is proportional to $\hat{\psi}(k, 0)$ where $p = hk/2\pi$. The probability density of the position of the particle at $t = 0$ is proportional to $|\psi(x, 0)|^2$ which is proportional to $\exp(-x^2/a^2)$. The variance of the position corresponding to this density is $a^2/2$ and hence the width of the wave packet in position space can be taken as $a/\sqrt{2}$. Note that

$$|\psi(\pm a/\sqrt{2},0)|^2 = |\psi(0,0)|^2 .\exp(-1/2)$$

The probability density of the momentum of the particle is proportional to $\exp(-a^2(k-k_0)^2)$. Thus, the variance of the normalized momentum k is given by $\frac{1}{2a^2}$. The width of the momentum packet can be thus taken as $\frac{1}{a\sqrt{2}} \times h/2\pi$. We find that

$$|\hat{\psi}(k_0 \pm 1/a\sqrt{2},0)|^2 = |\hat{\psi}(k_0,0)|^2 .\exp(-1/2)$$

The product of the position and momentum widths is given by

$$\frac{a}{\sqrt{2}} \times \frac{h}{2\pi a\sqrt{2}} = \frac{h}{4\pi}$$

which is the uncertainty principle.

(c) A wave having wave number k has energy given by $E = h^2k^2/8\pi^2 m$. Its angular frequency is then $\omega = 2\pi E/h = hk^2/4\pi m$. The wave amplitude as a function of space and time is thus given by

$$\psi(x,t) = \int_{-\infty}^{\infty} \hat{\psi}(k,0)\exp(i(kx - hk^2 t/4\pi m))$$

$$= \frac{Aa}{\sqrt{2\pi}} \int_{-\infty}^{\infty} \exp(-a^2(k-k_0)^2/2)\exp(i(kx - hk^2t/4\pi m))dk$$

$$= \frac{Aa}{\sqrt{2\pi}} \int_{-\infty}^{\infty} \exp(-a^2k_0^2/2)\exp(a^2kk_0)\exp(ikx)\exp(-0.5k^2(a^2 + iht/2\pi m))dk$$

$$= \frac{Aa}{\sqrt{a^2 + iht/2\pi m}} \exp(-a^2k_0^2/2).\exp(-(x - ia^2k_0)^2/2(a^2 + iht/2\pi m))$$

$$= \frac{Aa}{\sqrt{a^2 + iht/2\pi m}} \exp(-a^2k_0^2/2).\exp(-(x^2 - a^4k_0^2 - 2ixa^2k_0)(a^2 - iht/ 2\pi m)/2(a^4 + h^2t^2/4\pi^2m^2))$$

$$= \frac{Aa}{\sqrt{a^2 + iht/2\pi m}} \exp(-a^2k_0^2/2).\exp(-((a^2x^2 - a^6k_0^2 - 2xa^2k_0ht/2\pi m) - i(2xa^4k_0 - ht(x^2 - a^4k_0^2)/2\pi m))/2\,(a^4 + h^2t^2/4\pi^2m^2))$$

$$= \frac{Aa}{\sqrt{a^2 + iht/2\pi m}} \exp(-(a^2x^2 - 2xa^2k_0ht/2\pi m + h^2t^2a^2k_0^2/4\pi^2m^2) /2(a^4 + h^2t^2/4\pi^2m^2)) \times \exp(-i(2xa^4k_0 - ht(x^2 - a^4k_0^2)/2\pi m)/2$$

$$(a^4 + h^2t^2/4\pi^2m^2)) = \frac{Aa}{\sqrt{a^2 + iht/2\pi m}} \exp(-a^2(x - htk_0/2\pi m)^2/2 (a^4 + h^2t^2/4\pi^2m^2)).\exp((i\phi)$$

where ϕ is phase factor.

(d) From the above, we get

$$|\psi(x,t)|^2 = \frac{A^2a^2}{\sqrt{a^4 + h^2t^2/4\pi^2m^2}} \exp(-a^2(x - htk_0/2\pi m)^2/(a^4 + h^2t^2/4\pi^2m^2))$$

This shows that the mean position of the particle corresponding to the wave is $htk_0/2\pi m$. This corresponds to a momentum of value $hk_0/2\pi$ in accordance with De-Broglie's theory. The dispersion of the particle position varies with time as $(\frac{a^4 + h^2t^2/4\pi^2m^2}{2a^2})^{1/2}$. To determine the normalization constant, we simply use the value of the Gaussian integral:

$$1 = \int_{-\infty}^{\infty} |\psi(x, 0)|^2 \, dx = A^2 \int_{-\infty}^{\infty} \exp(-x^2 / a^2) dx = A^2 \sqrt{\pi a}$$

so that $A = (\pi a)^{-1/4}$. Note that the evolution of the wave function is unitary? (Prove this).

3. Consider the problem of determining the energy levels $E < U_1$ in a one-dimensional potential field given by the expression $V(x) = U_1$, $x < 0$, $V(x) = 0$, $0 < x < a$, $V(x) = U_2$, $x > a$ where $U_1 < U_2$. Solve this problem by first writing down the stationary Schrodinger equation.

Ans. The stationary Schrodinger equation is given by

$$-\frac{h^2}{8\pi^2 m}\psi''(x) + U_1\psi(x) = E\psi(x), \; x < 0$$

$$-\frac{h^2}{8\pi^2 m}\psi''(x) = E\psi(x), \; 0 < x < a$$

$$-\frac{h^2}{8\pi^2 m}\psi''(x) + U_2\psi(x) = E\psi(x), \; x > a$$

Define the constants α, β and k by the equations

$$\alpha^2 = \frac{8\pi^2 m(U_1 - E)}{h^2}, \quad \beta^2 = \frac{8\pi^2 m(U_2 - E)}{h^2}, \quad k^2 = \frac{8\pi^2 mE}{h^2}$$

Then the solution to the wave function in the three regions assuming that $E < U_1 < U_2$ and taking into account that the wave function cannot have any exponentially increasing term is given by

$$\psi(x) = A.\exp(\alpha x), \; x < 0$$

$$\psi(x) = B_1.\exp(ikx) + B_2\exp(-ikx), \; 0 < x < a$$

$$\psi(x) = C.\exp(-\beta x), \; x > a$$

Since ψ and ψ' must be continuous at a boundary, we must have the relations

$$A = B_1 + B_2, \; \alpha.A = ik(B_1 - B_2)$$

$$B_1\exp(ika) + B_2\exp(-ika) = C\exp(-\beta a), \; ik(B_1\exp(ika) - B_2\exp(-ika))$$

$$= -\beta.C\exp(-\beta a)$$

The first two equations give

$$B_1 = \frac{A}{2}(1 - i\alpha / k), \quad B_2 = \frac{A}{2}(1 + i\alpha / k)$$

The last two equations give

$$B_1 \exp(ika) = \frac{C}{2}\exp(-\beta a)(1 - i\beta / k), B_1 \exp(ika)$$
$$\doteq \frac{C}{2}\exp(-\beta a)(1 - i\beta / k)$$

It therefore follows that

$$A(1 - i\alpha/k) = C\exp(-(ik + \beta)a)(1 + i\beta/k)$$
$$A(1 + i\alpha/k) = C\exp((ik - \beta)a)(1 - i\beta/k)$$

Taking the ratio, we get

$$\frac{1 - i\alpha / k}{1 + i\alpha / k} = \exp(-2ika)\frac{1 + i\beta / k}{1 - i\beta / k}$$

The magnitude of both sides is unity and hence this reduces to just a single real equation for the phase. It is given by

$$-2\tan^{-1}(\alpha/k) = -2ka + 2\tan^{-1}(\beta/k)$$

or

$$\alpha / k = \frac{\tan(ka) - \beta / k}{1 + \beta \tan(ka) / k}$$

or

$$\alpha = \frac{k \tan(ka) - \beta}{k + \beta \tan(ka)}$$

4. The generating function of the Legendre polynomials is given by

$$F(t, x) = (1 - 2xt + t^2)^{-1/2} = \sum_{n=0}^{\infty} P_n(x)t^n$$

Derive recurrence formulae for $\{P_n\}$ by differentiating F with respect to t as well as with respect to x.

Ans. Differentiating with respect to t gives

$$\frac{\partial F}{\partial t} = \frac{x - t}{(1 - 2xt + t^2)^{3/2}} = \sum_{n=1}^{\infty} nP_n(x)t^{n-1}$$

Thus,

$$(x-t)\sum_{n=0}^{\infty} P_n(x)t^n = (1-2xt+t^2)\sum_{n=1}^{\infty} nP_n(x)t^{n-1}$$

Equating coefficients of t^n on both sides gives

$$xP_n(x) - P_{n-1}(x) = (n+1)P_{n+1}(x) - 2nxP_n(x) + (n-1)P_{n-1}(x)$$

or

$$(n+1)P_{n+1}(x) - (2n+1)xP_n(x) + nP_{n-1}(x) = 0$$

This is the first recurrence relation. The second is obtained from

$$\frac{\partial F}{\partial x} = t(1-2xt+t^2)^{-3/2} = \sum_{n=0}^{\infty} P_n'(x)t^n$$

or

$$t\sum_{n=0}^{\infty} P_n(x)t^n = (1-2xt+t^2)\sum_{n=0}^{\infty} P_n'(x)t^n$$

so that equating coefficients of t^n results in the recursion

$$P_{n-1}(x) = P_n'(x) - 2xP_{n-1}'(x) + P_{n-2}'(x)$$

5. Explain how one arrives at the Legendre differential equations and spherical harmonics functions starting from the differential expression for the angular momentum operator.

Ans. We have

$$-\frac{4\pi^2 L^2}{h^2} = \frac{1}{\sin\theta}\frac{\partial}{\partial\theta}\sin\theta\frac{\partial}{\partial\theta} + \frac{1}{\sin^2\theta}\frac{\partial^2}{\partial\phi^2}$$

We shall look at the eigenvalue problem for this operator, i.e. well behaved solutions of the eigen-equation

$$\frac{1}{\sin\theta}\frac{\partial}{\partial\theta}\sin\theta\frac{\partial\psi}{\partial\theta} + \frac{1}{\sin^2\theta}\frac{\partial^2\psi(\theta,\phi)}{\partial\phi^2} = \lambda.\psi(\theta,\phi)$$

We perform a separation of variables

$$\psi(\theta,\phi) = F(\theta)G(\phi)$$

which results in

$$\frac{1}{F(\theta)}\sin\theta\frac{d}{d\theta}(\sin\theta F'(\theta)) - \lambda.\sin^2\theta = -\frac{G''(\phi)}{G(\phi)}$$

The left side here is a function of θ only and the right side of ϕ only. Hence, both sides must equal a constant, say m^2. This gives $G(\phi) = exp(-\pm im\phi)$ and from the single-valuedness of ϕ, m must be an integer. the θ equation gives

$$\frac{1}{\sin\theta}\frac{d}{d\theta}\sin\theta F'(\theta) - \frac{m^2}{\sin^2\theta}F(\theta) = \lambda.F(\theta)$$

This equation we transform by a change of variable $\cos\theta = u$. It gives

$$\frac{1}{\sin\theta}\frac{d}{d\theta} = -\frac{d}{du}$$

and

$$\sin\theta\frac{d}{d\theta} = -\sin^2\theta\frac{d}{du} = -(1-u^2)\frac{d}{du}$$

so that the θ equation reduces to

$$(1-u^2)\frac{d}{du}(1-u^2)\frac{dF}{du} - \lambda(1-u^2)F - m^2F = 0$$

This is Legendre's differential equation. For $m = 0$, it reduces to

$$(1-u^2)F''(u) - 2uF'(u) - \lambda.F = 0$$

The values of λ for which we have well behaved solutions are given by $\lambda = -l\,(l + 1)$ where l assumes non-negative integral values.

ABOUT THE AUTHOR

The author Harish Parthasarathy completed his B.Tech. in Electrical Engineering from the Indian Institute of Technology, Kanpur in 1990 and his Ph.D from the Indian Institute of Technology, Delhi in 1994. His interests have been in signal processing, electromagnetic field theory and principles of quantum mechanics. For a brief period, the author was involved in a project on simulation of galaxy formation. The author has taught in the Indian Institute of Technology, Bombay and the Indian Institute of Technology, Kanpur. At present, he is working as Assistant Professor (Associate Head), Department of Electronics & Communication in the Netaji Subhash Institute of Technology, New Delhi where he teaches courses on basic signal theory, electromagnetic field theory.